Bob Schon

P9-AZW-378

NON-VERBAL COMMUNICATION

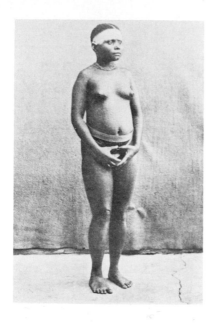

Andaman Islander
(from A. R. Radcliffe-Brown,
The Andaman Islanders, C.U.P.)

Mme. Proust with her sons
(from G. D. Painter, *Marcel
Proust*, Chatto & Windus).

(frontispiece)

NON-VERBAL
COMMUNICATION

edited by

R. A. HINDE

Royal Society Research Professor in the
University of Cambridge
Honorary Director of the MRC Unit
on the Development and Integration
of Behaviour

CAMBRIDGE
at the University Press, 1972

Published by the Syndics of the Cambridge University Press
Bentley House, 200 Euston Road, London NW1 2DB
American Branch: 32 East 57th Street, New York, N.Y.10022

© Cambridge University Press 1972

Library of Congress Catalogue Card Number: 75-171675

ISBN: 0 521 08370 2

Printed in Great Britain
at the University Printing House, Cambridge
(Brooke Crutchley, University Printer)

CONTENTS

PREFACE

In 1965 Sir Julian Huxley organized, and the Royal Society sponsored, a three-day meeting on the 'Ritualization of behaviour in man and animals'. The proceedings were lively and fertile, and the papers submitted were subsequently published in the Philosophical Transactions of the Royal Society (1966, **251**, 247–526).

It was apparent to those who attended the meeting that, although many of the individual contributions were of great interest, the terms 'ritual' and 'ritualization' were being used in quite different senses by students of animal behaviour, by psychologists, and by anthropologists. Although relatively little contact was achieved between these groups, many felt that contact could perhaps be achieved at a finer level of analysis – or, if it could not, the reasons why should be made explicit. Spurred on by Sir Julian Huxley's enthusiasm, the Royal Society set up a study group to explore further under the chairmanship of W. H. Thorpe.

This study group set itself more limited aims – the nature of 'non-verbal communication' in animals and man. It met thirteen times between 1965 and 1970, discussing contributions by its own members and invited guests. Its composition changed somewhat with time. In 1969 it was decided that some of the members should produce papers, based primarily on their own interests and taking into account the products of the discussions. A final two day meeting was held in the Ciba Foundation in September, 1970, at which pre-circulated drafts of most of the chapters in this book were discussed. In the light of these discussions, and numerous letters which followed them, chapters were revised and sent to section editors, and then to me.

In performing my task as editor, I had the advantage of copies of the correspondence which had passed between the contributors. I was able to see how far the final chapters had been influenced by the points raised, and what matters were still outstanding. From that point, I have attempted to synthesize and to link the contributions, bringing up issues not resolved in the discussions or in the correspondence, as well as other points which appeared when the material was collected together. No one editorial method was uniformly suitable. I have used notes, included amongst the author's own, for minor matters of fact; comments following the individual contributions for discussion of outstanding issues and for highlighting areas of disagreement; and these comments, prefaces and postscripts for linking the contributions. Where a particular individual was responsible for a point or opinion I have named him, but many of the issues arose from general interchange among the group. These editorial contributions were circulated amongst the authors, and were improved in the light of their comments, but at that point a halt was called to the interchange and the authors were not permitted further reply – though even here I have made some exceptions. Many of the individual contributions changed markedly between drafts as a result of our meetings and discussions, and the book as a whole was a collective enterprise. For defects in the comments and linking sections I alone am responsible.

The pleasantest of my editorial duties is to express the gratitude of all who took part to Sir Julian Huxley for setting us going and for his enthusiasm throughout; to the Royal Society for their generous support and for accommodating some of our meetings; to the trustees of the Ciba Foundation and to Ruth Porter, its medical director, for their hospitality and kindness, especially over the final meeting; to W. H. Thorpe for his patient and wise chairmanship; to M. Cullen and E. Leach for putting in much hard work as secretaries; and to D. M. MacKay, W. H. Thorpe and M. Argyle for acting as section editors.

R. A. H.

MEMBERSHIP OF
ROYAL SOCIETY STUDY GROUP ON
NON-VERBAL COMMUNICATION

* Attended final meeting.

*Professor W. H. Thorpe, Sub-Department of Animal Behaviour, Cambridge.

Dr E. R. Leach, King's College, Cambridge.

*Dr J. M. Cullen, Department of Experimental Psychology, Oxford.

*Mr M. Argyle, Department of Experimental Psychology, Oxford.

*Professor R. J. Andrew, School of Biological Sciences, University of Sussex.

Dr N. G. Blurton Jones, Institute Child Health, University of London.

Dr J. Bowlby, Tavistock Clinic, London.

Miss M. Brearley, Froebel Institute College of Education, London.

Professor R. Firth, Department of Social Anthropology, London School of Economics and Political Science.

*Mr J. A. W. Forge, Department of Social Anthropology, London School of Economics and Political Science.

Dr R. Fox, Department of Social Anthropology, London School of Economics and Political Science.

*Professor E. H. Gombrich, Warburg Institute, London.

*Dr G. A. Harrison, Anthropology Laboratory, Department of Human Anatomy, Oxford.

*Professor R. A. Hinde, M.R.C. Unit on the Development and Integration of Behaviour, Madingley, Cambridge.

*Mr F. J. H. Huxley, St Catherine's College, Oxford.

*Sir Julian Huxley, London.

*Professor D. M. MacKay, Department of Communication, Keele University.

Dr Jonathan Miller, London.

Dr D. Morris, Zoological Society of London.

Professor R. C. Oldfield, Speech and Communications Research Unit, Oxford.

Professor R. H. Robins, Department of General Linguistics, University of London.

Professor J. M. Tanner, Institute Child Health, University of London.

Professor N. Tinbergen, Department of Zoology, Oxford.

Professor J. Tizard, Institute of Education, University of London.

In addition the following attended meetings of the group to read papers or as discussants:

Dr A. Ambrose, Tavistock Clinic, London.

Professor R. Exline, Department of Psychology, University of Delaware, Newark, USA.

Dr E. Grant, Department of Psychiatry, Queen Elizabeth Hospital, Birmingham. (Now at Chelsea College, London.)

*Professor I. Eibl-Eibesfeldt, Max Planck Institüt, Seewiesen, Germany.

*Dr J. A. R. A. M. van Hooff, Laboratory for Comparative Physiology, Utrecht, Holland.

*Professor J. Lyons, Department of Linguistics, Edinburgh.

Professor D. Ploog, Max Planck Institute, Munich, Germany.

Professor D. Vowles, Department of Psychology, Edinburgh.

GENERAL INTRODUCTION

When we meet a stranger, we form an immediate impression of the sort of man he is. Without thinking we take in small details of how he stands, the way he uses his hands, his eye movements and facial expressions. The conclusions we draw may or may not be correct, but they certainly affect our actions. If we start to talk with him, our words are accompanied by other gestures which may elucidate, emphasize, enhance or even contradict what we say. At the beginning and end of our conversation we may use culturally determined signals, such as hand-shaking or hand-waving, which symbolize our readiness to enter into friendly acquaintanceship or our impending departure. Throughout this episode, much information passes between us which is never put in the form of words. Indeed a third party watching from a distance but unable to hear our conversation would be able to tell a great deal about us and about our relationship. It is signals such as these which form the subject matter of this book.

As discussed in later chapters, the importance of these signals has long been appreciated in the arts. The playwright can provide only the words to be spoken – the meaning which these convey to the audience must depend in no small way on the interpretation placed on them by the director and the actors, and this is carried by the particular emphases, facial expressions, gestures, postures and movements which accompany them. Novelists are able to supply their readers with more information about the non-verbal signals emitted by the characters they create, but still can cover only a small fraction of those which would be present in real life. The painter, in depicting human face or form, depends in part for his success on representing these non-verbal signals – and on their recognition by the beholder.

The painter's skill depends on long and patient objective observation of how people behave, but sometimes one must wonder just how far even the great masters consciously analysed all the signals which contribute to the impressions they convey. Indeed we use these signals ourselves nearly all the time without consciously recognizing that we are doing so. This was dramatically brought home to the participants at a recent conference at which Eibl-Eibesfeldt described the eyebrow flash – a rapid raising and lowering of the eyebrows which, though it appears in many and perhaps all human cultures, had not previously been described in the scientific literature. Suddenly the members of the conference were made conscious of a gesture which they were using many times a day without realizing it. From that moment on it became impossible to greet a colleague without a momentary mutual embarrassment as each realized that he had used this commonplace gesture without meaning to.

Outside the arts, many of these signals were documented by biologists and anthropologists in the nineteenth century (e.g. Charles Darwin, *The Expression of the Emotions in Man and Animals*, 1872) but, with a few exceptions, their detailed study was neglected for over fifty years. Recently, however, their importance has been more fully recognized and knowledge about them greatly increased. The reasons for this are multiple. In part, the descriptions by ethologists of signals in animals called attention to the functionally comparable, but more complex, signals in man. Independently psychologists realized their importance in human inter-personal relations, and anthropologists became increasingly conscious and objective about the role of non-verbal communication in the integration of societies. Also the practical importance of these signals became apparent – for instance by their absence in long-distance telephone and radio communication, and during interchanges between individuals from different cultures with dissimilar conventions about their use.

The problems which arise are numerous, and it may be helpful at this stage to illustrate them by a particular example which was used by Dr Leach in one of the earlier discussions of the group. The frontispiece shows members of two different cultures – an Andaman islander and Mme Proust – coming from opposite extremes of our normal scale of sophistication, but holding their hands in a similar way. The hand position is a striking one, but has no intended or explicit meaning. Is the resemblance accidental? If not, how is it to be explained? Does it represent a deep-rooted response which has some independence of culture? In any case, how is its appearance in the individual to be explained? In Western Europe, at least, girls are not often taught to stand in that way. Does it arise through imitation, or through some subtle social pressure common to both cultures, or is it independent of social experience?

We may also ask about the emotion or motivation associated with the posture, and its communicatory significance. In Mme Proust we might well call it a modest pose, but with the Andaman islander this seems more doubtful. Is it that our interpretation changes with the context in the picture, and depends on our own background? Or does the posture really mean different things in different cultures? If so, is the underlying motivation different in kind? Or different only in degree, with both women being ambivalent, but to differing extents? After all, we are familiar enough with the extent to which dress itself, ostensibly worn in the interests of decency, may in fact emphasize what it is supposed to obscure.

These are only a few of the many problems arising from the study of non-verbal communication between individuals. It has been the purpose of the contributors to this book to explore some of them. We have not attempted a complete review, but selected some issues for detailed discussion and merely sketched others, while attempting to highlight some of the threads which tie the whole field together and to underline some areas of controversy.

The chapters are grouped into three parts. Part A is concerned with establishing some basic distinctions, and with the relations between verbal and non-verbal communication. These are of necessity largely theoretical, and although they are fundamental to the serious student, the reader more interested in the empirical material may prefer to pass straight to the brief discussion at the end of the Section and thence to Part B. This is concerned with the problems raised by, and the diversity of, signalling systems in animals. Finally, Part C is concerned with our own species, except for one chapter which describes the way in which certain human expressive movements are related to those in sub-human primates.

PART A

The Nature of Communication

INTRODUCTION TO PART A

What is meant by communication? In summarizing a recent conference
on (non-verbal) primate communication, Ploog and Melnechuk (1969)
reported that some participants saw no differences between social behaviour
and communicative behaviour, while others thought that a distinction
'ought to be possible'. Some workers prefer a broad definition, using
communication to refer to the giving off by one organism of a signal which
influences the behaviour of another (e.g. Frings and Frings, 1964), but
others feel that this conceals fundamental differences of complexity between
different phyletic levels (e.g. Tavolga, 1968). When there is difficulty in
obtaining agreement over a dull matter of definition, one can be sure that
unresolved conceptual issues lie concealed. The important goal then
becomes, not the defining of the term, but specifying the sorts of distinctions which are useful. In the following chapter, MacKay, an information
theorist, analyses various concepts often confused under the label of
'communication', and provides a framework which makes it possible to
relate the different usages employed by other contributors. Thorpe, a
biologist, takes up the matter of species differences in systems of communication: he compares species in the extent to which their communication
systems possess the 'design features' of human language (Hockett and
Altmann, 1968). This leads him to consider how far the complex communicatory abilities, which have recently been taught to chimpanzees, do, in
fact, approach human language. In the third chapter, Lyons, a linguist,
takes up in more detail the relations of non-verbal communication to
verbal language, summarizes some of the basic principles involved in the
study of language, and discusses current views on its evolution.

1. FORMAL ANALYSIS OF COMMUNICATIVE PROCESSES

D. M. MacKAY

Department of Communication, University of Keele

I. INTRODUCTION

Every field of science, old or new, is haunted by the perennial question: how can we avoid foreclosing empirical issues, and missing essential points of the situations we study, by our choice of conceptual apparatus and working distinctions? There is, of course, no static answer. A restless Athenian eagerness to run after 'some new thing' is generally counterproductive and damaging to scientific standards. On the other hand, especially in a new field, we must expect our concepts to need constant refinement, and must be alert for signs that something important is escaping us because our customary ways of looking at the phenomena have a significant 'blind spot'.

Given the enormous variety of ways in which animals and men can influence one another without relying on words as such, it is not easy to see how such a vast field can profitably be studied at all as a unity. Clearly, we need some way of abstracting formally what these diverse situations have in common, and of systematically comparing and contrasting the mechanisms at work in them. On the other hand, any formal description of something as familiar as communication runs the risk of seeming both pretentious and banal. Unless it is clear to what end we wish to formalize, the effort were better not made, or at least not made public! The purpose of this chapter is not, then, to offer an armament of neologisms for general use, still less to lay down the law in matters of definition; it is rather to consider what problems, if any, in the study of non-verbal communication create a need for analysis in other than commonsense terms, and what the prospects are of meeting that need without defining out of existence some of the questions we must be ready to ask.

2. WHAT COUNTS AS 'COMMUNICATION'?

At the outset we need some kind of convention, however arbitrary, to define what will count for our purposes as an instance of 'communication'. The etymological root of the term (*communicatio*) means *sharing* or *distributing*. Thus we can speak of 'communicating' rooms, and the Apostle (in the classical English of the Authorized Version) can admonish us to 'forget not to do good and to communicate'. In this general sense, *A* communicates with B if anything is shared between A and B or transferred from A to B.

For scientific purposes we need a more restricted usage if the term is not to become trivial. (Otherwise, the study of 'non-verbal communication' covers every interaction in the universe except the use of words!) For the purpose of the present symposium, we might agree to begin by restricting ourselves to interactions between organisms – though even here some communication engineers would be left out of step.

By them, communication is used loosely to mean the transmission of information regardless of its origin or destination. They will happily speak of a rock on a hillside as communicating with an observer, if sunlight reflected from the rock reaches his eyes. Worse still, the definition of a 'communication channel' in some mathematical textbooks does not even require a causal connection between the two points in question! Provided that the sequence of events at A shows some degree of correlation with a sequence at B, their authors are ready to define a 'channel capacity' between A and B, regardless of the possibility that the correlation is due to a third common cause, and not at all to any interactions between A and B.

Already we see that to speak of 'communication' between A and B can have a multiple ambiguity. As used by different people it may imply:

(a) mere *correlation between events* at A and B,
(b) any *causal interaction* between A and B,
(c) *transmission of information* between A and B regardless of the presence of a sender or recipient, and/or
(d) a particular kind of action by an organism A on another, B, or
(e) a transaction between organisms A and B.

Since we already have perfectly good terms (those I have italicized) for (a) to (c), we might perhaps agree to use them instead of 'communication' in those cases. But what of (d) and (e)? Are all actions by an organism upon another, or all transactions between organisms, instances of 'communication'? In one sense, no doubt, the answer could be affirmative. Organisms in interaction can hardly fail to *receive information* about one another; and it has often been emphasized that such information can be conveyed by inaction as much as by action. All behaviour is potentially *informative* – even non-behaviour. Our main question is whether for our purposes we need to draw any further distinctions. Is it good enough to say bluntly that therefore 'all behaviour is communication'; or do we need sharper conceptual tools?

It is my impression that when we in the present context speak of 'non-verbal communication' between A and B we do want to distinguish a certain class of non-verbal behaviour from others. Broadly speaking, our concern is with behavioural and structural features of A that affect the *internal organization* of behaviour – the control centre – in B and vice versa. We look upon the recipient as a system with a certain repertoire of possible modes of action, both internal and external, which must have some central 'organizing system' to set targets for, and control, the running selection from the repertoire which we call its behaviour. In higher organisms the organizing system, (which we shall discuss more fully in Section 6) must determine not only the behaviour of the moment, but also a vast complex of 'conditional readinesses for action' – what kind of behaviour *would* result *if* such and such circumstances arose. To be more specific, it must be able to adjust behaviour to take account of *facts*, acquire and preserve *skills*, and adjust the *priorities* of different possible courses of action.

What interests us here, then, is the kind of interaction between organisms in which signals from A influence this central organizing system – B's internal representation of facts, skills or priorities. The ordinary 'Newtonian' effects of A on B, such as the mechanical vibrations of the body produced by a handshake, or even the reflex 'startle' response to a shout or a pinprick, are in a different and relatively uninteresting class. It is the 'informative', 'releasing', or 'coordinating' functions of behaviour and structure that raise the problems we want to solve. For our present purpose, an event is not communicative (in the relevant sense) unless it has some *internal organizing function* in a recipient. We shall take up this point also in Section 6.

This is a necessary condition – but is it sufficient? Here we reach the point where usages have tended to diverge. All would agree, for example, that a measly face can be *informative* to a qualified onlooker. But is it useful to speak of the sufferer himself (who may be unaware of it) as *communicating* this information? Is there no distinction to be made between the passive manifestation of a symptom and the deliberate (even if instinctive) production of words or non-verbal behaviour (including perhaps pointing to the spots) *calculated* to inform the observer? Again, shifts of gaze or posture may play a subtle part in coordinating the behaviour of two persons (Argyle, 1969); but may it not be useful to have different terms for those acts that are expressive of the originator's purpose and *perceived* or *interpreted* as such, and those that are not? For reasons to be elaborated in Sections 5 and 6, some of us interested in the internal organization of behaviour find such distinctions both operational and essential. Others concerned with purely external descriptions of behaviour may find the distinctions elusive and unnecessary for their purpose. This is understandable. But if we ourselves are to communicate without ambiguity, presumably some neutral terms will be needed which do not beg questions regarded by some of us at least as open.

Fortunately general terms are already available which leave room for most of the distinctions we want. *Signalling* is widely used as a neutral word for the activity of transmitting information, regardless of whether or not the activity is goal-directed, what impact if any it has on a recipient, or even whether the source is animate or not (we speak of 'signals from radio stars'). Its use would allow us to say, for example, that '*A* is signalling but not communicating' in circumstances where information is being transmitted from *A* but not affecting the organizing system of any recipient. Even here, care may be needed to preserve the distinction between signalling in a passive, impersonal sense – the sense in which a rock signals its presence by the light reflected from it – and signalling in the active, goal-directed sense of 'trying to communicate' – i.e. trying to establish a link with another agent – as when a mother calls 'dinner's ready' but the family is out of earshot. If this ambiguity proves serious, we may have to fall back on a still more neutral expression such as *emitting information. Control* can be used to denote regulative function by means of signals without either implying or denying that the regulation is consciously intended or consciously perceived. (We discuss the question of 'conscious intention' in Section 6.) *Coupling, interaction, correlation* are other terms with well-established functions that might with advantage take the place of 'communication' in appropriate contexts.

Similarly, it might help to sharpen our thinking if we were content to speak of structures or events as *symptomatic* or *informative* where one of those terms strictly and adequately describes their function, and allow the over-worked word 'communicative' to rest except when we are clear that it would add something both justifiable and essential.

3. THE TOOLS OF COMMUNICATION ENGINEERING

It is natural to ask how far the study of non-verbal 'communication' might be able to utilize the range of equipment and concepts already developed for the purposes of communication engineering. The answer, I think, depends very much on the kind of understanding we want, and on the extent to which we are prepared to adapt our questions to the tools that happen to be available.

(*a*) At the *taxonomic* level, the challenge has been to develop the equivalent of an 'Identikit' with which to classify the varieties of behaviour or structure that produce invariant (or rather, observationally indistinguishable) effects on the recipient of the resulting signals, and conversely. In the study of bird and mammal calls, for example, the sound spectrograph and its inverse the spectral synthesizer have played a valuable part in articulating the phenomena to be understood (Thorpe, 1967).

There are, of course, dangers in adopting the descriptive categories of the communication engineer, which can easily lead to biologically insensitive classifications. (Even with human speech sounds, for instance, the sound spectrograph can be notoriously ambiguous or misleading – see Fant, 1962; Lindblom, 1962.) There is perhaps a need here for the development of new automatic devices whose classifying categories are more biologically-oriented; but the work described by Thorpe and others in the present symposium shows that even the standard equipment used by communication engineers can be surprisingly effective.

(*b*) At the level of *mathematical* analysis rather less can be said for the quantitative tools developed in the past two decades under the name of Communication Theory. This may seem perplexing; a theory that purports to measure the traffic taking place between a sender and receiver separated by a communication channel might surely have been expected to throw useful light on the processes going on in the terminals themselves as well as in the channel? The reasons for the disappointment (by and large) of such hopes deserve detailed discussion in the following section. Meanwhile we may simply note that relatively little of the puzzlement actually felt by observers of animal communication has been of the right kind to be relieved by numerical answers to numerical questions. As we shall see, it is to a handful of qualitative concepts made precise by the theory, rather than to the great bulk of its mathematical formulae, that the study of non-verbal communication is likely to be most indebted in the immediate future.

(*c*) This brings us to the third level of understanding, that of *causal* analysis, at which engineering ideas have a crucial part to play. *How* do non-verbal signals elicit the response they do? What is going on inside the organisms concerned? What is the relation between the structure of a signal and the internal structure of the receiving organism, by virtue of which, presumably, the signal has the function it does? Answers to such questions can be sought in terms closely analogous to those of information-system engineering and the theory of automata; and the latter part of this chapter will be devoted largely to outlining the area of profitable contact between the two (though not to answering the questions themselves in any detail!).

4. THE MEASUREMENT OF INFORMATION-FLOW

The chief architect of the mathematical Theory of Communication, Claude Shannon, was at the time a communication engineer in Bell

Telephone Laboratories. Shannon (1948) is sometimes credited with having produced a mathematical definition of the concept of 'information', with the implication that if one now wants to use the word scientifically it must be strictly in the context of Shannon's theory. This impression is mistaken. As long ago as 1951, Shannon himself disclaimed any intention of defining 'information' *per se* (Shannon, 1951 a). What he defined was one particular measure of '*amount-of-information*', which as we shall see was especially appropriate for assessing the capacity of a communication channel to transmit code-signals, but which was designed to be indifferent to the *information* that those code-signals might represent.

The concept of information itself is most readily defined operationally, in terms of what it does (MacKay, 1969). Subjectively, we say that an event provides us with information when it causes us to know or believe something that we did not know or believe before. In other words, information-about-X determines the form of our readiness-to-reckon-with-X in appropriate circumstances. Objectively, information is said to be transmitted from A to B when the form of an event or structure at B is determined by the form of one at A, regardless of the source of the necessary energy. For example, if a heavy machine in a factory is suddenly switched on to a power line, the resulting flick of the ammeters back in the power station provides 'information' to the attendant. We can speak of this as the 'transmission of information' from the factory to the power station, even though the direction of energy-flow was from the power station to the factory.

Thus the general notion common to both subjective and objective uses of 'information' is *that which determines form*. In order to make it precise, we have always to ask: the form of what? In the case of an organism 'receiving information', the answer must be in terms of those internal features of its organizing system that represent 'that which is the case' for the organism – in a general sense, its field of action – the 'world' it is prepared to reckon with. Information-for-an-organism is operationally definable as that which confirms or changes its internal representation of its world. (This clearly leaves open, as it should, the possibility that such information may be true, doubtful, false or even illusory.)

In this context, the communication engineer regards himself as a go-between whose job is to determine the form of certain events at a 'receiver' in obedience to instructions given by a 'sender'. Instead of *constructing* each form *ab initio* (for example, by tracing out each letter of the alphabet in a run of text), it is often economical for the engineer to keep a pre-fabricated range of standard forms (such as the letters in a teleprinter) at the receiving end and to *select* the forms required in response to the sender's instructions. The set of rules prescribing what must be selected in response to each possible signal is known as a *code system*. On this basis, the cost of transmission is measured by the number of basic instructions needed to define the total *selective operation* required – quite regardless of what it is that has been selected.

At this point non-mathematical readers may prefer to skip to the next section; but for our later discussion a few further details will be relevant. If we take as the basic instruction (and as our unit of cost) the answer to a simple two-valued question (yes/no, left/right, 1/0), then the most economical selection process (as in the game of '20 Questions') will be one in which each instruction halves the range of equally-likely possibilities. One out of two items can be identified in one step; one out of four, in two; one out of eight, in three; and so on. The cost (in this logical sense) of identifying one out of n equally likely possibilities is thus measurable by the logarithm of n to base 2, $\log_2 n$. Since the prior probability of each item will in this case be $p = 1/n$, we can write $\log_2 n$ as $\log_2 (1/p)$. This is defined as the *selective information content* of an identification whose probability was p. It is measured in 'bits' or binary digits.

Where some forms are selected much oftener than others, the average cost (in basic instructions or 'bits' per selection) is reduced if the more frequent items are encoded so that they can be selected in fewer steps than the less common ones. It was shown by Shannon (1948) that where the relative frequencies of selection are $p_1, p_2, \ldots, p_i, \ldots$, the minimum average (logical) cost per selection achievable by encoding in this way is $H = \Sigma p_i \log_2 (1/p_i)$ bits; i.e. the weighted mean of $\log_2 1/p_i$. This is the expression loosely referred to by some textbook writers as 'information'; but as Shannon emphasized, what it makes mathematically precise is not the concept of information at all, but only a particular *property* of information – its prior uncertainty or statistical unexpectedness.

The expression H has its maximum value ($H_{\max}$) when all the probabilities p_i are equal – i.e. when all items in the repertoire are used equally often. ($1 - H/H_{\max}$) is known as the *redundancy* of the signalling process. It can be regarded as a measure of the extent to which the repertoire is under-utilized. Despite the pejorative flavour of the name, redundancy is of great value if signals are liable to distortion or corruption by 'noise', since it makes possible in principle the detection and correction of errors (Hamming, 1950). Our ability to detect misprints, for example, depends entirely on the redundancy of typical English text. Shannon (1948, 1951b) was able to show that in a strict sense a communication system's tolerance of transmission errors is directly proportional to the signalling redundancy.

5. COMMUNICATION THEORY IN BIOLOGY

The usefulness of Shannon's measures to people concerned with the economics of signalling systems is self-evident. What may be less obvious are the presuppositions that must be satisfied if they are to be applied in other contexts such as that of biological communication. At first sight it is all deceptively straightforward. An animal uses items from its repertoire of signals with different relative frequencies. By prolonged counting we can derive estimates for the probabilities (such as p_i) in Shannon's formula and compute an 'average selective-information-content' per item, H. So far, so good. But now, what does this mean? It means strictly that over a long run of items, an engineer who knew the probabilities (p_i) could specify the animal's behaviour by using a code sequence of about H basic instructions per item. The question is whether anything of biological interest is likely to correspond to or co-vary with this figure.

If the receiving animal's brain were organized to take account of relative frequencies on the same basis, then conceivably the brain might develop an optimal code-system whereby the items were identified 'economically' in the sense of Shannon's theory, and our estimate of H would then give some idea of the magnitude of the cerebral processing operation. But there are two snags. In the first place, for a biological system that *grows* its parts, the criteria of 'cost' may be very different from those used by the engineer. Logical economy and biological efficiency do not necessarily coincide. Large-scale parallel processing, for example, involving enormous numbers of signals per item, may be relatively cheap as a biological solution where it would be prohibitively costly in a wired automaton.

But more serious still is the problem of finding appropriate probabilities (p_i) for Shannon's formula. The whole computation presupposes a 'statistically stationary' situation: in other words, the relative frequencies must not change significantly as time goes on. But the behaviour patterns of greatest biological interest are often those that do change as communication proceeds; and if we confine ourselves to samples short enough to be 'stationary', we must then accept corresponding uncertainties in our estimates of the probabilities. Furthermore, the more labile the situation, the less confidence we can have that any probabilities we estimate are those reflected at a given time in the brain-state of the animal receiving the communication.

In short, any attempt to use Shannon's H as a measure of information-flow *between* organisms raises prior questions that are all too often unanswerable. As a measure of the variability or unexpectedness of behaviour for the scientific *observer*, H can, of course, have a precise significance; and the present symposium contains valid examples of its use as an overall statistical measure of the degree to which an organism's behaviour is coupled or correlated with that of another organism influenced by it. But unless one knows that the receiving organism has, so to say, a 'Shannon encoder' inside his head, one has no prior reason to regard H as a more biologically significant measure of the *information received* by that organism than, say, the total number or duration of the signals exchanged.

All this, however, does not mean that the engineering theory of communication can be ignored for biological purposes. On the contrary, since its basic notion is that of *selection from a repertoire*, it has an immediate link at the *qualitative* level with the characteristic habit of thought of the biologist. By viewing an organism as a system with a repertoire, and its environment as imposing a running pattern of demand for actions selected from this repertoire, we find ourselves talking essentially the same language as the communication engineer (MacKay, 1956, 1966). Terms such as 'redundancy', 'noise', and 'channel capacity' can be used quite rigorously to meet conceptual needs already recognized by biologists in the articulation of problems of behavioural organization. As we shall see in the next section, the related notions of 'feedback', 'feedforward' and 'evaluation of mismatch' have a still more direct (and long-recognized) application to the kinds of behaviour that most interest us in the present context. Finally, I hope it is now clear that ignorance of mathematical Communication Theory is no barrier to the use of 'information' in its ordinary sense, as long as we are not tempted to try to measure it. We are in much greater danger of bringing disrepute on our theorizing if we over-use the term 'communication' in such a way as to obscure necessary distinctions.

6. ORGANIZATION OF ACTION

The term 'action' in physical science has a very broad sense. When we speak, for example, of the action of frost upon rock, there is no implication that the frost is goal-directed, or governed to achieve a particular effect. In the present context, however, we are concerned with a more restricted usage. It is characteristic of organisms that they do act to achieve ends or maintain a required state.

Directed action by an organism is distinguished from mere undirected activity by an element of *evaluation*: a process whereby some indication of the current or predicted outcome is compared against some internal 'target criterion' so that certain kinds of discrepancy or 'mismatch' (those 'negatively valued') would evoke activity calculated to reduce that discrepancy (in the short or long term).

Directed action and undirected activity may sometimes bring about the same end-state. A criminal, for example, may repeatedly find himself in jail as a result of actions with quite a different aim! Mere observation of end-states is thus quite inadequate to answer the question whether an activity had a goal, and if so, which goal. From the outside, a man who acts *in such a way as to* bring about a particular event may sometimes be indistinguishable from one who acts *in order to* bring it about.

A little thought shows that the criterion we need in order to distinguish such cases is a *variational* one. We must ask not only what does or did happen, but what *would* happen *if* certain conditions varied. This is rather like the physicist's criterion of 'stability': if we imagine the system slightly displaced, would a restoring force be developed? If you wonder whether your criminal is trying to commit crimes with impunity, or (as may sometimes be the case) is really trying to get himself jailed, try *offering* him a prison sentence!

Such a criterion may seem too hypothetical to be useful; but it can readily be expressed in structural terms by the device of 'information-flow mapping'. Fig. 1, for example, shows the basic flow system used in automata to bring about goal-directed action (MacKay, 1951). Items selected from the repertoire of the effector system E bring about changes in the field of action F. A receptor system R generates an *indication* (I_F) of the current state of the field F. I_F is subjected to the process we have called *evaluation* in the 'comparator' C, which generates a mismatch signal if I_F fails to satisfy an internal goal-criterion I_G. This mismatch signal in turn activates an *organizing* system O in such a way as to change the pattern of E's activity. If the system as a whole is to pursue the goal G represented by I_G, the change thus brought about in I_F must be such as to reduce the mismatch: i.e. feedback must be 'negative'.

Given this condition, the structure represented by Fig. 1 embodies our variational requirement. Any change in the field F (and hence in I_F) that would increase the mismatch would find itself opposed by a corresponding modification of the action of E. Equally, any change in I_G would set the effector system moving so as to make I_F follow it. In other words, the system is 'conditionally ready to act' in pursuit of the goal G as need arises.

The point is that in order to establish these facts about the system, it is unnecessary actually to make such experiments. The facts could in principle be checked by detailed functional examination of the present flow-structure of the automaton (or the organism), and the knowledge so gained would serve the same diagnostic purpose as the hypothetical tests suggested by the variational criterion. (Note, however, that the functional knowledge must be sufficiently *detailed*. Merely superficial similarities to the flow-map of Fig. 1, with no functional information as to what goes on in the 'black boxes', would be quite inadequate.)

So, to return to our criminal, the hypothesis that he is trying to get himself a stretch in prison can be turned into a diagnostically equivalent hypothesis about the information-flow structure of his brain. The basic question is whether within this structure his evaluative system is set to give a positive or negative value to 'prospect of prison'. The answer, one way or the other, is a fact about him *as he is now*, which we could in principle imagine being settled by a sufficiently detailed functional inspection of his brain, much as the inspection of a burglar alarm could determine

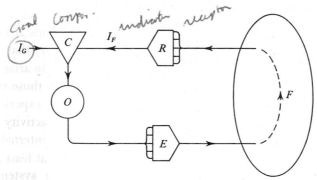

Fig. 1. Basic flow-map for goal-directed activity. The action of the effector system *E* in the field *F* is monitored by the receptor system *R*, which provides an indication I_F of the state of *F*. This is compared with the goal criterion I_G in the evaluator *C*, which informs the organizing system *O* of any mismatch. *O* selects from the repertoire of *E* action calculated to reduce the mismatch.

its 'state of conditional readiness' (to ring *if* certain events occurred) without actually setting it off. Normally, of course, we might be restricted to observing supporting evidence of a behavioural kind, including the criminal's response to suitable questions; but since we can have no guarantee that during the collection of such evidence his goal will have remained unchanged, it would be unwise as well as unnecessary to *define* the concept in terms of observed behaviour rather than the functional structure giving rise to it.

Unfortunately there are few organisms simple enough to have only one evaluatory process going on in their information system at any given time. In the human body, for example, we have many 'feedback loops' of the general form of Fig. 1 which operate without any *conscious* evaluation on our part. One example is the automatic mechanism that keeps the pupils of our eyes adjusted to the brightness of the light entering them. (The reader can verify that this process is quite unconscious, by holding before his eye a card with two pinholes $\frac{1}{8}$ inch apart. Two images of the pupil will be observed, and on opening and closing the other eye adaptive changes in their diameter can be clearly seen to happen without any conscious intention on his part.)

Where different evaluative subsystems have elements in common, any conflict between them must be resolved by hierarchic ordering of priorities. For example, at a given time an animal may be prepared to abandon feeding in favour of sexual activity, and to abandon either in favour of escape from a predator. In lower animals, as far as behavioural evidence goes, the rules governing such priorities are relatively few and simple. In man, however, they have a flexibility and lability that have scarcely begun to yield to systematic analysis. Our most complex conscious experience, indeed, is of events and processes that affect the relations between our conditional priorities, i.e. the target criteria (I_G in Fig. 1) that *would* be given precedence *if* various different circumstances were to arise.

Does this offer a clue to the elusive difference between those evaluative brain processes that correlate directly with our conscious experience, and those (like the pupillary reflex) that do not? The brain activity required for the evaluation of competing criteria must involve internal actions, including calculations and trial-and-error, which in man at least are quite as complex as any performed by the external effector system. These internal actions affecting priorities and criteria must themselves be evaluated according to higher-level criteria, so that we are left with the need for a complete hierarchic 'system within a system' to take care of this aspect of conscious agency. For short we may refer to this as the *meta-organizing* system (Fig. 2).

Just as information from the external field F has to be supplied to the evaluator C in order that effector E can be properly controlled by the organizing system O, so information from O itself, as well as from receptor system R, has to be supplied to one or more meta-evaluators (MC). These in turn can guide the meta-organizer MO to prepare trial subroutines and keep the programme of O updated to match the changing field F.

In the role we have sketched for it, the activity of this system seems well suited to be the physical correlate of conscious experience (MacKay, 1966). If it were, then *intentional* action would be identifiable in principle as action originated and evaluated within the meta-organizing system. Action not evaluated there would not be intentional, even when it was clearly under evaluation (as the pupillary reflex is) at some 'lower' level outside the meta-organizing system.

For our present purpose the details of this speculative suggestion are not important. What matters is that we recognize the unjustifiability of regarding all action-under-evaluation as *ipso facto* 'intended' to bring about its goal – a trap from which we could escape only by qualifying some 'intentions' as 'unconscious'. Flow maps on the lines of Fig. 1 must be used critically. Evidence that an evaluative criterion is operative is a necessary but not a sufficient condition for describing an individual as acting intentionally, i.e. acting *in order to* bring about particular consequences. We need also more detailed evidence as to the level in his information system at which the evaluation is taking place (MacKay, 1966; see also Hinde and Stevenson, 1970: 233).

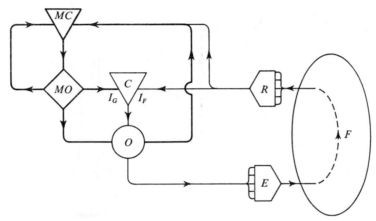

Fig. 2. Addition of 'meta-organizing system' *MO* (heavy lines) to determine goal priorities and govern internal action upon the organizing system *O*. *MC* represents one member of a hierarchy of evaluators.

One further elaboration of Fig. 1 is needed if we are to use it in discussing communication. As described thus far, the system takes no account of the contents of its field of action F, except as they affect the outcome of ongoing goal pursuit. In a reasonably stable environment, however, the pattern of demand represented by I_F will be 'highly redundant' in Shannon's sense, and the efficiency of goal-pursuit would be enormously increased if selections from the repertoire of action were made in the light of the constraints offered by permanent or temporarily stable features of the field. It would pay the organizing system to operate on a coding principle, whereby much of its activity was made up of components selected from a prefabricated internal repertoire of control patterns suitably matched to the current spatio-temporal features of the field of action. To this end it will need detailed 'feedforward' from sense organs, supplying information about the relevant features, if the repertoire is to be kept up to date. The structure of the organizing system amounts in this sense to an internal representation of the field of action – though not necessarily topographically similar to it. The correlate of perception may be regarded in these terms as the updating of this internal representation by the meta-organizing system in matching response to sensory signals (MacKay, 1956, 1967).

This leads us to the skeleton map of Fig. 3, in which the sensory input from R is analysed in a series of 'feature filter' systems FF to provide 'feedforward' (i.e. information in advance of any external action) to maintain the constantly changing 'state of conditional readiness' embodied in the organizing system O. One known filter system, for example, is

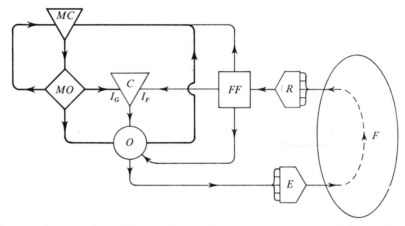

Fig. 3. Addition of 'feedforward' into the organizing system O from 'feature filters' (FF) sensitive to particular features of F.

believed to signal the *velocity* of the retinal image at different points in the visual field, so providing information as to the speed and direction of an animal's motion through its environment, and making it easier to detect other moving objects. Although many such filters are 'preset' at an early stage in development, it is possible that others may be 'tunable' to features of particular significance to an animal at a given time, as for example when looking or listening for something specific.

7. COMMUNICATION BETWEEN ORGANISMS

We are now in a position to think of communication between organisms in terms of its internal effects on the information system of each. For simplicity we may combine the elements of Figs. 1–3 in the skeleton map of Fig. 4. The three distinct (if sometimes overlapping) functions of the central system are represented in three compartments:

   (*a*) internal representation of the structure of the field of action, under 'feedforward' from sense organs,

(*b*) formation and storage of a hierarchy of 'subroutines' conditionally _facta_
ready to elicit patterns of appropriate effector action as complex
'wholes',

(*c*) determination of priorities among the hierarchically ordered set of _Spolya_
possible goals and norms.

For short, let us call these the representations of (*a*) 'facts', (*b*) 'skills'
and (*c*) 'priorities'.

The target of a communication may be any or all of these aspects of the
recipient's information system, and statements, instructions, commands,
questions and the like can be readily distinguished operationally in these
terms (MacKay, 1969). In the act of communication, one organism can be
thought of as wielding a tool (verbal or otherwise) in order to mould the
representation of 'facts', 'skills' or 'priorities' in another. The action of

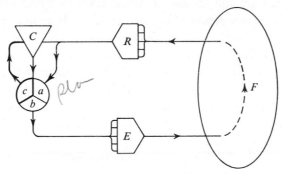

Fig. 4. Summary of essentials of Fig. 3 (see text).

the initiator to this end is generally called 'communicating', 'sending a
message' or 'generating a signal'; but of course it is open to us to regard
the whole causal sequence thus set in motion by the initiator, up to its
point of impact on the recipient, as the operation of one and the same tool.
What matters is that we distinguish sharply between

  (i) the originator's goal in wielding the tool and

  (ii) the actual effect in the recipient.

This relates closely to the controversial question of *meaning*. In everyday
language, as Professor Lyons points out in his chapter, we can distinguish
clearly between

  (i) the *intended* meaning of a message or signal

  (ii) the meaning as *understood* by its recipient, and

  (iii) the *conventionally* understood meaning.

These distinctions also have direct correlates in our present operational
terms (MacKay, 1969) if we take as the operational correlate of 'meaning'
the *selective function* (of the message) on the range of possible states of a

recipient's organizing system. This leads immediately to parallel distinctions between

   (i) the selective function *intended* by the originator;
   (ii) the *actual* selective function for the recipient, and
   (iii) the selective function for a 'standard' recipient.

In the light of the preceding section, I hope that the reference to 'intention' here need cause no alarm. As its etymology shows (*intendere arcum in*: to draw a bow at), the original notion behind the word was the selection of a target. By intended selective function we mean simply the aim implicitly represented by the criteria of evaluation in the originator's meta-organizing system. In these terms the notion can be made as operational as any other for scientific purposes, and there is no need to rob ourselves of any insight we can gain by using it.

Note that we are not in any of these cases defining meaning or its correlate as the *change produced* by the message. (It would be odd to have to call a 'No Smoking' sign *meaningless* for a given individual unless he actually read it!) Like the 'opening power' of a key, the 'selective function' of a message is a relational property, which will depend in obvious ways upon the target system in question. This property would doubtless be *demonstrated* if the key, or the message, were allowed to operate on the target system; but it could be no less well-defined even if the tool in question were never used.

As pointed out in Section 2, it is quite possible for some feature of an organism A (such as a measly face) to bring about such changes in another, B, without A's consciously having this as a goal. The distinction can now be spelt out in terms of Fig. 4. It turns upon whether or not the effect of this feature of A upon B is under *evaluation* in A's information system in the appropriate way.

Consider a few test cases.

(*a*) Schoolboy A appears with a measly face. His classmates run from him in alarm. Obviously, his appearance *informed* them that he had measles. But does it make sense (or is it useful) to say that *he communicated* this information? The answer is quite clear. If the 'I_G' of his evaluator would be matched only by their noticing the spots, so that any indication of failure to produce this effect would set up a negative feedback signal to his meta-organizing system, then his appearing before them was a communicative act in our present sense. He was, we might say, 'parading' his spots. If, however, his evaluator was not so set, the episode would not count as an instance of *communication by A to B*, but only one of perception of A by B.

(*b*) Again, suppose that A and B are in conversation. A stares B in the eye so unwinkingly that B is embarrassed and made to feel vaguely 'one down'. What is going on here? The same criterion enables us to sort out the possible cases. How is A's evaluator set? Is evidence of B's discomfort evaluated positively or negatively by A? If B showed signs of perceiving A as trying to make him uncomfortable by staring, would these signs evoke a positive or negative evaluation in A? and so on.

(*c*) A is blushing. B perceives his embarrassment and leaves A in no doubt of this, thereby reinforcing A's blushes. Is A communicating, and if so, what? Here we may presume that A's goal is *not* to give the embarrassing information that B has received. Blushing is not, alas, an item in the repertoire under the control of A's central evaluator. We may describe the situation by saying that A is 'unintentionally communicating' his embarrassment to B; but this is a paradoxical way of putting it, for from A's standpoint the blushing is not something he is *doing*, but something that is *happening to* him despite the negative value attached to it.

It would seem to do full justice to the facts on both sides to say that here A is simply *exposing* his embarrassment to B (or still better, that B is perceiving A's embarrassment). The blushing is a *symptom*, not a *message*. The words we use for this distinction are less important than the distinction itself; the foregoing suggestion is only in line with our general plea for mercy towards the overworked term 'communication'.

(*d*) A is sailing with the international distress signal at his mast head. He is besieged by offers of help. He replies 'I wasn't communicating at all – it was the first flag I found in the locker and I thought it would look nice up there'. Is it useful to reply: 'Oh but of course you were communicating, even if you didn't know it'? Is this purely a matter of verbal usage?

Once again, we can agree that A was *conveying information* (in this case false information) and that he was flying a recognized *communication signal*. But if he is telling the truth we would be mistaken to regard the outcome as his *goal*; and what we would be mistaken about is the setting of A's 'I_G' – a genuine empirical matter of fact, and not an empty matter of words. An accurate way of putting it might be to say that A *seemed* to be communicating and was *perceived* as communicating, but in fact was not doing so. He was only exposing, not *using* the communication signal. In other words, the claim that A is *communicating* is a claim not only about the form and the results of A's action, but also about its goal; it implies something about A's internal organization. Experimentally, the latter may of course be hard to determine with any finality; but this does not make it any less essential to an understanding, as distinct from a mere description, of what is going on.

Can we then make more explicit what it means for A to be perceived by B as communicating? A full answer would take us deeply into the question of perception in general; but we can pick out some key features without going into too much detail.

Anything perceived by B must presumably find itself represented in the 'facts' department of B's organizing system. A's measly face, for example, will have its representation there whether or not B perceives A as *communicating* the fact that he has measles. What is required, then, over and above this representation, if B does so perceive A?

The answer in general terms is that B's organizing system must be capable of representing A as a *goal-directed agent*. For B to perceive A as 'parading' his spots requires B to have an internal representation not only of A's visible structure and movements, but also (at least implicitly) of his internal criteria of evaluation. It may also require a representation in B of the perceiver himself (B) as *the target of A*. Indeed, as Oldfield has emphasized in our group discussions, the first exchange in a genuinely communicative situation is often devoted to establishing the *perception of each as the target of the other*,* either by a process of experiment and observation or (as in radio signalling) by the use of call-signs in a recognized ritual. For our present purpose, however, it is enough to note that our functional–structural criteria suffice in principle to define the distinction (in the recipient) between perceiving an individual as communicating, and merely perceiving information transmitted from that individual.

To take a simple example, suppose that in the Boy Scout tent A, poor fellow, turns out to have sweaty feet. B's internal state of readiness is likely to be very different according to whether he perceives A as an unsuspecting sufferer or as one who knows his olfactory armament and has the aim of stimulating B with it. The mapping of this difference on to their respective flow-maps can be carried out directly along the lines we have discussed.

The point of analysis in these terms is essentially methodological. An information-flow structure of this sort is meant to serve as a tool of research. Because it is framed in terms that have correlates at both the physiological and behavioural levels, it is able to generate or at least to suggest experimental questions at both levels, and so to help the two to interact more fruitfully. For example, in infants and in lower animals there must presumably be a level of brain development below which a reciprocal flow structure of mutual perception in the foregoing sense cannot be set up; so the flow-map can be used to guide a search for particular kinds of evidence from the

* In a sufficiently symmetrical situation of 'mutual perception' this requirement for self-representation can set up a logical regress, with intriguing consequences for the kind of knowledge that each can (logically can) have of himself and of the other (MacKay, 1957, 1966).

functional anatomist and physiologist. The answers to the experimental questions should lead to a refinement of the flow-map; this in turn to fresh and more sharply formulated experimental questions for the next round; and so on.

8. 'FEEDBACK' BY NATURAL SELECTION

At this stage some enthusiast for borderline cases may raise an objection. Suppose that a particular genetically-determined feature of an animal, such as a red spot on a conspicuous part of its anatomy, is known to function as a signal to other members of the species in such a way as to favour the survival of the species. Surely here we have something produced under 'negative feedback', in the evolutionary sense that mutants deficient in this feature would breed less successfully. Can we not then say that the signal is goal-directed (towards survival) and hence 'communicative'?

What this objection brings out is the importance of specifying explicitly the *agent* who is supposed to be communicating. The essential element in the initiation of a communication, we noted, was that of *evaluation*. If A is communicating with B, then not only does some feature or action of A have the effect of a signal upon B, but also A must be set to evaluate that effect appropriately. It is of no use our inventing a metaphysical pseudo-agent called Evolution (with capital E) or Nature (with capital N) or even the Species, and attributing to it the evaluatory role; for even if this made philosophical sense, it would not make the individual A the communicator in the case. And if A is not the communicator, who then *is* communicating with B?

A persistent source of trouble here is the ambiguity between A-as-a-material-structure and A-as-a-goal-directed-system. The question 'What is A doing?' obviously has quite different answers in the two senses. As a material structure, A is reflecting light and so transmitting information in all directions, including incidentally that of B. As a goal-directed-system, A may not even have B represented within it, let alone as the target of its present action. Any plausibility we may feel in still saying 'A communicates with B' in this case results, I think, from taking 'A' and 'B' in the first (material) sense, equating 'communication' with 'transmission of information', and then jumping to the 'system' level at which 'A' and 'B' are agents.

By recognizing all this we are not of course ruling out the possibility that an animal might actively *use* a genetically determined feature (such as a red spot) as a means of communication – for example, by 'parading' it in the sense discussed earlier. But in this case, if it occurred, the relevant negative feedback would be generated within the animal's own evaluatory system; its goal would be that of influencing the other animal. Our recognition of this action as communicative would be independent of explanations of the mechanism (whether selective or otherwise) whereby the particular item came to be included in the animal's repertoire.

At the risk of labouring the obvious, let us look at a few more examples. A sandwich-board man walks up and down the sea-front, his boards carrying inflammatory slogans. A visitor from an alien land wants to understand this phenomenon. Is it an instance of communication? Yes, we say: this is one of the more bizarre channels of communication evolved in our society. So this man can be understood as *communicating* with those who read his boards? Wait a moment – it turns out that in fact he, too, is an alien. He neither speaks nor understands our language, and is simply performing for a pittance a task whose communicative function means nothing to him. To describe *him* as communicating, as if he were indistinguishable for scientific purposes from the passionate devotee of the cause parading his boards on the next beat, would be to frustrate rather than further the understanding of this social phenomenon. Merely to be the bearer of an acknowledged symbol of communication, even one with a successful communicative function, does not make one *ipso facto* the communicator of its message. In the case of the innocent sandwich-board man, the communicator – the evaluator of the function of the message – is elsewhere.

What then of the case in which a genetically determined action, rather than a passive structural feature, has an adaptive effect on the internal organization of an observer? Here the temptation may be strongest to describe the agent without more ado as 'communicating'. A male robin sings in such a way as to deter other males from entering a 'territory' (Thorpe, Chapter 6). Are we justified *ipso facto* in concluding that he does so *in order to* deter them? Consider a counter-example. At a College of Music, anyone looking for a vacant practice room will be deterred from entering one from which he can hear the sounds of voice or music. The occupant acts *in such a way as* to establish a territorial claim; but can we necessarily conclude that he emits the sounds *in order* to do so? Suppose what we heard outside the door were an extremely private soliloquy? Would it be sensible to describe the occupant as 'communicating a territorial claim' by his words?

Here again, the distinction we need is between the agent and the signal emitted. Elliptically and loosely, it may be justifiable to say '*the sound informs* (i.e. conveys information to) me that the room is occupied'. We can correctly take the sound as a *symptom* of territorial occupation, and even of the occupant's probable 'readiness to defend the territory against all rivals'. But in order to justify a description of *the occupant* as communicating a territorial claim by the sound (as he would by hanging a card on the door saying 'OCCUPIED'), we would need a great deal more evidence of a 'variational' kind (see Section 6).

I am not at all suggesting that evidence of this kind is lacking in the case of the robin. I only want to emphasize the complexity of the 'evidential debt' incurred when we describe any action as a communication by the agent to his neighbours *with a view to* its observed effect on them. In particular the claim one sometimes hears, that 'birds don't sing for any of the old-fashioned reasons: they are *really* doing it in order to establish territorial claims', incurs a weight of both scientific and philosophical obligation that I think it would be healthy to recognize explicitly.

A similar point applies to deceptive behaviour (including the setting of lures) and to what some ethologists refer to as 'accidental communication' – where a feature or action with a recognized adaptive signalling function (in terms of species survival) happens to trigger off the 'wrong' response in an animal of the same species, or evokes a non-adaptive response in one of a different species. In the last case especially it would be an interesting empirical question – which ought not to be foreclosed by the terminology we use – whether the originator of the signal was in fact communicating or trying to communicate: whether the anomalous outcome was negatively evaluated within his meta-organizing system, evoking signals of dissatisfaction, and so forth. There must be many non-committal terms (such as accidental signalling, or mis-triggering) that would serve to describe such phenomena accurately enough while leaving this question open.

9. CONCLUSION

In trying to understand what is going on in non-verbal communication, we are usually (for good or ill!) deprived of the possibility of asking the animal concerned what it thinks it is doing, or trying or intending to do. It is tempting to conclude (with some of the older 'stimulus–response' behaviourists) that the right scientific course is to abandon all such concepts, and to restrict ourselves to a 'neutral' reporting of correlations between items of observable behaviour.

From the brief survey we have made of the 'information–system approach', it would seem that any such restriction is both unnecessary and potentially obstructive of a full scientific insight into the processes of non-verbal communication. In order to build up a functional understanding of what goes on inside the originator and the recipient of non-verbal signals, as well as of what is exchanged between them, it is not necessary to wait for the day when (if ever) we can describe these at the level of full physiological detail. The level of information-flow analysis, even in its crudest terms, offers clear functional correlates for some of the concepts and distinctions of most importance in the causal explanation of communication between organisms.

Once this is appreciated, it thrusts upon the observer of non-verbal communication a whole set of questions that he might otherwise pass by or even regard as scientifically 'meaningless'. For example, a new-born baby cries *in such a way as* to get attention. Later on, it may learn to cry *in order to* get attention. What are the stages by which the second develops out of or on top of the first? What kinds of behavioural situation might be diagnostic of the presence and nature of evaluative feedback upon the action concerned? . . . and so on.

Or again, a bird carries plumage coloured *in such a way as* to release particular responses from another. What sort of evidence would we look for to discover whether, or when, it *uses* its plumage *in order to* release these? I do not pretend that the best answer should be easy to find or to implement; but an information-engineer's-eye view in terms of Figs. 1–4 at least translates the question into concrete variational terms that can immediately suggest relevant experiments, and disposes of any suspicion that the distinction is an empty one.

When we add to this the distinction between signals perceived by their recipient as communicatively goal-directed, and those not so perceived, this presents us at the outset with four basically different categories of situation covered by our title, which may be summarized in the following diagram.

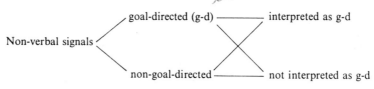

Situations of these four types may be expected to differ radically both in their dynamics and in the categories of scientific explanation that they will demand in order to be fully understood. It seems important that experiments on non-verbal 'communication' should be designed as far as possible to distinguish between them.

It should perhaps be emphasized that such conceptual distinctions are not at all of the kind that merely reflect 'gaps in our present knowledge'. On the contrary, the fuller our knowledge of the detailed mechanisms concerned, the easier it should be to make such distinctions operational and definitive in a given case.

The self-denying ordinance I have suggested regarding the use of the term 'communication' may be felt to be too strict by some of my colleagues; but I think that much would be gained and nothing lost if we were to say that *A communicates with B* only when *A*'s action is goal-directed towards *B* (in the sense of Section 6), and otherwise simply say that *B* perceives whatever he does about *A*, or that information flows from *A* to *B*, or whatever neutral expression says more precisely what we mean.

It is always tiresome to have to maintain distinctions, with their entail of technical terms. The argument of this chapter has been that although much of the mathematical apparatus of the communication engineer would be ill-adapted to the elucidation of non-verbal communication, an essential level at which to seek causal understanding is that of the information-flow structure of the participating organisms. At this level, the distinctions I have been labouring stand out sharply of their own accord; and the qualitative thinking of the communication engineer seems to have a natural application both in the design of relevant experiments and in the insightful interpretation of their results.

2. THE COMPARISON OF VOCAL COMMUNICATION IN ANIMALS AND MAN

W. H. THORPE

Most of those whose studies are represented in the first two sections of this book will have felt the need for some basis or scheme comparing the communicative abilities of animals and men. The most complete system so far developed for attempting such comparisons is that of Hockett (1960a, b). This scheme has been elaborated and developed over a number of years, not only by Hockett himself but also in collaboration with several of his colleagues. I have adapted this system to compare, on the basis of modern knowledge, animal 'languages' with those of men. The results are summarised in Table 1. The first column on the left indicates the 'design features', all of which are to be found in human verbal language. The other columns deal with the communicative abilities, as compared with the peak example of man, in a number of different groups of animals.

In addition to this, it is to be noted that the columns numbered 1 and 10 refer to other aspects of human communication. The first is designated 'human paralinguistics', with the implication that this is also exclusively human. A word is necessary at the outset as to the meaning of 'paralinguistics' (see Lyons, Chapter 3). Not all the sounds, and features of sound, produced by articulatory motions in the human being are part of language. As Hockett himself has pointed out (1960a: 393), the activity of speaking produces also, besides visible gestures, a variety of sound effects which may perhaps be conveniently termed vocal gestures. Neither the visible signals, perceivable at a distance in the form of gestures or attitudes, nor the sounds which sometimes may accompany these, are part of human language itself in the usually accepted sense. They are, rather, 'paralinguistic phenomena' and are of great importance when we are considering the relationship between animal and human communication. For this reason I have included human paralinguistics in the first column. Except for the last two columns, the table refers to all sensory modalities used by the species in question. In column 9 I have, for convenience sake, omitted consideration of vocal elements at all and so am merely speaking about the gestural or postural language of which these animals are capable. Column 10 includes the gestural ('deaf and dumb') language which the current work of Gardner and Gardner (1970, 1971) has shown to be within the powers of a chimpanzee which is daily exposed to the use of these gestures by human associates: this is discussed later in this chapter.

[27]

Table 1. *A comparison of the communication systems of animals and men based on the design features of Hockett*

Design Features (All of which are found in verbal human language)	1 Human paralin-guistics	2 Crickets, grass-hoppers	3 Honey bee dancing	4 Doves
1. Vocal–auditory channel	Yes (in part)	Auditory but non-vocal	No	Yes
2. Broadcast transmission and directional reception	Yes	Yes	Yes	Yes
3. Rapid fading	Yes	Yes	?	Yes
4. Interchangeability (adults can be both transmitters and receivers)	Largely yes	Partial	Partial	Yes
5. Complete feedback ('speaker' able to perceive everything relevant to his signal production)	Partial	Yes	No?	Yes
6. Specialisation (energy unimportant, trigger effect important)	Yes?	Yes?	?	Yes
7. Semanticity (association ties between signals and features in the world)	Yes?	No?	Yes	Yes (in part)
8. Arbitrariness (symbols abstract)	In part	?	No	Yes
9. Discreteness (repertoire discrete not continuous)	Largely no	Yes	No	Yes
10. Displacement (can refer to things remote in time and space)	In part	—	Yes	No
11. Openness (new messages easily coined)	Yes	No	Yes	No
12. Tradition (conventions passed on by teaching and learning)	Yes	Yes?	No?	No
13. Duality of patterning (signal elements meaningless, pattern combinations meaningful)	No	?	No	No
14. Prevarication (ability to lie or talk nonsense)	Yes	No	No	No
15. Reflectiveness (ability to communicate about the system itself)	No	No	No	No
16. Learnability (speaker of one language learns another)	Yes	No(?)	No(?)	No

5 Buntings, finches, thrushes, crows etc.	6 Mynah	7 Colony nesting sea birds	8 Primates (vocal)	9 Canidae non-vocal communica- tion	10 Primates– chimps, e.g. Washoe
Yes	Yes	Yes	Yes	No	No
Yes	Yes	Yes	Yes	Partly Yes	Partly Yes
Yes	Yes	Yes	Yes	No	No
Partial (Yes if same sex)	Yes	Partial	Yes	Yes	Yes
Yes	Yes	Yes	Yes	No	Yes
Yes	Yes	Yes	Yes	Yes	Yes
Yes	Yes	Yes	Yes	Yes	Yes
Yes	Yes	Yes	Yes	No	Yes
Yes	Yes	Yes	Partial	Partial	Partial
Time No Space Yes	Time No Space Yes	No	Yes	No	Yes
Yes	Yes	No?	Partial	No?	Yes?
Yes	Yes	In part?	No?	?	Yes
Yes	Yes	No?	Yes	Yes	Yes
No	No(?)	No	No	Yes	Yes
No	No	No	No	No	No
Yes (in part)	Yes	No	No?	No	Yes

Now let us consider one by one the design features themselves. It must be emphasised again that all these are shared by all human languages, but that each is lacking in one or another of the animal communication systems which have so far been studied.

1. VOCAL–AUDITORY CHANNEL

This design feature is perhaps the most obvious and hardly needs comment. It has two supremely important features: (*a*) that the production of sound requires very little physical energy, and (*b*) it leaves much of the body free for other activities that can then be carried on at the same time.

2. BROADCAST TRANSMISSION AND DIRECTIONAL RECEPTION

This feature and, similarly, design feature 3 which follows, are the almost inevitable result of the physics of sound. Nevertheless, they are so obvious that they can be easily overlooked and their significance, and the benefits derived from them, ignored or under-estimated. Broadcast transmission is, of course, inevitable with sound, but directional reception depends upon the structure and design of the auditory organ: this, in animals generally, and obviously in mammals and birds, is clearly an adaptive result of evolutionary development.

3. RAPID FADING

This indicates, of course, that the message does not linger for reception at the hearer's convenience, but quickly vanishes, leaving the communication system free for further messages. This is in strong contrast to such things as animal tracks and spoors, and also to the production of messages by means of those scent glands which secrete persistent chemical substances; likewise urine and excreta, which may persist, sometimes for a very long while.

4. INTERCHANGEABILITY

This is meant to imply that the adult members of any speech community are interchangeably transmitters and receivers of linguistic signals. In general, a speaker of a language can reproduce any linguistic message he can understand, whereas the mating displays of, for instance, the spiders, are usually confined to the male sex only and cannot be executed by the female who receives them. The same lack of interchangeability is often characteristic of those vocal displays which are known generally as 'song', female song being the exception rather than the rule in the birds. In contrast, however, directly we come to the 'call notes' of birds (which are used for maintaining contact and co-ordinating movements amongst flocks and families not in breeding condition) the signals may be interchangeable in Hockett's sense.

5. COMPLETE FEEDBACK

This simply means that the speaker hears everything relevant of what he himself says. The important fact that we ourselves, and presumably, most animals, can hear ourselves as others hear us but cannot see ourselves as others see us, is a feature often of great importance in the development of communication systems.

6. SPECIALISATION

Specialisation implies that the direct energy consequences of linguistic signals are biologically unimportant; only the response-eliciting consequences are significant.

7. SEMANTICITY

This refers to the fact that in man language is meaningful. One way in which meaning may be defined (and this is the way it is defined by Hockett) is in terms of the associative ties between signal elements and features in the outside world.* If meaning is defined in this way, it is obvious that many signalling systems other than language have the property of semanticity, as a glance at Table 1 will show.

8. ARBITRARINESS

This simply refers to the fact that the symbols are abstract in the sense that the relation between a meaningful element in a language and its denotation is independent of any physical or geometrical resemblance between the two. For instance, the word 'three' is in no sense triple nor is the word 'four' quadruple.

9. DISCRETENESS

This implies that the repertoire is discrete, not continuous; and that the possible messages in any language constitute a discrete repertoire rather than a continuous one. This distinction between discrete units and graded units can also be regarded as the difference between digital and analog communication. Human verbal language is characterised by its digital information coding system. Thus we may have two words which are

---

* Lyons (Chapter 3 below) cogently criticises this broad use of the term with reference to human language; indeed he goes so far as to doubt 'whether any single precise unequivocal interpretation can be given to "meaning" – and hence to semanticity'. Nevertheless it has decided advantages for comparative purposes, as in this chapter – as I think Lyons will agree.

acoustically very similar but with entirely different meanings and with no possibility of bridging the gap; e.g. the difference between the English words 'smock' and 'smog' is discontinuous. By contrast, the facial displays and the gestures of human beings and many animals often show continuous gradations between 'types'. Although the two ends of the continuum may carry different messages the cues by which we decide what is meant are not discrete but are graded – the information coding is analog.

10. DISPLACEMENT

Displacement implies that signals can refer to things remote in time and space. Human language can refer to things millions of light years distant and aeons removed in time. They can also refer to events here and now taking place in our heads. Here, the difference between human and animal languages is far from concrete and very often may be merely a matter of the length of the time or space gap to be bridged.

Obviously a blackbird (*Turdus merula*) can give its 'flying-predator' alarm call even though the 'danger' to which it is responding (e.g. a hawk in the sky) is still many hundreds of meters away in space and many seconds away in time. Similarly birds of many species (e.g. the domestic fowl), having been frightened by a fox or a stalking cat in the undergrowth, may continue, in a 'jittery' state, to utter alarm calls long after the actual danger has vanished. In other words its alarm calls are the effect of a past event and can be interpreted as such. The same applies, *mutatis mutandis*, to the conveyance of information by the 'dance language' of the Honey bee (*Apis mellifica*) (see Chapter 5).

11. OPENNESS

Openness implies that new linguistic messages are coined freely and easily and, in context, can be immediately understood. This is sometimes referred to as 'productivity' in the sense that a communicative system in which new messages can be coined and understood is open or productive. Obviously human communication is open in this sense; we can talk about things never talked about before. But so can a bee by means of its dances, since a worker may report a location which has never been reported before. In principle, openness is possible in any continuous, as opposed to a discrete, system of communication. It will become clear that some bird systems, as also primate gestural systems, can be open even though discrete.

12. TRADITION

This indicates that conventions can be passed on by teaching and learning from one group or generation to another.

13. DUALITY OF PATTERNING

This essentially implies that though the signal elements themselves may be meaningless, patterned combinations of them are meaningful. This is a feature of the sounds of some birds and also of the paralinguistic communications of animals such as the Primates. Thus a bird's song may, depending on the species, be made up of anything from half-a-dozen to several hundred 'notes' – which are the signal elements. Most of these 'notes' are quite meaningless if sounded alone; but grouped in the correct pattern of the song they can convey a great deal of information both as to the species and the individual involved. The summary of the work of Gardner and Gardner on teaching American Sign Language to a chimpanzee, at the end of this chapter, will provide many instances of duality of patterning in this learning situation.

Number 13 completes the design features originally elaborated by Hockett in his papers in the early 1960s. However, Hockett and Altmann (1968: 61–72) have recently given three more which are numbers 14–16 in my Table 1.

14. PREVARICATION

This connotes the ability to lie or talk nonsense with deliberate intent. It is highly characteristic of the human species and hardly found at all in animals. Possible exceptions occur in the play of some mammals and a few birds, where we see what appear to be gestures, feints, and ruses designed to mislead.

15. REFLECTIVENESS

Reflectiveness or reflexiveness simply indicates the ability to communicate about the communication system itself. This is undoubtedly peculiar to human speech and not found anywhere else, as far as we know, in the animal kingdom.

16. LEARNABILITY

This implies that the speaker of one language can learn another language. This is most obviously true for human beings; for there is no known human language which cannot be learnt by all normal members of the human race. It is much more difficult to decide how to evaluate this in animals. Certainly animals can learn new signals and the meaning of new signals; and learning plays a vital part in many communicative systems of birds and mammals. In birds, of course, 'imitativeness', i.e. the ability and tendency to imitate the sounds produced by congeneric associates (or foster parents, as in mynahs) without specific reward, is an outstandingly important feature. Thus it will be seen that we can give a fairly confident 'yes' to this question in columns 1, 5, 6 and 10, and a rather less confident 'no' in some of the other columns.

From what has been said it will be evident that, admirable as this outline scheme may be, it raises a number of problems and it is difficult – as the table shows – to decide what answer to give in some of the columns. Hockett and Altmann (1968) are themselves very conscious of this, emphasising two points of difficulty. The first is the one to which I have just referred – that each feature seems to be set forth in an all-or-none manner whereas, in fact, they are surely matters of degree (e.g. the difficulty with displacement referred to above). The second point is that the sixteen features tie into communicative behaviour in different ways. Some have to do with the channel, others with the repertoire of messages, some with the mechanisms by means of which the system is passed from generation to generation and group to group, and so on. Accordingly Hockett and his co-workers have suggested some superordinate groupings or 'frameworks' which suggest further things to be looked for in the communicative behaviour of any one species or society.

Framework A concerns features relating to the channel or channels and the points which are suitably inquired into in these circumstances. This will include the type and capacity of the emitting and receiving apparatus; the medium by which the signal is transmitted; the rate of travel and the rate of decay of such signals; the source of energy and the sources and kinds of channel noise; particulars about the directionality of the transmission, for example, is it broadcast or tight beam? Similarly the orientation of the receptive mechanism: whether, for instance, the receiver's posture has to be adjusted to achieve the best reception and how far distance is involved in determining the response, would suitably be inquired into here. This framework *A*, as will quickly be noted, is in fact particularly concerned with design features 1, 2 and 3 and, perhaps, also 5. If we were dealing with human communication here, under this heading we should ask questions such as 'Is it vocal–auditory?' For our present purposes it would be better to frame the question in a much more general fashion, namely 'What is its channel?'

Framework B is concerned with the social setting of the communication. The kinds of question suggested include the number and position of transmitters, for instance, the number and position of birds in the avian mobbing response; and the number and location of addressees, i.e. to a particular one or ones, or to all of the species in the neighbourhood. We should also want to inquire about differentiations resulting from age, sex, and, in some cases of mobbing for instance, even the species, for the alarm signals produced by many birds are understood by more than one species.

Framework C relates to the behavioural antecedents and to the consequences of communicative acts. Here the kind of questions that we ask are as follows: how far is the signal effective in itself in producing direct behavioural results in the receiver (as when the charge of a stag forcibly removes its rival), or does it produce a triggering effect? If the effect is trigger-like, is this due to an associative tie between the signal and other familiar things in the environment of the community? If such ties exist, do they rest on some similarity between the signal and what it stands for, or are they purely symbolic? Finally, having asked whether the signal causes a response even when its subject (e.g. food) cannot be seen or otherwise sensed by the receiver, we can go on to ask whether, in fact, the signal is ever used by the species in question to transmit an intentionally false message? (See design feature 14). In fact, this framework *C* may bring in everything that is known about the natural history and developmental state of the animals in question. It affects particularly design features 12, 14 and 16.

Framework D covers continuity and change in communication systems, raising points such as the variations from group to group and community to community with age and sex, and differences between successive generations.

Framework E comprises features of repertoire and of messages for a single system, asking, for example, is the repertoire open or closed, continuous or discrete? Are new messages sometimes transmitted, and if so are they received and interpreted appropriately? And do the messages of the repertoire differ from one another as total 'Gestalten' or are they built out of recurrent partials in some patterned way?

It would be quite out of the question to go into these five frameworks in any more formal detail here. They comprise, in fact, circumstances which should at once spring to the mind of any good ethologist or ecologist in considering the significance of, and the differences between, different methods of communication.

In considering the relations of animal communication systems to human language, special interest attaches to the complexity of communicative behaviour which can be achieved by chimpanzees. There have been, in the past, several attempts to train chimpanzees to use human language. It is not putting it too strongly to say that they have been monumental failures. The best known was that of Hayes and Hayes (1952, 1955) with the young chimpanzee 'Viki'. I was fortunate to be able to spend some days with Dr and Mrs Hayes and Viki in 1952, and have vivid memories

of their experiments and the immense care and the sophisticated psychological technique which they employed in their attempts to teach Viki to
talk. The result was that in six years Viki learned only four sounds that
approximated to English words, and even then the approximation was
somewhat tenuous. As Gardner and Gardner (1969) point out, human
speech sounds are unsuitable as a medium for communication for the
chimpanzee, whose vocal apparatus and vocal behaviour are different from
ours. These animals do make many different sounds but generally vocalisation occurs in situations of high excitement and tends to be specific to the
exciting situation. Undisturbed, chimpanzees are usually rather silent
when in captivity. It is consequently improbable that a chimpanzee could
be trained to make refined use of its vocalisations.

With this problem in mind, the Drs Gardner adopted an infant female
chimpanzee at an age between eight and fourteen months and proceeded
to attempt to teach this animal a gesture language known as the American
Sign Language (ASL). This is the language widely used in the United
States for communication between deaf human beings, and it is systematically taught to deaf children. It is entirely different from the deaf and dumb
language of Great Britain, which essentially is a method of spelling – an
alphabetical means of communication. Such systems of finger-spelling are
of course widely used by the deaf and dumb in conjunction with and to
supplement sign languages.

ASL is in fact composed of manually produced visual symbols, called
signs, which 'are strictly analogous to words as used in spoken languages'
(Gardner and Gardner, 1971). The Gardners point out that as words can
be analysed into phonemes, so signs can be analysed into what have been
called by Stokoe 'cheremes'. A system of fifty-five cheremes has been
devised for ASL. Nineteen of them identify the configuration of the hand
or hands making the sign, twelve the place where the sign is made and
twenty-four the action of the hand or hands. Thus the configuration of
pointing the hand yields one sign near the forehead, another near the cheek,
another near the chin, yet another near the shoulder and so forth. At any
given place the pointing hand signal yields one sign if moved towards the
signer, another if moved away, another if moved vertically and yet another
if moved horizontally, and so on. But if the 'tapered hand signal' is used
instead of the pointing hand a whole new family of signs is generated.
This summary shows that ASL satisfies the criteria for Hockett's design
feature 13 in that there are arbitrary, but stable, meaningless signal elements
and that these are arranged in a series of patterns which constitute minimum
meaningful combinations of those elements. The formal analysis of the
ASL language by linguists is still in a relatively elementary state. But it is

already clear that ASL has a syntactical structure of its own that is different from English. It will surprise non-linguists to learn that the complete linguistic analysis of spoken English is even yet incomplete, so as the Gardners point out, it will be a long time before a precise comparison of ASL with English in regard to syntax can be established. But while such precise comparison is not yet possible, it is nevertheless clear that ASL is a language in essentially the same sense that English is a language, and that if a chimpanzee can be taught ASL to such a degree that it can carry on some or most of the normal communicative activities of its life by that means, then that chimpanzee has learned a language. The process of teaching ASL to a chimpanzee has the further very great advantage that the animal's achievements can be compared directly with the achievements of normal young children in their acquisition of English and with deaf children of the same ages in their acquisition of ASL.

The basic situation employed was as follows. A young chimp (named 'Washoe') was kept in a room that contained most of the usual items of a modern human dwelling; that is to say the environment was as interesting as possible and the training programme was made an integral part of this environment throughout the waking hours of the ape. There was, however, no attempt to make the chimpanzee into a normal member of a human family. The work was planned so that during the whole of her waking hours Washoe was in the presence of one or more human companions with whom she took part in the routine activities of the day – feeding, bathing, dressing, etc., with whom she played games, examined new objects, was shown picture books and magazines, and so on. However the only form of verbal communication that was used in Washoe's presence was ASL: all her human companions had to master ASL sufficiently well for them not to need to use any other form of verbal communication, with the exception of finger-spelling for unusual or technical words. But apart from the prohibition on the use of spoken language, there was no rule of silence. Vocalisation was permitted if it was not verbal or if the sounds were sounds that Washoe could imitate, such as laughter and cries of pleasure and displeasure. Washoe's attention was obtained by clapping the hands and many other noises, and there was also music. Burglar alarms and boat horns were used to frighten Washoe away from forbidden places. After the first thirty-nine months of this work the results are briefly as follows.

One of the most widespread hypotheses invoked to explain the development of language in the human infant may be called, for convenience, the 'babbling hypothesis'. This assumes that the 'random' babbling and jabbering of the human baby comprises a great variety of sounds and so will include most if not all of the sounds which are employed in the language of whatever culture it happens to be reared in. Mothers often talk or croon to their children when attending to them, and so the sound of the mother's voice will become associated with comfort-giving measures. From this it is to be expected that when the child, alone and uncomfortable, hears his own voice, this will likewise have a consoling, comforting effect. So it is supposed (e.g. by Mowrer, 1950) that the infant will be rewarded for his own first babbling and jabbering without any necessary reference to the effects produced upon others. Before long, however, the infant will learn that if he succeeds in making the kind of sound his mother makes, he will get more interest, affection and attention in return; so the stage is set for the learning of language. There are, in fact, a good many difficulties in the way of this theory: perhaps the major one of these is the difficulty, if not impossibility, of characterising the kind of reinforcing event which could adequately account for the acquisition of language in such a learning situation. In fact, as will be made clear in other sections of this book, Chomsky (1957) has shown, with seemingly incontrovertible argument, that such a theory cannot possibly account for the basic events in the child's acquisition of language. Nevertheless at the commencement of the work with Washoe it was assumed that the infant chimpanzee would in fact show a great deal of 'manual babbling' in the form of random or accidental gestures which would come so close to the cheremes of ASL that they could be readily enhanced and shaped by the methods planned for the experiment. However, during the earliest months of the project the amount of manual babbling was very small. The ape would manipulate objects, poke at them with her fingers and so forth, but seldom engaged in the kind of manual play that had been expected – such as wiggling her fingers before her eyes or poking the fingers of one hand at the other hand. As time went on cheremic and incipiently cheremic gestures seemed to increase in frequency, and during this period the ape acquired several signs which she could use appropriately. Moreover she seemed to be paying more and more attention to the signs made by her human companions, and showed that she could comprehend the meaning of many of these signs. Manual babbling was encouraged by the experimenters as much as possible, and one sign which did appear to be learned largely in this way was the ASL 'word' for funny. Washoe was fond of touching her nose or her friends' noses with her index finger, and in the 'funny' sign the extended index and second fingers are brushed against the sides of the nose. Washoe herself introduced a variation which consisted of snorting as the nose was touched and gradually she came to make the 'funny' sign in 'funny' situations without any prompting. Although manual babbling did appear to play its part at this stage, it declined as time went on and its function in the total process of acquiring ASL seems to have been comparatively slight and rather doubtful. Particularly in the later stages of the learning, imitation was playing a greater and greater part in the process (a fact that in any case could have been anticipated by the striking powers of manual imitation shown by Viki). One of the signs which was clearly corrected by imitation was that for 'flower'. In the acceleration of ASL acquisition observational learning, as distinct from imitation, was almost certainly the prime factor. But Gardner and Gardner remark (1971), 'However, if Washoe could reach a point at which she asks us for the names of things, then imitative guidance might become the most practical method of introducing new signs'.

Some of the signs developed by Washoe seem to be best described as straight inventions. They are quite different from the signs which had, until then, been modelled for her, and their occurrence throws a particularly strong light upon the animal's mental processes. One example will suffice. The experimenters sometimes could not find an ASL equivalent for an English word which they wished to use. In such cases they would adapt a sign of ASL for the purpose: the sign for 'bib' was one of these cases. They happened to use the ASL sign for 'napkin' or 'wiper' to refer to bibs as well. This sign is made by touching the mouth region with an open hand and a wiping movement. During the eighteenth experimental month Washoe had begun to use this sign appropriately for bibs, but it was still unreliable. 'One evening at dinner time, a human companion was holding up a bib and asking Washoe to name it. She tried "Come–gimme" and "Please", but did not seem to be able to remember the "bib" sign that we had taught her. Then, she did something very interesting; with the index fingers of both hands she drew an outline of a bib on her chest – starting from behind her neck where the bib should be tied, moving the index fingers down along the outer edge of her chest, and bringing them together again just above her navel.' The authors remark, 'A high level of cognitive ability must be possessed by a creature that can represent the concept of a bib by drawing an outline on its chest with its fingers'.

A number of instances have recently come to light which demonstrate that many animals, at least including the birds and the mammals, must possess greater powers of conceptualisation than hitherto seemed credible. Among the pioneers in this field were Herrnstein and Loveland (1964), whose work showed quite dramatically that the domestic pigeon is capable of forming a broad and complex concept when placed in a situation that demands one. The pigeons used by these workers were in fact taught to respond to the presence or absence of human beings in photographs. They were trained to peck one disc if there was any sign of a human being in the photograph, and another if there was no sign. It was found that the most fragmentary, and presumably unfamiliar, aspects of the human being, or parts of the human being, were sufficient to cause the birds to give a positive response. (For similar recent work on monkeys, see Lehr, 1967.) Accordingly it was not surprising that the ape Washoe often showed this kind of ability. Many of the items for which Washoe learned the names and the appropriate ASL signals could of course be classified into meaningful groups. Thus there were six items, 'brush', 'clean soap', 'comb', 'oil', 'lotion' and 'toothbrush' that could be classified as grooming articles; five, namely 'bird', 'bug', 'cat', 'cow' and 'dog' which could be classified as animals; and five, 'banana', 'drink', 'fruit', 'meat' and 'sweet' that could be classified as foods, and so on. When Washoe made errors it was possible to show that these were governed by the category of item that was actually presented. Thus in one set of experiments in which Washoe made errors in twelve trials, seven of the twelve were the signs for other items in the appropriate category – in this case the category of grooming articles. Similar results were found for other categories of items. In fact these errors revealed just the kind of conceptualisation referred to above.

This was further displayed in Washoe's ability to learn the use of pronouns. When she was being taught, the experimenters often had to use a few specific instances of the appropriate referent: a few items of clothing for 'pants', a few examples of her companions' possessions for 'yours'. When Washoe transferred the sign to another referent or when she used it in a new combination she produced examples of the usage that went beyond the ones that had been taught to her. It was from these that it is possible to infer and evaluate her meaning. It would never have been possible to show Washoe all the persons she should designate as 'you', nor all the actions that 'you' was capable of performing. Yet it appeared that Washoe used the pronouns appropriately, that is, for any companion, and in combination with a wide variety of signs for action and for attributes. She began to use the two pronouns 'you' and 'me' in January, 1968. In spring, 1968, she had signed 'you me out' in a doorway situation and later produced many variants such as 'you Roger Washoe out', 'you me go out', 'you me go out hurry'.

At the time of the latest report Washoe does not have signs for words that can be used to join members of combinations – e.g. 'and', 'for', 'with', 'to'. But it is significant perhaps that words of this type are noticeably absent from the early sentences of young children (Brown and Bellugi, 1964; see also Brown, 1970).

According to the information available at the time of going to press Washoe, in the approximately three years since the experiment commenced, has reliably mastered the use of eighty-seven signs included in, or of a type similar to, those which make up ASL. Let us remind ourselves that in a comparable period the chimpanzee Viki, belonging to Dr and Mrs Hayes, acquired a vocabulary of four vocalizations, one of which was still rather doubtful. Nothing further needs saying to stress the impact and significance of this new approach to the understanding of the communicative abilities of the higher animals.

We must now come to what is in many respects the key question in this whole topic, namely the question of syntax. From what was said above it will be already clear that the popular notion, derivable from any dictionary, that syntax is a set of rules governing sentence construction, will not carry us very far at the present day. In fact Chomsky's *Syntactic Structures* (1957) was so to speak a signpost marking the direction of the new road. The problems raised by the advances and developments resulting from the work of Professor Chomsky have been discussed elsewhere in this book and need not be entered into here. But it is still necessary to inquire, as far as we can, whether and in what senses of the word the achievements of Washoe can be described as syntactic. Certainly Washoe combines words meaningfully. In the thirty-eight months following April 1967, the experimenters recorded 294 different two-sign combinations used by this animal. In arriving at this total different sequences of the same signs, e.g. 'open please' and 'please open', and again 'gimme food' and 'gimme food gimme', were tabulated as the same two-sign combinations. It was soon noticed that the vocabulary seemed to be divided into a small group of words called 'pivots', which are found in the bulk of the two-word utterances, and a larger group which are seldom found in two-word utterances unless they are paired with one of the pivots. All the 87 word-signs were scored according to the number of different two-sign combinations in which they appeared. It was found that at least one of the twelve highest ranking signs occurred in 240 of the 294 different two-sign combinations. These twelve word-signs, in descending order of their scores, were 'come gimme', 'please', 'you', 'go', 'me', 'hurry', 'more', 'up', 'open', 'out', 'in' and 'food'. As Gardner and Gardner point out, if the tabulation had been based on the total number of two-sign combinations that Washoe produced it would only indicate that she had some favourite signs and perhaps some favourite combinations. But in this tabulation each two-sign combination was counted only once no matter how many times it was used by Washoe, and moreover, reversals of order and repetitions of the same sign were not counted. This indicates that the sample is a sample of types, and the finding that at least one out of a small number of signs appeared in most of the combinations, indicates that certain types of combinations were more permissible (according to Washoe's own 'rules') than others. This fact suggests, to a non-specialist, the beginnings of an approach to 'syntax'.

Since Washoe's achievements, significant though they are, are so far removed from the achievements of a fully articulate adult human being, Gardner and Gardner very wisely proceeded by comparing Washoe's two-sign combinations with children's two-word combinations in a way that follows the scheme proposed by Brown (1970). First they listed the eighty-seven signs and grouped them into six categories. This analysis is shown in Table 2 herewith. It will be seen that the six categories, which are 'Appeal', 'Location', 'Action', 'Object', 'Agent' and 'Attribute', overlap to a considerable extent. But for Washoe it was found necessary to establish an additional category, namely the first one, 'Appeal', which does not appear in Brown's scheme for children's two-word combinations. As Brown suggests, further analysis may make it possible to fit these four signs, 'gimme', 'hurry', 'more' and 'please', into the other five categories.

A further point of interest appears. Apart from the pronouns 'me' and 'you', most readily combined signs are all to be found amongst the 'Appeals', 'Locations' and 'Actions'. This suggests that these more readily combined signs serve certain constructive functions. It is quite evident, from a study of the records of characteristic dialogues between Washoe and one or other of her human associates, that, if in response to an initial word from Washoe, for instance the sign 'please' or the sign 'come', the human partner asked the question 'What you want?', Washoe elaborates by adding a second word in the form of an answer; and this answer expresses considerably more than the initial one-sign utterance. Thus in response to these questions the commonly formed two-sign combinations that Washoe uses are as follows: 'please out', 'come open', 'more tickle'. Leading questions indicate what the human partners in this project, Washoe's linguistic community, would accept as completed ASL utterances, and it is clear that she gives them.

Just because infant human beings and infant chimpanzees cannot act as 'informants' in the way that native speakers serve as informants when an unknown human language is being studied, lists of the attempts so far made to analyse children's early language for structure have used the utterances themselves as raw data. And this must obviously be the method employed in the study of the communications between Washoe and her associates. Table 3 compares Washoe's achievements with the structure of the earliest two-word combinations of children. The essence of the scheme for children is the notion that the conjunction of two words can express a relationship that is independent of the particular words in the

Table 2. *Washoe's signs grouped into six overlapping categories* (from Gardner and Gardner, 1971)

Appeal	Action	Object	Agent	Attribute
gimme	in come bed look	baby cheese hurt spoon	me	black
hurry	out go brush oil	banana climb key string	you	green
more	up help catch tooth-brush	berry clothes leaf sweet	Washoe	red
please	*Location* there down			
	hug clean ride	bib cow light tree	Dr G.	white
	kiss comb	bird dog meat window	Mrs G.	enough
	open cover	book flower pants	Greg	funny
	peek-aboo dirty	bug fruit pencil	Naomi	good
	spin drink	car grass shoes	Roger	quiet
	tickle food	cat hammer smell	Susan	sorry
	listen	chair hat smoke	Wende	mine

Object

Table 3. *Parallel descriptive schemes for the earliest combinations of children and Washoe* (from Gardner and Gardner, 1971)

Brown's (1970) scheme for children		The scheme for Washoe	
Types	Examples	Types	Examples
Attributive Ad+N	big train, red book	*Object–Attributable* *Agent–Attribute*	drink red, comb black Washoe sorry, Naomi good
Possessive: N+N	Adam checker, mommy lunch	*Agent–Object* *Object–Attribute*	clothes Mrs G., you hat baby mine, clothes yours
N+V	walk street, go store	*Action–Location*	go in, look out
Locative N+N	sweater chair, book table	*Action–Object* *Object–Location*	go flower, pants tickle baby down, in hat
Agent–Action: N+V	Adam put, Eve read	*Agent–Action*	Roger tickle, you drink
Action–Object: V+N	put book, hit ball	*Action–Object*	tickle Washoe, open blanket
Agent–Object: N+N	mommy sock, mommy lunch	*Appeal–Action* *Appeal–Object*	please tickle, hug hurry gimme flower, more fruit

pair. The comparison set forth in Table 3 will be discussed in detail by B. T. and R. A. Gardner (1971), and readers who wish for a more detailed discussion of the problems and difficulties it raises, must seek it there.

It seems safe, however, to say that Brown's scheme shows that the types of construction which can be distinguished in the earliest combinations of young children do express relationships that are somewhat independent of the specific words that appear in the combinations. That is to say, to some extent children's utterances are characterised by structure at this level of development. From the data available to Gardner and Gardner at the time of writing it is estimated that between 70 and 85 per cent of the two-word combinations in the children's samples can be accounted for by the Brown scheme. In the sample of Washoe's 294 two-sign combinations 228, or 78 per cent, can be accounted for by the scheme that the Gardners have derived from Brown's schemes. 'At this level of analysis, then, Washoe's earliest two-sign combinations were comparable to the earliest two-word combinations of children.' (Gardner and Gardner, 1971.) There remains the question whether this structure represents the first emergence of syntax or whether it is a stage of semantic structure that precedes the first stage of truly syntactical construction. The present view of the Gardners is that this question remains open.

We may now turn to another recent series of experiments with a chimpanzee using a quite different technique. It will be obvious that the ability to develop an association between two objects or events in contiguity is part of the basic learning repertoire of every animal that can at least establish a conditioned response. Every dog can learn to associate a sound or a gesture with an object which it desires or fears. Premack (1970*a*, *b*) in a remarkable series of experiments with another young female chimpanzee 'Sarah', has developed an ingenious technique for teaching language by means of plastic 'words'. Each 'word' is a metal-backed piece of plastic of a given and unique combination of shape, size, texture and colour which adheres lightly to a magnetised slate. Association between the plastic 'word' and the object (e.g. an apple) which it is to represent, is readily taught to the animal by simple reward. From this simple beginning Premack, by an elegant series of steps, has been able to teach Sarah to 'speak', by tokens, to a degree not inferior to Washoe and in some respects superior. The work has not yet been reported in sufficient detail to enable a full comparative assessment to be made, nor is it possible here to describe Premack's method in detail. But not only does Sarah's vocabulary now exceed that of Washoe, her use of it is in many respects more sophisticated and complex. Thus Sarah has a vocabulary of 112 plastic 'words' which comprise eight names of persons (or chimps), twenty-one verbs, six colours, twenty-one foods, twenty-six miscellaneous objects and thirty 'concepts, adjectives and adverbs'.

Exact comparison with Washoe is of course misleading in that the 'words' are made for Sarah; she does not have to learn to make them herself. Nor does she have the free choice of actions open to Washoe; she can manipulate what is given to her but not invent or choose others. Nevertheless, the work with Sarah gives even stronger suggestions of syntax than does that with Washoe. Premack says 'We feel justified in concluding that Sarah can understand some symmetrical and hierarchical sentence structures, and is therefore competent to some degree in the sentence function of language'.

Another achievement of Sarah that is worthy of mention is her ability to answer questions about her classification of objects. It is possible for instance to ask her '*A* is what to *A*?' or '*A* is what to *B*?'. Two object words are put on the board with the question symbol between them, Sarah's task is to replace the question mark with the right word 'same' or 'different'. A second question may be translated '*A* is the same as what?' or '*A* is not the same as what?'

Finally Sarah can be asked a variety of yes–no questions thus: 'Is *A* the same as *A*?', 'Is *A* not the same as *A*?' and so on. Sarah, it is claimed, learned to answer all three types of questions for an essentially unlimited variety of items, words and concepts. There are indeed objections which can be made as to how far these are true questions; this cannot be discussed here but there is interesting evidence that the chimp thinks of the word not as its literal form (say, a piece of blue plastic) but as the thing it represents (e.g. a red apple).

Finally we come to the thorny question of language and its relation to speech. The general colloquial use of the term 'language' is so vague as to be of little value. But there has been for many years a widespread tendency both among students of animal behaviour and a good many physiological psychologists (e.g. Teuber, 1967) to use the term language for many of the more elaborate examples of communication amongst animals; especially for the transfer of social information and particularly when the transfer is vocal or auditory. While there is much to be said for this, the usage in the past has perhaps been too naive (Thorpe, 1968). Hebb and Thompson (1954) proposed that the minimal criterion of language is twofold. First, language combines two or more representative gestures or noises purposefully for a single effect, and secondly it uses the same gestures in different combinations for different effects, changing readily with circumstances. It will be quite obvious from this book that not only the communication of Washoe with her associates but also the communication of many birds and some other animals comes within this definition of language. There remains, however, the problem of intent or purpose, a topic which is discussed elsewhere in the present book (e.g. Chapter 1 and the postscript to Part A). To express my own view, and I do not wish to in any way saddle the Drs Gardner with it, I would say that no one who has worked for a long period with a higher animal such as a chimpanzee, particularly in the circumstances of the Gardners' work, is justified in doubting the purposiveness of such communication. I believe such purposiveness is also clear to the experienced and open-minded observer with many of the Canidae, with some, probably many, other mammals, and with certain birds (Thorpe, 1966, 1969). Some of the philosophical depth and significance of this question of purposiveness, both for students of animal behaviour and for philosophers such as Price and Whitehead, will be found in Pantin (1968).

It would no doubt be easy to devise definitions of language such that no examples of animal communication could readily find inclusion therein. There have always been, and no doubt there will continue to be, those who resist with great vigour any conclusions which seem to break down what they regard as one of the most important lines of demarcation between animals and men. We must surely be justified in accepting such preconceived definitions only with the utmost caution. One of the tasks of the scientific student of animal behaviour is to attempt to establish whether or not there are such hard and fast dividing lines and if so what and where they are. Of one thing we can be certain: it is that work such as that of the Gardners and of Premack is only the beginning of the application of an important and powerful new technique from which we stand to learn much in years to come. I believe that no one should have anything to fear from its cautious and objective application.

3. HUMAN LANGUAGE

JOHN LYONS
Department of Linguistics, University of Edinburgh

I. INTRODUCTORY

Many people would say that the epithet 'human' in the title of this chapter
is redundant, on the grounds that the capacity for language, properly so
called, is unique to man; and this is probably the view of most linguists.[1]
Whether anything that might be properly called language is found in other
species is a central question which has been discussed by Thorpe in relation
to Hockett's 'design features' in the previous chapter, and I will return to it
in the final section. Meanwhile, since I will be concerned with human
communication by means of language and will be making only cursory and
indirect reference to communication in animals on the one hand, or to
non-linguistic communication in man on the other, I will henceforth use
the term 'language' without qualification to refer to 'human communication
by means of language'.

The purpose of this chapter is to give some account of the concepts and
techniques that have been developed by linguists for the analysis and
discussion of language. Since the account is intended primarily for students
of non-verbal communication, my selection of topics, and the emphasis I
have chosen to give them, has been determined by my judgement of their
relevance to the investigation of systems of animal communication and of
non-verbal systems in man. The reader will note that I have deliberately
raised in his mind the suspicion that others might judge the relationship
between verbal and non-verbal communication somewhat differently. In
the present state of linguistic theory, it would be misleading for me to
suggest that I am presenting *the* linguist's, rather than merely *a* linguist's
point of view. I shall of course endeavour to give as balanced and as
objective an evaluation as I can: but the reader should be warned at
the outset that much of what I have to say is to a greater or less degree
controversial.

The first point I wish to make is that 'language', in my usage at least,
is not synonymous with 'verbal communication'. As I will explain later,
the term 'language' is itself ambiguous; and I will distinguish the two
senses by means of the terms *language-behaviour* and *language-system*. I
will also suggest that many of the important functions of language (and
not merely of its non-verbal component) are not in fact communicative, in
the strict sense of this term.

4   [49]

But first I wish to distinguish the 'verbal' from the 'non-verbal' components of language. Unfortunately the term 'verbal' is not used at all by most linguists in anything like the required sense (i.e. as relating to words, rather than as relating to verbs). Outside linguistics it has been used in a variety of different senses. For example, Skinner (1957) uses the term 'verbal behavior', in the title of his book and throughout, to include not only the whole of language, but much else besides (in fact anything that can be described as 'behavior reinforced through the mediation of other persons' needs'). But this is an extremely idiosyncratic use of the term 'verbal' and, once noted, is not likely to lead to confusion. More serious is the inconsistency and imprecision with which the distinction between 'verbal' and 'non-verbal' has been drawn in relation to the different components or aspects of language. The same inconsistency and imprecision, it may be added, attaches to the term 'paralinguistic', which we shall be distinguishing from 'non-verbal'; and also to 'vocal'. For a comprehensive discussion of the relevant literature from a linguistic point of view, the reader is referred to Crystal (1969), whose terminology I will myself for the most part adopt.

We may begin by distinguishing between *vocal* and *non-vocal* signals, according to whether the signals are transmitted in the vocal–auditory channel or not. The vocal–auditory 'channel' is here defined, it will be observed, in terms of its two end-points and the manner and mechanisms by means of which the signals are produced at the source and received at the destination, rather than simply in terms of the properties of the channel itself, which links the terminals and along which the signal travels. This point in itself is worth noting, since there are alternative definitions to be found in the literature; and the choice between them may imply, whether this is explicitly avowed or not by the author, a set of assumptions and presuppositions about the whole field of human signalling. By making our primary classification one of vocal *vs.* non-vocal signals (rather than for example acoustic *vs.* non-acoustic, or auditory *vs.* non-auditory signals) we are deliberately taking the view that the so-called speech-organs enjoy a position of pre-eminence among the signal-transmitting systems employed by human beings. It is because the vocal–auditory channel serves for the transmission of language and because, by common consent, language is the most important and most highly developed signalling system employed by human beings, that we start by distinguishing vocal from non-vocal signals. (I am for the present concerned solely with speech, which we will assume, with most linguists, is in some sense more 'basic' or more 'natural' than written language. See Section 3 below.) Among the non-vocal systems used by man there are the 'paralinguistic' systems of head-nods, gestures, eye-movements, etc. of the kind described by Argyle in Chapter 9; and these play an important role in the 'punctuation' and 'modulation' of speech (see below).

Not all vocal signals are linguistic. First of all, we must exclude what Crystal (1969: 99, 131) calls *vocal reflexes*: sneezing, yawning, coughing, snoring, etc. Usually they are physiologically determined; and, although they are signals, in the sense that they are transmitted (for the most part involuntarily) and can be interpreted by the receiver, no-one would wish to regard them as being other than external to language. When they occur 'normally' (i.e. as physiological reflexes) during speech, they merely introduce noise into the channel; and when, by prior individual or cultural convention, they are deliberately produced 'abnormally' for the purpose of communication (when, for example, we cough to warn a speaker that he might be overheard by someone approaching), they operate outside and independently of language.

Rather more debatable, from a linguistic point of view, is *voice-quality* (also called *voice-set*) by which term is meant 'the permanent background vocal invariable for an individual's speech' (Crystal, 1969: 163; cf. Abercrombie, 1967: 91; Laver, 1968). Unlike the vocal reflexes, voice-quality is a necessary concomitant of speaking. Furthermore, it plays an important role in signalling the identity of the speaker, both as a particular individual to those who know him, and more generally, as having certain characteristics which may correlate with membership of particular social groups within the community (being of a certain age, of a particular sex, of a certain physical build and personality, etc.). Voice-quality, which may have both a physiological and a cultural component, is very relevant to the phenomenon known as 'self-presentation' (see Chapter 9). Whether we call voice-quality a part of language or not is a matter of dispute. I will here follow the majority of linguists in describing it as 'non-linguistic', without thereby wishing to suggest that it should not be studied by linguists or regarded as irrelevant to the investigation of 'language proper'.

I am using 'voice-quality' in a sense which will allow us, when necessary, to distinguish between 'individual-identifying voice-quality' and 'group-identifying voice-quality'. One might also wish to recognise, as distinct from these two, 'personality-identifying voice-quality'. The speaker's own distinctive (i.e. 'personal') voice-quality, as the (normally) invariant 'background' against which linguistic variation is evaluated, is a complex of all three. The terms 'individual-identifying', 'group-identifying' and perhaps also 'personality-identifying', it should be noted, need not be restricted to voice-quality, but may be applied, where appropriate, to paralinguistic, prosodic and verbal variation. The whole area, however, is in need of more than just terminological clarification. Not only must we be careful to distinguish between 'real' and 'stereotyped' qualities (cf. Crystal, 1971; Laver, 1968), but we must give recognition to the fact that, for certain listeners, a particular feature may be merely individual-identifying, whereas for the linguist describing the language, and for other listeners of wider experience, it will be group-identifying (and simultaneously, of course, individual-identifying, if representatives of the group, or conjunction of groups, in question are rare in the community in which the speaker is at present living).[2] For all of these source-identifying features (which have an obvious parallel in animal signalling, as Cullen and Thorpe make clear in their chapters in Part B) I will follow Abercrombie (1967) and Laver (1968) in using the term *indexical* (see below Section 6): they *indicate* or are indicative of (or identify) their source.

We now come to the distinction of *verbal*, *prosodic* and *paralinguistic* features or components. (What is meant by prosodic and paralinguistic will be explained presently.) It will be convenient to recognise two separate dichotomies; and for this purpose, I will use the terms *linguistic* (subsuming verbal and prosodic) and *non-segmental* (subsuming prosodic and paralinguistic). For linguistic and paralinguistic features taken together I will introduce the term *locutional*. Note that I use the term 'segment' for both phonological units (phonemes) and the grammatical units (words). My reason for wishing to establish these two different dichotomies is that, from one point of view, the verbal and the prosodic components go together; they would definitely be regarded as part of language by almost all linguists, whereas the situation with respect to paralinguistic features is far less clear. From another point of view, however, prosodic and paralinguistic features go together: they are 'superimposed', as it were, upon the segments (phonemes, syllables, words, etc.) which constitute the verbal component of the utterance.

We cannot go into all the details of the various prosodic and paralinguistic features of utterance. As instances of *prosodic* features I will mention just two: *intonation* and *stress.* Every normal English utterance is produced with a particular intonation 'pattern' or 'contour', which is determined partly by the grammatical structure of the utterance and partly by the 'attitude' of the speaker (as dubious, ironical, surprised, etc.). If two utterances of the same sequence of words differ with respect to their intonation they would probably be described by linguists as two 'different' utterances, rather than as two instances of the 'same' utterance (i.e. they would be regarded as being different in type, rather than as being tokens of the same type: see Section 2). In an English utterance each word will be pronounced with a certain degree of stress (or 'emphasis') according to its grammatical function and a variety of other factors including the contextual presuppositions of the utterance, the 'attitude' of the speaker, and so on. For example, if I stress *seen* in *I haven't seen her* (in reply, let us say, to *Have you seen Mary?*) it might be taken to imply that, although I haven't seen her, I have some news of her. By contrast, if I stress *Mary* it will imply that, although I haven't seen Mary, I have seen someone else. Again, if two utterances of the same sequence of words differ in stress they will probably be described as two different utterances. Intonation and stress (as well as certain other prosodic features I have not mentioned) are almost universally regarded as being part of language.[3] They are non-verbal: they do not identify or form part of the words of which the utterance is composed.[4] And yet they are an essential part of what are commonly referred to as 'verbal signals'. If I have laboured this point unduly, it is because it is not clear to me whether the term 'non-verbal communication', as it is used by many authors, is intended to include the essential linguistic non-verbal component in verbal communication or not. Terms like 'verbal language', which occur in the literature of semiotics, psychology and communication theory, are equally obscure. One obvious source of confusion is the ambiguity of 'verbal' with respect to duality of patterning (see Section 4): phonemes are 'verbal' in the sense that they identify and form part of words, whereas words, phrases, etc. are 'verbal' in the sense that they are composed of words.

The term *paralinguistics* is particularly troublesome. As Crystal (1969: 140) says: 'There is substantial disagreement . . . in the literature, and the tendency has been to broaden its sense to a point where it becomes almost useless.' Crystal himself describes paralinguistic features as 'combinations of physiologically grounded parameters with pitch, loudness, duration and silence being variable in relation to their identification (for example, different degrees of loudness distinguish "ordinary" from "stage" whisper)'. This restricts the term to features of vocal signals. However, a case can be made for applying it, as certain authors do (e.g. Abercrombie, 1968), to those gestures, facial expressions, eye-movements, etc. which play a 'supporting' role in normal communication by means of spoken language (cf. Chapter 8). The important point about paralinguistic features is that they differ from prosodic features in not being so closely integrated with the grammatical structure of utterances.[5]

I have been at some pains in this section to emphasise the point that there is room for considerable disagreement as to where the boundary should be drawn between language and non-language. We can of course decide, as a matter of methodological fiat, that the boundary should be drawn as I have drawn it above; and this is where most linguists do in fact draw it (although they might differ as to what vocal features apart from stress and intonation should count as prosodic, and therefore as linguistic, and what as paralinguistic). The fact that there is such a complete and intimate interpenetration of language and non-language should always be borne in mind in considering the relationship between verbal and non-verbal communication.

Having established this point and having given it due emphasis, I can now attempt to make certain generalisations about language on the basis of our dichotomous classification of the various components. First, as Crystal (1969: 128 ff.) puts it (in a somewhat different sense of the term 'linguistic'), there is a scale of 'linguisticness' within what I am calling the locutional component of language, such that the verbal component is the 'most linguistic', prosodic features are more 'linguistic' than paralinguistic features, and within the set of prosodic features some, like stress and intonation, are more 'linguistic' than others; and paralinguistic features are more 'linguistic' than the non-linguistic components, voice quality and vocal reflexes. By 'more linguistic', in this sense, we mean 'taken as more central by the linguist'. The interesting question is why the linguist should evaluate 'centrality' in the way that he does. And the answer would seem to be that the more 'central' a particular component is the more specific it is to human language by contrast with the signalling systems of other species and human signalling systems other than language.

It is noteworthy in this connexion that the sixteen 'design features' of language identified by Hockett (1960, etc.) and discussed by Thorpe in the previous chapter are all incontrovertibly applicable to the verbal component. Fewer of them apply to the prosodic features; yet fewer to the paralinguistic features; and so on. For example, intonation almost certainly lacks the property of 'displacement' in the sense of being able to make reference to 'things remote in space and time'. Whether intonation has the property of 'reflectiveness' (the 'ability to communicate about the system itself') is more debatable, since the repetition by the listener of some part of what the speaker has just said with a different and distinctive intonation (and stress) may have the effect of querying, not what has been said, but the appropriateness of the words chosen to describe it. Whether intonation is characterised by 'duality of patterning' and the degree to which it is 'arbitrary' and 'discrete' – these are matters of dispute among linguists (a fact that is perhaps not sufficiently reflected in Hockett's own discussion of language). I will give more detailed consideration below to what I consider to be the more important 'design features' of language. Here let us be content with the generalisation that the more 'linguistic' a component is, the more of the 'design features' of language it possesses; and the most 'linguistic', from this point of view, is the verbal component. This, then, is the first generalisation.

The second has to do more particularly with the communicative function of the different components of language. Argyle suggests in his chapter below, not only that non-verbal communication sustains and supports verbal communication (a point to which I have referred earlier and would here reiterate) but also that non-verbal and verbal communication 'normally play two contrasted roles', the former 'to manage the immediate social relationship – as in animals', the latter to convey factual information, give orders and instructions, etc. If 'non-verbal and verbal communication' is interpreted as 'the less linguistic and the more linguistic components of human signalling' respectively, Argyle's generalisation would seem to apply satisfactorily to language. It would, however, be erroneous to suggest that a very sharp distinction can be drawn between the two roles. After all, the choice of one word rather than another might be as much determined by the desire to placate (or antagonise) the listener as it is by the desire to give him a particular piece of information. The two roles, or functions, of language would seem to interpenetrate as intimately as do the verbal and the non-verbal components; and they are complementary, rather than contrasting. The most we can say, perhaps, is that the verbal component is more closely associated with the 'cognitive' and the non-verbal with the 'attitudinal' or 'social' function of language (see Section 6).

2. THE NATURE OF LINGUISTIC THEORY

It is not generally realised by non-linguists how indirect is the relationship between observed or observable utterances and the description that the linguist makes of these utterances. This is a complex and controversial topic; and naturally enough, linguists tend not to dwell upon it in the less technical presentations of their subject. However, in the present context it seems to me essential that we should go at least some little way into the question. For the very fact that a set of rather complicated terminological and conceptual distinctions have been found necessary, or at least useful, in the description of language and yet have not been drawn, as far as I know, in the investigation of other kinds of human and animal signalling suggests that there may be an important difference between language and these other kinds of signalling. Alternatively, it might be the case that the same distinctions should be drawn in the investigation of other kinds of signalling.

I will begin by developing the distinction I hinted at earlier between *language-behaviour* and the *language-system* which underlies it. When we say that someone is speaking a particular language, English for example, we imply that he is engaged in some kind of behaviour, or activity, in the course of which he produces vocal signals, as well as various non-vocal paralinguistic signals which punctuate and modulate the vocal signals (as explained in the previous section). Under the term 'language-behaviour' I include both the vocal signals and the associated non-vocal paralinguistic signals. The vocal signals we will call *utterances*. The term 'utterance' is ambiguous in that it may refer to a piece of behaviour, an 'act of uttering', as well as to the vocal signal which is a product of that behaviour. These two senses may be distinguished, when this is necessary, by means of the terms 'utterance-act' and 'utterance signal'. So far I have used the term 'utterance' in the sense of 'utterance signal'; and this is the sense in which it seems to be most commonly used by linguists. (In passing, it may be worth observing that this distinction between the activity and the signal that is produced by that activity does not appear to be so clear in the case of certain systems of non-verbal signalling.) Both the speaker's behaviour and the utterances he produces are observable and can be described in purely physical terms. So too can the totality of the *speech-event* of which the utterance is a part. All these terms – 'language-behaviour', 'utterance', 'speech-event' – belong to the pre-theoretical, *observational* vocabulary of linguistics. They are terms that the linguist can employ to talk about the 'raw data' prior to and independently of its description within a particular theoretical framework.[6]

But the linguist also has another set of terms at his disposal, which, as we shall see, relate less directly to the primary data. These we will call *theoretical* terms, since, in contrast with the observational terms, their definition and interpretation is fixed within a particular linguistic theory. The necessity of distinguishing clearly between the observational and the theoretical vocabulary of linguistics has been emphasised, in a number of publications, by Bar-Hillel (especially 1957, 1967, 1969).[7]

The linguist is not generally concerned with utterances as unique physical events. He is interested in *types*, not in *tokens*, and the identification of utterance-tokens as instances of the same utterance-type raises important theoretical questions. When we say that two utterances produced on different occasions are instances of the 'same' utterance (that they are two tokens of the same type), we are implying that they have some structural or functional identity in terms of which native speakers will recognise their 'sameness'. One might expect that it should be possible to account for their structural identity in terms of an acoustic analysis of the two signals; and to account for their functional identity (i.e. what 'meaning' they have or what communicative purpose they serve) in terms of a purely behavioural analysis of the two acts of utterance, as psychologists and linguists have occasionally suggested. Without going any further into the details, let me say that linguists would now seem to be agreed that the token–type identification of utterances cannot be carried out in terms solely of what one might call 'external' or purely observational criteria. Two utterance-tokens might differ quite grossly acoustically and yet count as structurally identical for the native speaker, and they will be described as such by the linguist.

Nor does it seem possible to identify utterances as tokens of the same type, as far as the vast majority of utterances are concerned at least, in terms of their functional identity as 'responses' to the same 'stimuli' (when the 'stimuli' have been independently and appropriately grouped together as tokens of a particular type). We can of course postulate, if we wish, that every utterance is determined by the internal psychological or physiological state of the speaker, such that two utterance-tokens are identical in type if they are the (observable) effects of the same (unobservable) internal state.[8] This may be a useful and productive hypothesis for the investigation of the psychological and physiological mechanisms involved in speech-production. But we cannot invoke it as an explanation of token-type identity nor can we use it to derive a set of empirical procedures for the establishment of token-type identity, since we have no independent knowledge of the internal state of the speaker. In the present state of research at least, it would appear that the relationship between the psychological or physiological hypothesis I have referred to, and linguistic description, is the converse of what it is commonly supposed to be by those who advocate a purely observational account of language-behaviour. The prior establishment of token–type identity by the linguist serves as evidence for the psychologist or physiologist in his investigation of the 'black box' of which utterances are the observable output. Chomsky, in his famous review of Skinner's theory of operant conditioning and elsewhere (1968), has perhaps exaggerated the degree to which language-behaviour is 'stimulus-free'.[9] But it does seem clear that any account of language-behaviour that rests content with correlating features of utterances with features of the external and observable environment or with unobservable, but postulated, features of the internal state of the speakers is, at present at least, incapable of accounting for the structure and function of the majority of utterances.

What the linguist does when he describes a language, English for example, is to construct a model, not of actual language-behaviour, but of the system of regularities which underlie that behaviour (more precisely of that part of language-behaviour which the linguist defines by methodological decision to be 'linguistic', rather than 'non-linguistic', along the lines suggested in the previous section) – a model of what I am calling the *language-system.* When we say that someone 'speaks English', or 'can speak English' we imply that he has acquired (normally as a young child in the process of first-language acquisition) the mastery of a system of rules or regularities underlying that kind of language-behaviour which we call 'speaking English': to use Chomsky's terms (1965), he has acquired a certain *competence*, and it is this which makes possible, and is manifest in, his *performance* (or language behaviour).

By the 'structural and functional identity' of utterance-tokens the linguist will mean 'having the same phonological, grammatical and semantic structure'. We have just touched on two of the important 'design features' of language that I will discuss in more detail later: duality of patterning (having both a phonological and a grammatical structure) and semanticity (having 'meaning'). Here it will be sufficient to say that by 'having the same grammatical structure' is meant 'being composed of the same words in the same relationship to one another'. The important point to note at this stage is that *phonological*, *grammatical* and *semantic* (as well as *word* which, as we shall see, is especially complex) are theoretical terms: they refer to units and relationships defined within the linguist's model of the language-system. In most models of linguistic description, including Chomsky's highly formalised model of generative grammar, the main unit of linguistic description is the *sentence*; and I will now attempt to suggest how utterances are related to sentences.

We can usefully distinguish three stages of 'idealisation' (not necessarily successive, of course) in our description of the 'raw data'; and it is at the third of these that we bridge the gap between utterances and sentences. The first two stages are involved also in the token–type identification of utterances.

First of all, we discount all 'slips of the tongue', mispronunciations, hesitation pauses, stammering, stuttering, etc; in short, everything that can be described as a 'performance phenomenon'. At the first stage of idealisation, Chomsky's distinction of 'competence' and 'performance' is helpful, and, as far as I am aware, is not seriously questioned by any linguists. There are of course practical difficulties involved in identifying errors, but the principle is not in doubt. It may be worth mentioning in passing that performance errors and hesitation phenomena of various kinds are far more frequent in everyday conversation than is generally appreciated (cf. Quirk, 1970). The speaker and hearer may not even notice them during the conversation itself, since there is generally sufficient redundancy to compensate for the channel noise that performance errors introduce. But, in many instances at least, speakers will accept, or indeed readily volunteer the information themselves, that what the linguist identifies as performance errors are 'mistakes', if they are confronted with a recording or transcript of the conversation afterwards. It should also be stressed that the investigation of performance errors is by no means empty of theoretical interest. The incidence and nature of such errors, to the extent that they are non-random, provides important evidence for the study of the physiological and neurological mechanisms involved in speech-production (cf. Laver, 1970). They are not represented, however, in the linguist's model of the language-system. The first stage of idealisation, which consists in the elimination of performance phenomena in the primary data, I will refer to as the stage of *regularisation*.

The second stage of idealisation I will call *standardisation*. Here I have in mind the discounting of a certain amount of the systematic variation between utterances that can be attributed to personal and sociocultural factors. We have already seen that the linguist will generally discount differences of voice-quality, and probably paralinguistic differences also, in deciding that two utterances are structurally identical. Nothing more need be said about this, except that even this degree of idealization presupposes some knowledge of the language-system (cf. Ladefoged, 1964: 104; Laver, 1968). We are concerned with differences in the linguistic part of utterances. And here, we must admit, there is room for disagreement as to how much systematic variation should be included in the description of a language.

At one extreme, we have something like the degree of standardisation that has been deliberately imposed upon written English by the conventions of printers. At the other extreme, which is perhaps unattainable in practice, however desirable it might be in theory, one would incorporate in the model of the language-system all the systematic differences in the language-behaviour of individual speakers that are attributable to such factors as geographical and social origin, professional occupation, education and interests, the nature and occasion of the speech-event, and so on. Some of these differences are covered by the terms *dialect*, *accent* (in the sense in which we talk of a 'Lancashire accent' or an 'upper-class accent'), *style*, *register* (cf. Halliday, McIntosh and Stevens, 1964; Strang, 1968). The point I wish to emphasise here is that the degree to which one takes account of such factors in the description of a particular language is to some extent arbitrary. There is a sense in which it is true that everyone we normally describe as a 'speaker of English' speaks a different 'English'; he has his own *idiolect*, distinct to some degree in vocabulary, grammar and phonology. It would, however, be absurd to hope to describe all these differences; and it would be misleading to define the language-system as either the intersection or the union of the idiolects of the speech-community – the former definition would be too restrictive, the latter too inclusive. In practice, what the linguist will do is to discount all but the major systematic variations in the speech-community whose language he is describing. (For further discussion and references see Hymes, 1962, 1970; Labov, 1966; Pride, 1970.)

The third stage of idealisation is the one at which we need to invoke the distinction between sentences and utterances. It is a commonplace of traditional grammar that the utterances of everyday conversation tend to be, in some sense, 'grammatically incomplete' or 'elliptical'; and one of the things we are taught (or, perhaps I should say, used to be taught) at school in our 'composition' classes is to write in 'complete sentences'. I will assume that this sense of 'complete sentence', for written English at least, is reasonably clear; and that the term can be made applicable in roughly the same sense to the spoken language, with which we are primarily concerned.

Every act of utterance is made in a particular *context* and must be interpreted in the light of the relevant information 'given' in that context.[10] The context may be purely *situational*; it may be both situational and *linguistic* but it is never purely linguistic.[11] The first utterance of a conversation will normally have a purely situational context. Thereafter, very many, and possibly the majority of the utterances in the same conversation will be dependent, both for their form and for their interpretation, on the preceding utterances of the same speaker or the person with whom he is conversing. To repeat what I have said elsewhere: 'An example [of a context-dependent utterance] might be *John's, if he gets here in time*, which could hardly occur except immediately after a question which 'supplied' the words required to make it into what would traditionally be regarded as a 'complete' sentence. For example, it might occur after *Whose car are you going in?*, but not after *When are you going there?* (Lyons, 1968: 175). I have cited these invented, but I trust realistic, utterances in the normal orthography: I am assuming that they could be identified as tokens of particular utterance-types (with reference to some explicit or implicit description of English) and represented as such phonologically (after regularisation and standardisation). The same utterance (that is to say, what would be described as a token of the same type) might occur, in other conversations, as an answer to indefinitely many different questions other than the one expressed by *Whose car are you going in?* Suppose, for example, it had occurred after *Whose father is going to pay for the drinks?* Not only is its meaning now quite different. It is ambiguous, as long as we take into account only the previous utterance, in that *he* might refer either to John or his father. In all probability, yet more of the linguistic context or situational features (such as the fact that John was already present) would of course eliminate this ambiguity.

The point I have been making, then, is that utterances are typically *context dependent*, with respect both to their meaning and their grammatical structure. An utterance may be made context-independent by adding to it various elements 'given' in the particular linguistic and situational context in which it has occurred: by converting *John's, if he gets here in time*, for example, into *We are going in John's car if he gets here in time*. Now this sequence of words, with the appropriate prosodic features, is context-independent (provided that *he* refers to John and not to some other person not mentioned) and it has the properties normally thought of as being characteristic of 'complete' sentences. It is such entities as these, and not utterances, that linguists generally describe in their models of the language-system.[12]

It will be clear from what has been said so far that sentences never occur in speech (except in contexts of 'mention' – that is to say, in the 'reflective', or metalinguistic, use of language). There are, however, certain utterances or parts of utterances which stand in a particularly close relationship with sentences. The relationship can be described as a one-to-one, order-preserving, mapping of the words of the sentence on to the words of the utterance. As we have seen, most of the actual utterances of everyday conversation do not stand in this particularly close relationship with sentences, but they can be described as being derived from them by the deletion of certain contextually 'given' elements, by the substitution of pronouns for the names of persons and objects and phrases referring to them, and by various other devices.

This section is already longer than was intended; and there are many further points that would need to be made in a more complete account of the relationship between the linguist's description and the primary data. I will conclude by stressing the perhaps obvious point that the linguist's model of that part of the underlying system that he defines to be within his province is not to be taken as a model for the production or understanding of utterances as these processes are carried out in 'real time' by the speaker or hearer. It need not be supposed that there is any moment at which the sentence as such is represented in the speaker's brain. The linguist's model, as I have been emphasising throughout, rests upon a considerable idealisation.

One final point may be made with particular reference to the subject of this volume. It is sometimes said by investigators of non-verbal communication that 'something signalled in one modality may be contradicted in another' (Austin, 1967). This phenomenon of 'contradiction' can hold,

however, not only between vocal and non-vocal signals, or even between the
linguistic and paralinguistic components of an utterance, but also between
the verbal and the prosodic components. We have already seen that an
utterance may have the grammatical structure of a declarative sentence
(as far as its verbal component is concerned) and yet have 'superimposed'
upon this, as it were, the intonation more generally characteristic of a
question. It seems to be the case that, whenever there is a contradiction
between the overt form of the verbal part of an utterance and the associated
prosodic or paralinguistic features, it is the latter which determine the
semiotic or 'functional' classification of the utterance (as a question rather
than a statement, as a tentative suggestion rather than a question, and so
on). In this respect, the prosodic features of an utterance, like the vocal
and non-vocal paralinguistic features associated with it, may be thought
of as being 'selected' by the speaker as a consequence of his uttering a
particular sentence (or the context-dependent 'residue' of a particular
sentence) in the performance of one of the semiotic functions. It may even
be the case that we should recognize more than two levels in the selection
of the relevant features such that as prosodic overrides verbal, so paralinguistic overrides prosodic (when there is conflict between these). But
I am now letting speculation run ahead of any evidence that I can call
upon. It would be interesting to know whether there is any clear confirming
or disconfirming evidence.[13]

3. LANGUAGE AND SPEECH

Language, as we know it today in most parts of the world, exists in two
main forms: speech and writing. Of these, it is the former that the linguist
takes as 'basic' or 'primary'. According to a common formulation of the
relationship between the two, the spoken language is *prior* to the written
in the sense that the latter results from the transference of the former from
its 'natural', or primary, medium to a secondary medium. But what do
we mean when we say that the spoken language is prior to the written
language? We can distinguish at least four relevant kinds of 'priority'.

(i) *Phylogenetic priority*

Every community of men of which we have any direct knowledge, or any
historical knowledge, has, or had, a spoken language. It is reasonable to
suppose that in all cases written language is based upon (i.e. derived from)
speech; though, in the case of languages with a long literary or scribal
tradition, we may have to go back a long way before arriving at the point
of derivation.

(ii) *Ontogenetic priority*

Normally, a child learns his native language in the spoken form first; and he does so very young, and in a way that can be described as 'natural'. By this term I mean to imply that children are genetically 'programmed', not only for vocalisation, but for the vocalisation of language (cf. Lenneberg, 1967). He acquires the corresponding written language, in a literate community, by formal instruction based upon prior knowledge of the spoken language.

(iii) *Functional priority*

Even in highly literate communities, spoken language serves in a wider range of communicative functions than does written language. Writing is employed in some, though not all, of these as a substitute for speech (e.g. in informal letter writing); but it also has characteristic functions of its own. Technological developments (the invention of the telephone, radio, etc.) have had some effect upon the functional relationship of speech and writing; but this has not so far been changed in any fundamental way.

(iv) *Structural priority*

The basic units of the script can be put into correspondence (not necessarily one-to-one correspondence) with units of the spoken language, either phonological units (as in the case of alphabets and syllabaries), or grammatical units (as in the case of so-called 'ideographic' scripts); and, as far as alphabets and syllabaries are concerned, the patterns of combination into which the letters or characters enter, though arbitrary and inexplicable in terms of their shapes, can be accounted for by relating them to the patterns of combination into which corresponding phonological units enter.

For all these reasons speech is taken by most linguists to be primary and writing secondary. Sound (and more particularly that range of sound which can be produced by the human 'speech organs') is the 'natural' *medium* in which language is realised: written utterances or texts result from the transference of language from the *phonic* to the *graphic* medium. The distinction between channel and medium is debatable, and I do not propose to go into the question in detail. It is at least terminologically convenient for the linguist to draw the distinction. When he uses the term 'channel' he tends to be concerned with the transmission of signals (typically vocal signals); when he uses 'medium', he tends to be concerned with such systematic functional and structural differences between written and spoken language as can be attributed to physical differences in the *characteristically* distinct channels of transmission. (I say 'characteristically' because the spoken language can be transmitted in writing, though not in the normal orthographic script; and the written language can be transmitted, and commonly is nowadays, along a variety of channels, not excluding the vocal–auditory channel.)

Granted the priority of speech, it is important to give due recognition to the functional and structural differences between speech and writing. There may be important grammatical and lexical differences. The prosodic and paralinguistic features of speech are only crudely and very incompletely represented by punctuation and the use of italics, etc., in writing. Written texts may be composed, reflected upon and edited, before any part is transmitted. They are more enduring (or were until systems of sound recording were developed); they have therefore been used throughout history in literate communities for the codification, preservation and citation of important legal, cultural and religious documents. This fact has contributed to the greater formality and prestige of the 'written word'; and also to the development of the concepts of 'literature' and 'scripture'. Any account of language and its communicative function in modern society must recognise that written language, despite its derivative character, phylogenetically and ontogenetically, has a considerable degree of structural and functional independence.

Should we take vocalisation to be an essential feature of language, as many linguists do? I think not. Let us suppose that we did in fact discover a society making no use of vocalisation (except possibly for the emission of vocal reflexes and a limited set of indexical signals symptomatic of such emotional states as anger, fear, sexual arousal, etc.), but communicating by means of inscribing visible shapes on paper or some similar substance. Let us further suppose that, upon analysis, these texts were found to have the same kind of grammatical structure (or a grammatical structure of a similar degree of complexity) as our own written texts; that they were used for the same or a similar variety of communicative functions, and had the 'design features' of openness, semanticity, prevarication, reflectiveness, etc. (see Chapter 2). Faced with this hypothetical 'discovery', we would surely say that the society in question had a language. Is it not possible to conceive of a group separating itself from the main body of our own society and forswearing all use of the spoken language? In what sense would their written language still be dependent upon the spoken language that elsewhere in society was associated with it? Could it not develop independently? Could it not be taught to children without the previous development of their genetically 'programmed' disposition to acquire spoken language? These are difficult and perhaps unanswerable questions.[14]

The point is that language *can* be considered independently of the medium in which it is primarily and 'naturally' manifest; and written language already has some degree of independence as one of man's principal means of communication. At any rate, in any comparison of human signalling with animal signalling, or of verbal and non-verbal signalling in man, due weight should be given to the fact that people can learn, fairly easily and successfully for the most part, to transfer from one medium to the other, holding invariant much of the verbal part of language. It is doubtful whether this phenomenon is found in non-verbal signalling. Indeed, it is not clear how one would establish the invariance. I would suggest that the *medium-transferability* of language is as important as what Hockett calls 'learnability' (i.e. the ability a speaker of one language has to learn another).

4. DUALITY OF PATTERNING

Other terms for duality of patterning (see Chapter 2) are 'double articulation' (e.g. Martinet, 1960) and 'duality of structure' (Lyons, 1970*b*: 12). In all spoken language so far investigated, and presumably in all languages yet to be investigated, there are two *levels* of structural organisation: *phonological* and *syntactic*. At the syntactic level, utterances can be represented as sequences of *words* (some linguists say *morphemes*: I will come back to this); at the phonological level, as sequences of phonemes. (Words and phonemes are *segments*, according to the usage established in Section 1. In some languages, words are generally analysable into smaller, grammatically functioning segments. Phonemes, though minimal segments, are not the minimal units of phonology: they may be analysed into 'simultaneous' *features*. For simplicity, however, we will neglect 'subphonemic' analysis.)

As commonly discussed, the principle of 'duality' is made dependent upon the somewhat Procrustean model of linguistic analysis exemplified by Trager and Smith (1951), according to which there are only two kinds of basic units in language, phonemes and morphemes, and everything that is recognised as being part of language proper must be described as a combination of one or more of these units (whether it is segmental or not). Few linguists would now accept this viewpoint, though it is still taken for granted by some writers on non-verbal communication (e.g. Austin, 1967). The trouble with the morpheme is that, if it is defined to be a segment of an utterance, it is not a unit that is found in all languages. On the other hand, if it is defined in more 'abstract' terms, it loses its usefulness for the description of those languages which can otherwise be reasonably described as having a level of morphemic patterning. Let us consider a simple illustration of this point. If we define the morpheme as a segment (composed of phonemes) we can quite reasonably describe the word *walk-ed* as a combination of two morphemes (as indicated by the hyphen); but the grammatically parallel *went* cannot be described in these terms (without having recourse to some rather dubious supplementary constructs). If we take the alternative approach, defining the morpheme in more 'abstract' terms, we can say that both *walked* and *went* realise, or represent, combinations of two morphemes; but then we cannot say that the morphemes are composed of phonemes.

The relationship between the two levels, in many languages at least, is far more complex than any statement of the principle of duality in such terms as the following would suggest: 'signal elements meaningless, pattern combinations meaningful' (see Chapter 2). Consider a simple Latin utterance (the point can be made more clearly with respect to an 'inflecting' language): *trucidaverunt regem crudelem*, 'They killed the cruel King'. What is the meaning of the second word? The answer depends, not only upon what is meant by 'meaning' (which is a question we may postpone), but also upon what we mean by 'word'. But we do not normally mean the 'pattern combination' of phonemes (represented here by a sequence of five letters) which actually occurs in the utterance (i.e. *regem*), but rather the *vocabulary item* (or lexical item), of which *regem* is but one of the 'inflected' forms. We would want to say, in this sense of 'word', that the meaning of the second word in the above utterance is the same as the meaning of the first word in *Rex mortuus est* 'The King is dead'. However, as 'pattern combinations' of 'lower level' units these are in principle totally, and not just partially, distinct. In order to account for the grammatically acceptable combinations of actual *word-forms*, like *regem*, *rex*, *trucidaverunt*, etc., we might wish to recognise two 'higher' sub-levels of syntactic patterning within the grammar, at each of which the units are 'words' in a different sense of the term. There is the sense of 'vocabulary item', already explained. There is also the sense in which we refer to such entities (using traditional terminology) as 'the *accusative, singular* of REX' as 'words', independently of their *realisation* at the lowest sub-level as 'pattern-combinations' of phonemes. For a fuller and more formal treatment, see Matthews (1970). In short, the relationship between the meaningful units of language and the lower-level 'signal elements', in some languages at least, is not simply one of composition.

The point about duality is that there are systematic limitations upon the permissible combinations of phonological units and these limitations are largely independent of the grammatical function or meaning of the combinations; it is because of these limitations that we say language has a phonological *structure.* Suppose we took the total stock of written English word-forms and, having listed them in alphabetical order, we numbered them from 1 to *n*, so that every word-form was kept distinct and every number from 1 to *n* was matched with some word form. We could then write sequences of digits instead of sequences of letters; and there would no longer be anything corresponding to phonological structure. And yet there would probably still be sufficient redundancy to counteract channel noise despite the loss of redundancy at the level of the signal-elements. Surely we should still want to say that the new system was a language; moreover that it was, in some important sense, written English. However that may be, duality is found in all languages; whether we regard it as an essential feature or not is perhaps of little consequence.

5. SYNTAX

In this section I will, for simplicity, talk as if words in the sense of word-forms (i.e. those 'pattern combinations' of 'signal elements' which occur between spaces in written English and which are partly distinguished by phonologically suprasegmental features in spoken English)[15] were the minimal syntactically-functioning elements in all languages. It is possible to discuss the general principles of syntactic structure in these terms. But it will be clear from the previous section that the grammar of some languages must contain rules specifying the rather complex relationship which holds between the word-forms and the minimal meaningful elements in the language (in the normal sense of 'meaningful'). I will use the term 'word' throughout in the equivocal sense permitted by this declared simplification. It may be observed in passing that most non-linguists writing on language, as well as certain linguists, continually use the term 'word' in this way without apparently noticing the equivocation and its theoretical importance.

The simplest kind of syntax from a purely formal point of view would be a system where the grammatically well-formed combinations of words had the properties of *associativity* and *commutativity* (i.e. where $x+(y+z) = (x+y)+z$ and $x+y+z = y+x+z = \ldots$). Once we take account of the 'computing' mechanisms of the organism or the machine operating with the system we may of course rate simplicity differently. In particular, commutativity might be a complicating property from this point of view. The syntax of human language, as we shall see, is at least partially non-associative (or, to put it more cautiously, of all human languages that have so far been studied and reported upon in sufficient detail for it to be possible to discuss their syntax in such formal terms). Whether language is necessarily non-commutative is a controversial issue.

The question depends partly upon the way one interprets 'equivalence' (symbolised by ' = ' in the equations above) and partly upon whether one distinguishes between 'sentence' and 'utterance'. For example, in relation to languages with a so-called 'free word order', if one distinguishes between 'sentence' and 'utterance' as I have done above and defines syntactic equivalence to hold over sentences, it might be reasonable to say that *Brutus occidit Caesarem* and *Brutus Caesarem occidit* ('Brutus killed Caesar') are syntactically equivalent utterances derived from the same sentence and that the words in the sentence are sequentially unordered. (To a limited extent this phenomenon is found in English: *Mary I love* and *I love Mary*. But in the systematic description of English syntax it is probably preferable to describe *Mary I love* as a contextualised 'transform' of the sequentially ordered sentence.) Chomsky's formalisation of his 'three models for the description of language' in terms of the basic operation of *concatenation* (1956, 1957, etc.) presupposed non-commutativity; and he argued for it explicitly later (1965).

The importance of this point for our present purpose lies in the fact that psychologists and ethologists commonly discuss patterns of behaviour as having a 'sequential' structure, presumably implying thereby that these behaviour patterns lack the property of commutativity.[16] It is an associative, non-commutative system that can be satisfactorily accounted for by means of a *finite-state grammar*: the least powerful of Chomsky's three models. Granted that Chomsky has proved that such a model (with or without the introduction of some mechanism for computing the transitional probabilities between particular words or particular classes of words) is inadequate for the description of the syntax of human language (Chomsky, 1957: 21–4; Lyons, 1970a: 47–55), the interesting question arises whether the behaviour patterns of other species and the non-verbal signalling systems of man, in so far as they are syntactically structured, fall within the scope of finite-state grammars.[17]

I have said that non-associativity is a characteristic of the syntax of human language. Hence, the phenomenon of *structural ambiguity*, of the type illustrated by *old men and women, beautiful French teacher*, etc. (cf. Lyons, 1970a: 58, 1970b: 117). Non-associativity requires for its description a model of syntax which incorporates the notion of 'bracketing'. One such model is the second of Chomsky's 'three models'; *phrase-structure* grammar. And various other systems have been developed which have been shown (in most respects at least) to be formally equivalent.

It is not possible, in the space available, to give even a brief account of Chomsky's system of 'transformational grammar' in either of its two versions (Chomsky, 1957, 1965; cf. Lyons, 1970a: 66–82, 1970b: 121–39) or of proposed alternatives which, like Chomsky's system, are more powerful than phrase-structure grammars. It would probably be agreed by all linguists who have concerned themselves with this question that what is required for the description of the syntax of language is a model which is more powerful than phrase-structure grammars (or their equivalents), but less powerful than transformational grammar, as it is at present formalised. Whether the restriction in power is achieved by imposing limitations upon the operation of the rules within a grammar of essentially the same type as that formalised as the third of Chomsky's 'three models', or by constructing a radically different model, is an open question.

What I have tried to do in this section is to illustrate some of the simpler formal properties of syntax. Even with respect to such simple properties, it is interesting to compare various signalling systems; and the theory of generative grammar provides us with the tools for making such comparisons. It is here more than anywhere else that linguistics would seem to be in a position to contribute something of value to the study of other signalling systems and 'patterns' of behaviour.

Syntactic structure is a necessary, though not sufficient, condition of 'openness' (see Chapter 2). It is important, in evaluating systems in terms of this property, not to overlook the formal complexity of the rules by virtue of which many of the new 'messages' are 'coined' in human language; and the number of new messages coined in relation to the number of 'old' messages previously experienced and perhaps stored, either as wholes (as exemplars) or as analysed, will not necessarily reflect this complexity. For example, it might prove possible to teach an animal, or indeed a machine, to 'speak English' (whether in the spoken medium or not is irrelevant) in the sense of its learning to produce a very large number of utterances that it has not experienced before, all of these new utterances being correct and being coined by the relatively simple process of substituting a new word for one of the words in one of its stored utterances. If we were successful in an experiment of this kind (and psychologists are conducting such experiments at the present time: see the account of the Gardners' work given by Thorpe in Chapter 2), we could certainly claim that the animal or machine in question was capable of learning a signalling system with the property of 'openness'. This would not mean, however, that there is no difference in principle between the child's ability to acquire language and the ability demonstrated by the animal or machine. It is perhaps not the simple fact of 'openness' (or 'creativity' to use Chomsky's, 1968, term) – the ability to produce and understand utterances not previously encountered – that is so characteristic of human language, but the fact that the 'openness' of human language is determined by principles of great formal complexity.[18] This should be borne in mind in the evaluation of the results of language-learning experiments.

6. MEANINGFULNESS

Language is frequently described as a system of *signs* and thus brought within the scope of *semiotics*. Following Morris (1946), and ultimately Peirce (1931), many authors recognise three areas within semiotics: *syntactics*, *semantics* and *pragmatics*. How these three areas are related varies somewhat from one author to another. Fairly typical is the following: '. . . syntactics studies how signs are related to one another. Semantics studies . . . how these signs are related to things. And pragmatics studies how they are related to people' (Smith, 1966: 4–5). Although this scheme of classification has been adopted by many communication-theorists, and may have some value for the analysis of other systems of signalling, it is hopelessly vague for the study of human language. The inter-relationship of words (if words are taken to be signs) is governed by at least two distinguishable, though related, principles of well-formedness ('being grammatical' and 'being meaningful'); and both of these are dependent upon semantics. However, the tripartite scheme does not allow, in any obvious way, for combinatorial semantics. Further, if 'things' is taken in any reasonable sense, very many of the words and combinations of words in human language cannot be said to refer to or denote 'things' at all; and yet we shall wish to say that they are meaningful.

Similar criticisms can be made of any definition of 'meaning' for human language which describes this in terms of an 'association' between utterances and particular 'states of affairs' or 'situations', or between parts of utterances and 'features' of the external environment. What is the nature of this 'association'? Is the external 'feature' or 'situation' a necessary or sufficient cause of the signal? Is it a stimulus to which the signal is a response? If we define the nature of the putative association more closely, we will find that very many utterances have no discernible 'association' with any independently identifiable 'feature' or 'situation'. Hence Chomsky's insistence (1968) that language is free from stimulus control.[19] It is preferable to abandon such a general definition of 'meaning', which has the unfortunate consequence that it allows as 'meaningful' an enormous variety of behaviour both human and non-human, and yet excludes a considerable amount of language-behaviour. It is in fact doubtful whether any single precise and univocal interpretation can be given to 'meaning' (and hence to 'semanticity': see Chapter 2). However, there are various distinguishable functions reasonably called 'semantic' and various ways of classifying them.

Let me first introduce a distinction between the terms 'communicative' and 'informative' (cf. Marshall, 1970: 2). A signal is *communicative*, if it is intended by the sender to make the receiver aware (to *inform* him) of something of which he was not previously aware.[20] One sense of 'meaningful' is 'communicative'; and this rests upon the possibility of *choice* on the part of the *sender*. (If the sender cannot but behave in a certain way, then he cannot communicate anything by behaving in that way.) Another sense of 'meaningful' is 'informative'; and here we look at the principle of 'choice' from the *receiver's* point of view. A signal is *informative* if it informs the receiver of something of which he was not previously aware (regardless of whether the sender communicated this information or not). Under a fairly standard idealisation of the process of communication, what the sender communicates (the information put into the signal by choice) and the information derived from the signal by the receiver are assumed to be identical. But there are, in practice, frequent instances of misunderstanding.

It is probable that most, if not all, animal signals are not communicative, if we take 'choice' in a very strict sense. Marshall (1970: 235) suggests that the clearest evidence of the ability to communicate would be evidence of an 'intention to misinform'; and evidence of this so far in animals is merely anecdotal. One of the 'design features' of human language listed by Hockett, it will be recalled, was 'prevarication' (see Chapter 2).

The communicative component in human language, important though it is, should not be overemphasised to the neglect of the non-communicative, but nevertheless informative, component, which human language appears to share with various animal systems. Relatively few utterances in everyday language-behaviour are totally non-communicative.[21] But all utterances will contain a certain amount of information which, though signalled by the speaker, has not been 'put there' by the exercise of 'choice'.

Of particular interest in this connexion is information of the kind that is described as *indexical* by Abercrombie (1967) and Laver (1968): i.e. information about the sender. (The 'sender' is normally both the 'source' and the 'transmitter' – there are, not uncommonly, situations in language-behaviour where the source and transmitter are to be distinguished; and the signal when it reaches its 'destination' conveys information that is indexical with respect to both the 'source' and the 'transmitter'.)[22] Abercrombie (1968: 7) and Laver (1968: 48) each offer a different, though not necessarily conflicting, tripartite classification of indexical information: Abercrombie into '(a) [indices] that indicate membership of a group; (b) those that characterise the individual; (c) those that reveal changing states of the speaker': Laver into '*A*. Biological information, *B*. Psychological information, *C*. Social information'. Both Abercrombie and Laver are concerned primarily with voice quality, paralinguistic features and those aspects of 'pronunciation' that can be summarised as 'accent'. But 'indexical', on their definition of the term, is clearly and usefully applicable to every component, including the verbal component of utterances. A person's employment of a particular word may be *indicative* of (may *indicate*) his membership of a particular regional or sociocultural group, or who he is, or what his emotional state or 'attitude' is. The same is true of certain prosodic features; and doubtless of many of the non-vocal paralinguistic phenomena that 'punctuate' and 'modulate' language-behaviour. Very many animal signals are indexical; and I would hazard the opinion that in many species *all* signals are indexical.

Within the class of signals distinguished as 'indexical' there is one particular subclass worthy of special mention, and possibly of a separate term. This is the third of Abercrombie's subtypes ('those that reveal changing states of the speaker'), to which he gives the label 'affective'. Indexical information of this kind is what many linguists refer to as 'attitudinal'; and we have already noted that much of the non-verbal component of speech is characteristically attitudinal, or affective. But I would propose a somewhat wider definition of 'state of the speaker' than 'attitudinal' or 'affective' would suggest; and it is for the correspondingly wider class of indexical signals that I am putting forward the term *symptom* (cf. the use of this term by MacKay in Chapter 1). By a 'symptom' I mean a signal, or some part of a signal, which indicates to the receiver that the sender is in a particular state, whether this be an emotional state (anger, fear, etc.), a state of health (suffering from laryngitis, etc.), a state of intoxication, or whatever. In many cases, the state in question will be plausibly interpreted as the 'cause' of the symptom.

The notion of 'speaker's choice' has been given a rather narrow interpretation above: something like 'the exercise of voluntary control'. It can be widened, and perhaps with advantage, by counting as 'choices', not merely those decisions to act one way rather than another on any given occasion, but also the maintenance of some habitual 'setting' of the 'controls'. (For this use of the term 'setting', cf. Honikman, 1964; Laver, 1968.) This would allow as communicative a good deal of signalling, both in speech and in non-verbal behaviour, that would otherwise be described as non-communicative. The justification for this extension of the notion of 'choices' (i.e. taking it in the sense of 'options' in the system (Halliday, 1970: 142)) might be as follows: the signalling system permits at various points a number of options, one of which *must* be selected by the sender; in principle, senders can select every time afresh;

in practice, many of the options become habitual with individual senders (and therefore signal indexical information); and, although we cannot say that the sender continues to exercise any positive control, we can perhaps say that he confirms his original options by the fact of not changing the 'settings'. It is only by means of this extension of the notion of 'choice' that much of the behaviour called 'self-presentation' (Argyle, 1967; and Chapter 9 below) can be described as communicative. For the projection of one's self-image, like most of one's social behaviour, is very largely habitual.

Looking at language-behaviour from a somewhat different point of view, we can distinguish two broad classes of communicative and informative functions that it fulfils – functions that are complementary, rather than in contrast: *cognitive and social*. (The terms 'ideational' and 'interpersonal' have been used in a similar sense by Halliday (1970: 143).) The first of these refers to what many would consider to be the most distinctively human function of language, the transmission of propositional, or factual, information and discursive reasoning or 'cogitation'; the second to the establishment and maintenance of social rapport. It was for the latter function that Malinowski coined the felicitous phrase 'phatic communion' (1946). Due weight must be given to each of these functions: philosophers have frequently overemphasised the cognitive function, anthropologists and social psychologists have commonly overemphasised the social function. Correspondingly (though there is perhaps no necessary connexion), overemphasis of the cognitive aspect of language has frequently been associated with a typically 'rationalist' insistence upon 'freedom from stimulus control', and overemphasis of the social functions with an insistence that language-behaviour is very largely determined by the social situation.

Without attempting anything like a complete, or even representative, list of the more specific *semiotic functions* of language it might be worth mentioning a few of the more 'basic' ones: 'basic' in the sense that they are identifiable in the earliest linguistic development of the child (though not always unambiguously) and perhaps also in non-verbal communication both in men and in other species.[23]

(i) *Deixis*

Signals, or parts of signals, which draw the attention (of the receiver) to a particular object in the situation of utterance, not by naming it, but by locating it in relation to the source may be described as *deictic*. (*Look here! Over there!* etc.) The situation of utterance, and the deictic system (personal pronouns, demonstratives, etc.) which depends upon it in language, is typically egocentric (Lyons, 1968: 275). (A signal which draws attention to the source by means of information characterising the source as an individual, a member of a group, etc. is both deictic and indexical.) Deixis would seem to be the ontogenetic source of reference (identifying for the receiver the object about which one is making a predication); and the 'definite article' in English, whose historical and perhaps ontogenetic source is deictic, has a characteristically referential function. But by means of language we can of course refer to objects outside the situation of utterance; and it is this feature that Hockett calls 'displacement' (see Chapter 2).

(ii) *Vocative signals*

Signals which attract the attention of a particular person (organism, machine, etc.) whether by 'name' or otherwise. (*Look out! John!* etc.) Such signals are commonly used in language-behaviour (including the associated paralinguistic signalling) to invite the addressee to assume the role of receiver (i.e. to become the 'second person' in the communicative 'drama').

(iii) *Nomination*

By *nomination* I mean not the use, but the assignment, of a name. At its most explicit (combined, as typically, with deixis) this is exemplified by an utterance like *This is John Smith*. Explicit and implicit nomination plays a considerable role in language-acquisition (cf. Brown, 1970); and the learning of the denotation of concrete nouns (i.e. the extension of the class of objects to which they are applicable) may well develop on the basis of nomination, a word like *chair* first being learned as a proper name, as it were, and only subsequently being 'generalised' to a whole class of objects. Explicit nomination is one obvious aspect of what Hockett calls 'reflective-ness' (see Chapter 2).[24] Presumably nomination is not found in any signalling system other than language, though one can perhaps relate it to something more general and more 'basic' like 'associating a symbol with what it symbolises', as many semioticians and philosophers have tried to do (cf. Morris, 1946; Quine, 1960).

(iv) *Desiderative signals*

Under this heading I am referring to signals that indicate the desire of the organism for some object (whether communicatively or symptomatically). These have often been regarded as especially 'basic'.

(v) *Instrumental signals*

These are signals that serve to get something done, to bring about a particular state of affairs. The *instrumental* function is perhaps dominant within the broader class of functions that Halliday (1970) calls 'inter-personal', to which he relates the grammatical category of 'mood'. The differences between questions, requests, commands, etc., in language would be describable in terms of a finer classification of instrumental functions; and it is here that we would need to take note of the notions of 'illocu-tionary force' (Austin, 1962; Searle, 1969) and 'speaker's intention' (Grice, 1957; Strawson, 1970).

I have mentioned only a few of the semiotic functions which one needs to recognise in discussing the nature of verbal communication. It would be out of place, in the present context, to go into all the complexities that arise in the analysis of 'meaning' in language: the necessity of distinguishing between 'sense', 'reference' and 'denotation' in the analysis of 'cognitive meaning', between 'predication', 'presupposition' and various types of 'implication'. That such distinctions have been found necessary for the analysis of the semantic structure of language merely reinforces the point with which I began this section: that there are many different kinds of 'meaning' expressed by language and no single gloss or definition can be expected to capture all of these.

7. CONCLUSION

I now wish to summarise the more important of the points I have made in the previous sections and, in doing so, to relate them to the questions raised at the very beginning of the chapter: is the capacity for language, properly so called, unique to man? An associated question, and one which inevitably arises in the present context, is whether language, as we know it, can be shown or plausibly assumed to have evolved from some system of non-verbal communication. I will not give a direct answer to either of these questions. For, in my view (and once again I would emphasise that it is a personal view rather than one which all linguists would necessarily share), neither question, in the form in which I have just put them, admits of a direct affirmative or negative answer on the evidence available at the moment.

Men have been debating the origin of language from the very earliest times (cf. Lyons, 1970b). Long before the publication of Darwin's *Origin of Species*, scholars had been putting forward theories designed to account for the evolution of language from such systems of non-verbal communication as 'instinctive' emotional cries, gestures, rhythmic communal chanting, and so on. But it was the work of Darwin, including his own speculations on the origin of language (cf. Marshall, 1970: 229–31), that gave particular impetus to the attempt to construct an evolutionary theory of the origin of language in the late nineteenth century. At that time linguistics was very strongly influenced by the theory of evolution. Over the last fifty years or so, however, most linguists have refused even to discuss the origin of language. The reason is simply that no sign of evolution from a simpler to a more complex (from a more 'primitive' to a more 'advanced') stage of development can be found in any of the thousands of languages known to exist or to have existed in the past. If we had interpretable records of the forms of communication employed by earlier human or hominid species we might be in a better position to discuss the origin of language. As things are, most linguists would say that the question is totally irrelevant to their primary concerns: the construction of a general theory of the structure of language and the description of particular languages within the framework of this general theory. What one might call the 'official', or 'orthodox', professional attitude of linguists to evolutionary theories of the origin of language tends, therefore, to be one of agnosticism. Psychologists, biologists, ethologists and others might say, if they so wish, that language *must* have evolved from non-verbal communication; the fact remains that there is no actual evidence from language to support this belief.

It might be argued (as it has often been argued in the past) that, although there is no evidence of evolution from a more 'primitive' to a more 'advanced' form of structure in existing languages, there are two other kinds of evidence relevant, and possibly decisive, in any discussion of the relationship between language and non-verbal communication: evidence derived from the study of children's language, and evidence derived from a comparison of the structure and functions of language and non-verbal communication. The acquisition of language by children is a topic to which we will return presently. Evidence of the second kind has already been mentioned in earlier sections of this chapter and is presented in almost all the other chapters of this book. The problem is how to evaluate it. One of the points I particularly emphasised in the first section was that there is no clear distinction in language-behaviour between what is linguistic and what is non-linguistic (paradoxical though this statement may appear), and in Section 6 I stressed the fact that many of the semiotic functions one finds manifest in language-behaviour (in particular those which I distinguished as social, rather than cognitive) can be identified also in non-verbal communication, whether human or animal. Does it follow from the intimate interpenetration of the linguistic and non-linguistic features of language and from the fact that some aspects of meaning, or 'semanticity', are common to language and non-verbal communication that the former has 'evolved' from the latter? Clearly it does not. In default of any explanation of what is meant by 'evolve', and of the mechanism by means of which the 'evolution' of language from non-language is presumed to have operated, evidence based on the structural and functional continuity of verbal and non-verbal communication is purely circumstantial, compatible with, but not conclusively demonstrative of, the evolutionary development of the one from the other. The verbal component in language might have been of totally distinct origin and its interpenetration with the non-verbal component a matter of subsequent and gradual development. The one hypothesis is surely as plausible, *a priori*, as the other.

A more important conclusion to be drawn from our comparison of the verbal and non-verbal components of language is that the questions whether language is unique to man and whether it evolved from non-verbal communication are not formulated precisely enough to be answered with a 'yes' or a 'no'. Although there is no sharp distinction between language and non-language, there are certain properties of language having to do with grammatical complexity and the cognitive functions of language which, as we saw in Sections 5 and 6, appear to be unique to language and associated more particularly with its verbal component. If we decide to make the possession of these properties a defining characteristic of what we will call 'language', we will then say, and correctly, that non-verbal communication is 'radically', or 'fundamentally', or 'qualitatively', different from language. We might equally well have framed a definition of 'language', however, according to which one would be inclined to say that the difference between verbal and non-verbal communication is 'quantitative', and a matter of 'degree' rather than 'kind'. This purely definitional aspect of the question should be borne in mind when one considers the arguments of those who maintain that language is or is not unique to the human species.

Lenneberg (1964: 65–9, also 1967; for further discussion and references see Marshall, 1970) lists under five headings what he considers to be the principal reasons for believing that language is species-specific and depends upon particular biological 'properties' not found in animals other than man: (i) Anatomic and physiological correlates; (ii) Developmental schedule; (iii) Difficulty in suppressing language; (iv) Language cannot be taught; (v) Language universals. Not all of these reasons, however, are equally cogent. We will take each in turn (though not in the order in which they have been listed).

Under (i), Lenneberg says: 'There is increasing evidence that verbal behaviour is related to a great number of morphological and functional specializations', including the morphology of the vocal tract, cerebral dominance and the localisation of language functions in certain areas of the brain. (A more detailed treatment of this aspect of the 'biology' of language is given in Lenneberg, 1967.) Although the concept of localised brain functions, in its simplest form at least, appears to be less generally accepted today than it was until recently (Marshall, 1970: 230; Laver, 1970: 63–5), there still remains a considerable body of anatomical and physiological evidence to suggest that human beings are 'designed' for the production (and reception) of speech. At the same time, it must be remembered that language is in principle dissociable from speech (under the principle of medium-transferability discussed in Section 3); and none of the anatomical and physiological evidence that is cited in connexion with the claim that language is species-specific relates directly and solely to the kind of semanticity and the degree of grammatical complexity which, as we have seen, appears to distinguish verbal from non-verbal communication (cf. Section 1).

Under (iii), Lenneberg refers to the fact that 'the ability to learn language is so deeply rooted in many that children learn it even in the face of dramatic handicaps' (congenital deafness, blindness, etc.). This is well attested; and it is a fact which speaks in favour of the hypothesis that all children are very strongly motivated to acquire the principal system of communication employed in the society in which they are reared. The further fact, if it is a fact, that there is 'a "critical period" when the brain is specially "tuned" to language' (Marshall, 1970: 240) and that, if language is not acquired during that period, it may not be properly acquired at all, whilst it imposes some limitations upon the principle that language is 'difficult to suppress', would seem to suggest that the child's motivation and ability to learn society's principal means of communication are innate. Once again, however, it must be insisted that the facts referred to under this heading clearly do not prove that the child is biologically disposed to learn language as such.

Under (v), Lenneberg claims that 'every language, without exception, is based on the same *universal principles* of semantics, syntax, and phonology' and refers to a number of linguists (Chomsky, Greenberg, Hartmann, Hjelmslev) who, 'working quite independently of each another', have maintained that the 'syntax of every language shows some basic, formal properties, or, in other words, is always of a peculiar algebraic type'. The force of this claim is somewhat lessened, it should be realised, by the fact that the linguists referred to are far from being in agreement as to what these formal properties are. That the syntax of all languages is of considerable formal complexity, as was said to be the case in Section 5, is no doubt true. Just how universal the more particular aspects of this formal complexity are is a matter of controversy (see note 18 below). For a general discussion of the problem of language universals see House-holder (1970: Chapter 3).

Under the term 'developmental schedule' Lenneberg discusses the fact that 'the onset of speech is an extremely regular phenomenon, appearing at a certain time in the child's physical development and following a fixed sequence of events.' All the systematic research carried out so far on the acquisition of language by children (for references see Campbell and Wales, 1970) confirms that there is a relatively constant 'developmental schedule', the early stages through which the child passes being: (1) babbling, in association with which the prosodic patterns of the language are developed; (2) holophrastic speech, consisting of grammatically unstructured 'one-word' utterances; (3) simple two-word and three-word utterances of a 'telegraphic' character, lacking an overt marking of such distinctions as present v. past tense, singular v. plural, or 'a' v. 'the'. We are perhaps entitled to conclude from the fact that there is a relatively fixed 'developmental schedule' that 'innate schemata, determined by genetic and maturational factors, impose fairly strong constraints upon the mechanisms of language acquisition' (Marshall, 1970: 240). We are obviously not entitled, without further evidence, to conclude that the innate schemata referred to are specific to language acquisition rather than to the more general cognitive development of the child. In this connexion, it is interesting to observe that certain psychologists working in the field of language acquisition have suggested a correlation between the developmental schedule for speech and the stages of cognitive development recognised by Piaget (Campbell and Wales, 1970: 252 ff.; Brown, 1970).

I have left until the end what might, on the face of it, seem to be the most directly relevant argument for (or against) the species-specificity of language. When he says, under (iv), that 'language cannot be taught', Lenneberg means that 'there is no evidence that any non-human form has the capacity to acquire even the most primitive stages of development'. Marshall (1970: 231) refers to 'attempts to teach interesting subparts of natural languages to chimpanzees' and comments: 'Past efforts in this direction have been failures, although such studies continue unabated . . . and eventual success cannot be ruled out on *a priori* grounds.' The question which now arises is whether more recent evidence than that which was available at the time Lenneberg and Marshall were writing, such as that derived from the Gardners' experiment with the chimpanzee Washoe (described by Thorpe in the previous chapter), is sufficient to contradict the assertion that language cannot be taught to any non-human species.

Brown (1970), in his discussion of the Gardners' work and of its theoretical implications, makes it clear that Washoe's use of the sign language being taught to her was characterised at the age of 37 months by the properties of 'semanticity' and openness (but not displacement). If these properties are taken jointly as being sufficient to define language then we must of course conclude that Washoe had by the age of 37 months demonstrated an ability to learn language. But I have already suggested that openness as such should not be regarded as criterial (see Section 5). According to Brown (1970: 229) whether Washoe can be said to have acquired any 'real syntactic capacity' or not is still uncertain. Her 'utterances' can certainly not be described as formally complex; and it is not even clear whether they manifest the property of commutativity or not. So far, therefore, it seems to be simply a matter of how we define 'language' whether we say that Washoe's ability differs from the human capacity for language in 'degree' or in 'kind'. In saying this one is not of course denying that the Gardners have shown that a chimpanzee is capable of learning a system of communication with properties which many had thought previously were beyond the intellectual capacity of any non-human species.

Furthermore, the utterances of very young children have a very simple grammatical structure; and the particular interest of Brown's paper lies in his comparison of Washoe's 'utterances' with those of young children at an equivalent stage in the developmental schedule ('equivalence' being defined for the purpose, not in terms of age, but in terms of the average length of utterances). In common with a number of other scholars investigating the acquisition of language by children, Brown has come to the view that the utterances of very young children (at the earliest stage at which one can say that their utterances have any grammatical structure at all – that is to say, when they have passed through the holophrastic stage) can all be accounted for in terms of a small set of 'structural meanings' (many of which can be identified with semiotic functions of the kind that I described as 'basic' in Section 6). These 'structural meanings', it may be added, cannot be assigned to utterances solely on the basis of the occurrent words: the same combination of words may be given two or more different interpretations according to the context in which it is produced by the child and the investigator's judgement of the child's 'intention' (e.g. *Mommy lunch* might be described as an instance of a possessive utterance, 'Mommy's lunch', or an agent-object utterance, 'Mommy is having lunch'). In his comparison of Washoe's 'utterances' (i.e. sequences of signs) with the utterances of young children analysed in terms of their structural meanings, Brown (1970: 224) points out that many of Washoe's two-term and three-term sequences are similar to children's utterances both in grammatical structure and in meaning. He also relates the set of structural meanings that he recognises to the 'sensory–motor' intelligence postulated by Piaget: 'If the meanings of the first sentences are an extension of sensory–motor intelligence then they are probably universal in mankind. Universal in mankind but not limited to mankind and not innate. Animals may operate with sensory–motor intelligence, and Piaget's work shows that it develops in the infant, over many months, out of his commerce with the animate and inanimate world'. The implication here seems to be that the earliest stages of language development, but not the later stages, are under the control of the sensory–motor intelligence; and that as a consequence, we might expect certain species of animals to reach, but not go beyond, these earliest stages. In view of the structural and functional parallels that have been drawn between human non-verbal communication (including the non-verbal component in language) and animal signalling systems, one might perhaps go on to hypothesise that non-verbal communication, in general, is under the control of 'sensory–motor' intelligence, whereas language in its fully developed form (though it continues to make use of the sensory–motor basis) requires the higher modes of cognitive ability.

One further point should be made in connexion with Brown's reference to Piaget's work. This has to do with the suggestion that the meanings of the earliest utterances are 'universal in mankind . . . but not innate'. Campbell and Wales (1970: 257) in their reference to the work of Piaget, talk of 'certain, presumably innate, principles of internal reorganisation' – what Piaget calls 'processes of equilibration' – as co-operating with 'external environmental factors' (cf. Brown's 'commerce with the animate and inanimate world') in 'the development of the system of language in children'. Here we find adumbrated an alternative theory of language acquisition to the one presented by such scholars as Chomsky (1965, 1968) and Lenneberg (1964, 1967). According to Chomsky, the fact that children are able to learn successfully the language of the community in which they are brought up, in a relatively short time and on the basis of a very limited sample of utterances, cannot be satisfactorily explained other than by assuming that children are genetically endowed with a 'knowledge' of the structural principles common to all human languages (see note 18). According to the alternative theory of language-acquisition, there is no need to postulate any specifically linguistic 'knowledge' that is innate in the child and brought to bear upon the utterances he hears around him. It is assumed instead that the child gradually comes to an understanding of the formal principles underlying the construction of the utterances he hears by applying to this task abilities which, far from being specifically linguistic, are applicable in every aspect of his cognitive development. It would seem to follow from this that the inability of animals other than man to get beyond the earliest stages of language acquisition derives from their failure to reach the necessary level of cognitive, or intellectual, ability rather than from their lack of a specific 'language faculty'. At the present time the evidence in favour of the one or the other account of language-acquisition is inconclusive. Interestingly enough, both theories are compatible with the view that language, in its fully developed form, is unique to man; and neither has anything to say about the origin of language.

NOTES

1 I am indebted to the following for their comments on an earlier draft of this chapter: R. E. Asher, Michael Argyle, E. K. Brown, Gillian Brown, J. J. Christie, David Crystal, Michael Cullen, John Laver, D. M. MacKay, P. H. Matthews, J. P. Thorne, H. G. Widdowson.

2 'We all act, as listeners, as if we were experts in using information in voice quality to reach conclusions about biological, psychological and social characteristics of speakers. . . There is a good deal of evidence that in such subjective judgements we operate with stereotypes. . . Listeners, if they are from the same culture, tend to reach the same indexical conclusions from the same evidence, but the conclusions themselves may, on occasion, bear no reliable relation to the real characteristics of the speaker. Of the three types of indexical information in voice quality, biological, psychological, and social, it is the biological information which probably tends to lead to the most accurate conclusions, especially as to sex and age. . . Psychological and social conclusions are much more likely to be erroneous, because of their culturally relative nature, and because they derive from a more variable strand of the speaker's voice quality, the habitual settings of the larynx and vocal tract.' (Laver, 1968: 50–1.) In this article Laver puts forward a labelling system for voice quality according to which impressionistic terms such as 'sepulchral voice', 'reedy voice', 'harsh voice', would be replaced with more precise phonetically interpretable labels. He gives a number of examples of the indexical information signalled by voice quality.

3 More precisely, they will be taken account of within the linguist's model of the language-system; and this is what is implied by calling them 'linguistic'. I use the adjective 'linguistic' in this sense exclusively throughout this chapter: i.e. as 'pertaining to the language-system (as described by the linguist)', and not as 'pertaining to language-behaviour' (see Section 2). It should be added that there has been some necessary simplification in the exemplification of intonation and stress given here. Stress is not independent of, but interacts with and affects, the intonation 'pattern'. For discussion and references to the literature see Crystal (1969).

4 I am not here concerned with what is sometimes referred to as the 'accentuation' of words by means of such prosodic features as tone or stress. For example, there are many pairs of words in English distinguished by stress: e.g. *contrást* (verb) and *cóntrast* (noun): cf. Gimson (1970: 233–6). In many so-called 'tone languages' (of which the most familiar is perhaps Chinese) unrelated lexical items are distinguished by tone in much the same way as unrelated lexical items in English may be distinguished by their constituent vowels and consonants (e.g. *pet: pat, pet: pen*). Lexical tone is part of the phonological system of a language (see Section 2). Complications and difficulties arise, however, when we come to consider the full range of prosodic phenomena and their function in

various languages. It is not always easy to draw a sharp distinction between lexical and non-lexical 'accent'. This does not, however, affect the argument of the present chapter.

5 It should perhaps be added that there may be conflict between a phonetic and a functional classification of the non-segmental features of utterances. That is to say, what might be regarded as prosodic in terms of some general phonetic account of the phenomena could be treated as paralinguistic in a particular language by virtue of its function in that language (the kind of information signalled, the optionality of the features in question, their lack of integration with the grammatical system, etc.). This is yet another reason for saying that there is no clear-cut distinction between the prosodic and the paralinguistic, and consequently between language and non-language.

6 I am not of course implying that there can be no theoretical discussion of the 'raw data'. As I will argue in this section the identification of utterances as tokens of the same type must be made within some theoretical framework.

7 My own account of this distinction has been strongly influenced by Bar-Hillel's, but differs from his in certain respects.

8 They would be *symptoms* of the internal state of the speaker according to the terminology introduced below.

9 By 'stimulus-free' or 'free from stimulus control' Chomsky means 'free from the control of detectable stimuli, either external or internal' (1968: 11).

10 The process of utterance may be described as one by virtue of which sentences are *contextualised*.

11 I am using the term 'linguistic', as earlier, to include both the verbal and the prosodic features of utterances. To the extent that paralinguistic features, voice-quality and other signals in the vocal–auditory channel produced as part of previous utterances convey contextual information, this may be included in the admittedly vague notion of 'situation'. Many authors distinguish between 'situational' and 'verbal' context; but it is not always clear how much they intend to be covered by 'verbal'.

12 Many linguists would not accept the distinction between 'complete' and 'incomplete' sentences and would prefer to talk in terms of a distinction between 'favourite sentence types' and 'non-favourite sentence types' (cf. Robins, 1964: 231–4; see also Pike, 1967: 147; Halliday, 1970: 145). If the distinction between sentences and utterances is drawn as I have drawn it here, it seems to me that we can give a satisfactory interpretation to the traditional notion of grammatical 'completeness'. And it is my assumption that the rules one would need to set up in order to account for the contextualisation of sentences (as utterances) would be to a high degree independent of the more particular grammatical features of different languages and furthermore that the rules would require access to only a relatively small part of the information in the structural analysis assigned to the sentences by the grammar which generates them. If these assumptions are justified there would seem to be good reason for adopting the view I have put forward in this section (cf. Matthews, 1967: 119–20).

13 It has been suggested that one of the difficulties with the hypothesis I have just put forward is that non-linguistic behaviour inconsistent with an accompanying utterance might be simply ignored (i.e. taken as communicatively irrelevant) rather than interpreted as 'contradicting' the utterance. This is no doubt true in many instances, but not, I think, in all.

14 Cases such as those of Helen Keller, though sufficient to establish the point that it is possible to learn to read without having first acquired the corresponding spoken language, do not suffice to establish the complete independence of the

written language, since Helen Keller's teacher spoke English normally and made use of it in teaching her pupil, not only to read written English, but also to lip-read spoken English. For further discussion of this question, see House-holder (1970: Chapter 3).

15 I am not suggesting that the orthographic English word-form (i.e. 'pattern combination' of letters occurring between spaces) invariably corresponds to what one might wish to recognise as a word-form in the spoken language. For a discussion of the English word-form from the point of view of its phonemic and prosodic characteristics, see Gimson (1970: 222–303).

16 There are of course two senses in which they might be non-commutative: (1) by having $x+y$ as a permissible pattern, but not $y+x$; (2) by having both $x+y$ and $y+x$, but as non-equivalent patterns. It seems clear that a signalling system which permits $x+y$ and $y+x$ as non-equivalent patterns is a more complex system than one which permits $x+y$ but not $y+x$; and other things being equal, it should perhaps also be rated as more complex than a commutative system. It has been pointed out to me that the ethologist's appeal to 'sequence' does not necessarily imply an assumption of non-commutativity in the system underlying the behaviour-patterns they are studying.

17 It is perhaps debatable whether Chomsky's proof of the inadequacy of finite-state grammar is valid, as far as weak generative capacity is concerned, since it depends upon there being no upper limit to the embedding of non-adjacent interdependent words in sentences; and there are acknowledged limits upon this process in language-behaviour. Chomsky's view is that these limitations are matters of performance, not competence. From the point of view of strong generative capacity, in whatever sense this is interpreted, Chomsky is surely right. Finite state grammars are 'unnatural' and 'clumsy' for the description of the syntax of human language; and there is plenty of evidence to suggest that language is not, in fact, processed in a 'finite-state' manner.

18 Chomsky has argued that the formal principles underlying the structure of language are so specific and so highly articulated that they must be assumed to be biologically determined (cf. Chomsky, 1968; also Lenneberg, 1967). This is a very controversial issue. While I see nothing implausible in the suggestion that a 'knowledge' of the universal principles of language-structure should be genetically transmitted, I do not consider that the evidence from linguistics is as yet very convincing (cf. Matthews, 1967; Lyons, 1970a: 109–116). For that reason I have not gone into the question here.

19 See note 9 above. Freedom from stimulus control does not necessarily imply the denial of any causal connexion between a certain state of the organism (and, more particularly, of the brain) and the emission of a certain utterance. What is being denied is the assumption that language can be accounted for within the framework of a mechanistic theory of the kind put forward by behaviourist psychologists.

20 The notion of 'intention' is perhaps reducible to something with a less 'mentalistic' flavour about it (see Chapter 1). But it cannot be entirely eliminated.

21 By 'utterances' I here mean 'utterance-types'. Since tokens of many such utterance-types are of high frequency in recurrent, and independently-identifiable, situations, it may well be the case that a considerable portion of our everyday utterances counted as tokens is non-communicative.

22 This sense of 'indexical' is not to be confused with the sense of the term as it was introduced into philosophical discussion by Bar-Hillel (1954: cf. also Searle, 1969: 80). Both senses derive from Peirce's definition of an 'index' (e.g. 1940: 107). For Bar-Hillel's sense of 'indexical', the term *deictic* is now

quite commonly used by linguists (as in Lyons, 1968: 275) and would seem to be preferable (see below).

23 Many linguists now use Searle's (1969) term 'speech acts' to refer to the phenomena I am calling 'semiotic functions'. This is unfortunate in two respects. (i) 'Speech act' has long been employed in linguistics in the more obvious sense of 'act of (spoken) utterance': cf. Halliday (1970: 142). (ii) The functions in question are just as relevant to the interpretation of the written language as they are to the analysis of speech; and some of them at least would seem to be identifiable in other signalling systems than language.

24 An interesting subclass of nominative functions (defining, christening, etc.) which operate within a particular framework of cultural assumptions and practices may be described as 'performative' (Austin, 1962).

COMMENTS ON PART A

The chapters in this section have been concerned with classifying the subject matter, with relating its different aspects, and with drawing a number of distinctions which will be useful in its study.

(I) DEFINITION OF 'COMMUNICATION'

The first problem concerns the use of the term 'communication'. As MacKay emphasizes, it is important to use words in such a manner that essential distinctions are maintained. From the point of view of the biologist interested in the internal organization of the communicatory process, the distinction which stands out concerns the goal-directedness of the sender. MacKay therefore advocates that for present purposes the term communication be limited to those cases in which the behaviour of the sender is goal-directed with respect to the receiver. In the commonest form of goal-directed signalling the effect of the sender's behaviour on the receiver is monitored by the sender in such a way as to promote corrective action by the sender if the message appears to be ineffective. The criterion is thus a variational one, but MacKay (Chapter 1) dismisses the possible objection that this is too hypothetical to be useful on the grounds that it can readily be expressed in terms of the 'information flow map' of the sender, which is in principle open to examination.

MacKay's distinction between an animal which acts merely *in such a way as* to bring about a particular event (which is thus an accidental consequence of its behaviour), and an animal which acts *in order to* bring about an event (which is thus a goal), is clearly an important one. But as he points out it is by no means so easy to apply in practice as it might seem. From a control system approach we should expect that, in a goal-directed system, (*a*) there should be a repertoire of alternative courses of action; (*b*) that most appropriate to the goal should be chosen; (*c*) the intensity or frequency of corrective behaviour should be determined by discrepancy between the present situation and the goal situation and (*d*) the correction should average to zero when present and goal situation coincide. In fact each of these points presents difficulties.

(*a*) Criteria for the existence of alternative courses of action are difficult to specify. Does it imply that the sender should possess alternative effectors or central nervous mechanisms? What if it had these but never used them in the context in question – in what sense could an alternative course be said to exist if the animal never used it? Or does it mean the animal should have a general ability to develop new courses? Lyons faces a similar difficulty when he discusses the criterion of choice on the part of the sender: if he 'cannot but behave in a certain way, then he cannot communicate anything by behaving in that way'. But Lyons in fact finds it necessary to extend 'choice' to include habitual modes of behaviour signalling indexical information over which the sender is not exercising any immediate positive control.

(*b*) Goal-directed behaviour requires the postulation of an image or internal correlate of the goal. But the usefulness of this is not always clear-cut. In another context, the long history of dispute between stimulus–response (e.g. Hull, 1943) and cognitive (e.g. Tolman, 1932) theorists testifies to the difficulty of deciding whether or not an activity such as maze-running necessitates the postulation of cognitive processes which could involve comparison between present and goal situations (e.g. Osgood, 1953).

(*c*) If the intensity of corrective behaviour is governed by the difference between the goal situation and the present situation, we might expect it to decrease as the goal is approached. But sometimes activities which might be described as 'goal-directed' *increase* in frequency or intensity as a goal is approached: rats may run faster as they get near the goal (Miller, 1959) and human subjects may work harder as a task nears completion (Welford, 1962). Blurton Jones (Chapter 10) cites a case of 'goal-directed behaviour' where the intensity of the corrective behaviour is more closely related to time since initiation of the behaviour than to proximity to the goal.

(*d*) Responses may come to an end in a number of different ways, many of which do not involve the perception of new 'goal' stimuli – for instance through the removal of the eliciting stimuli, or by internal consequences of responding. Furthermore, and with reference to all these criteria, activities which meet any particular definition of goal-directedness may differ widely in complexity (Hinde and Stevenson, 1970).

Comparable difficulties arise with verbal behaviour. True, if we wish to cause another individual to behave in a particular way, we may select from our verbal repertoire those items most likely to be successful, and we cease talking when our aim is achieved. But how much more complex is the case of informative statements, where our criteria of reaching a goal depend on assessing a change in the 'internal organization' of the recipient. And what of performative statements, such as 'I name this ship . . .' (Austin, 1962)? Again, when the Italian 'prego' comes as an automatic response to 'grazie', a superficial application of MacKay's criteria might suggest that it is non-communicative. In such a case we can either (*a*) regard the utterance as mandatory in the context, and thus as having no meaning there: in that case this use of verbal language would not involve communication (Lyons, 1963). Or (*b*) we can consider what the alternative of silence might have conveyed to the receiver, and recognize the long-term consequences of the interchange in maintaining a good relationship between the speakers as part of the intention behind the utterance. We shall return to this point later (see comment by MacKay on Chapter 4).

Such difficulties may be thought of merely as presenting a challenge to the control systems approach. But they certainly indicate that a distinction between behaviour which merely affects others and behaviour which 'is intended to' may not be so easy as the engineer's diagrams imply.

In addition to this, MacKay makes a further distinction according to whether or not the sender's behaviour is perceived to be goal-directed by the receiver – a matter, as he emphasizes, of crucial importance in many human interchanges (see also Lyons in Chapter 3), but difficult to verify in sub-human species.

MacKay's dichotomies are not merely means of classification – they highlight new problems, such as the nature of the differences between the social behaviour of young and old organisms, or between animals at different phyletic levels, and indicate some ways in which they can be tackled. As an example, the usefulness of such distinctions in defining problems in the complicated case of the development of the human mother-child relationship is apparent from Bowlby's (1969) work on 'Attachment'. But students of social behaviour come from many different backgrounds and have diverse interests, and for some MacKay's dichotomies are not the most pressing ones. For instance biologists interested in the evolution of behaviour have tended to focus on the distinction between behaviour which appears to have become adapted in evolution for a signal function, and that which does not. The evidence for such adaptation is complex and is discussed by Cullen in Chapter 4: in brief, it comes in part from the nature of the behaviour itself and its relation to other functional requirements, in part from the context in which it appears, and in part from the comparative study of closely related species (see also e.g. Tinbergen, 1959; Hinde, 1970). As MacKay stresses, behaviour adapted for a signalling function may or may not be goal-directed to that end. Indeed, some such behaviour may be goal-directed in a sense, but towards broadcasting signals rather than towards affecting the behaviour of a particular individual. Furthermore, behaviour which is goal-directed towards affecting the behaviour of others may be idiosyncratic and not adapted through processes of natural selection to that end (see e.g. several of the chapters in Part C). Thus the distinction between behaviour which is adapted for a signal function and that which is not cuts across MacKay's dichotomies. It is, however, the focus of attention of all the contributors to Part B of this volume, and of some of those in Part C (especially van Hooff and Eibl-Eibesfeldt).

A third type of dichotomy comes from questions concerning the bases of the social organization in a community of individuals. If we wish to understand the full nature of their interactions, we must distinguish between individual behaviour which does and does not influence the behaviour of others. For example, Altmann (1962, 1965) attempted to make a catalogue of the 'social behaviour' of rhesus monkeys on the criterion that a pattern of behaviour would be 'considered to be social if, and only if, it affects the behaviour of other members of the society'. Such a distinction can be based on an analysis of sequential associations between the behaviour of senders and potential receivers. Altmann's catalogue includes many patterns for which there would seem to be no

Table 1. *Focus of interest of socio-biologist, evolutionist and 'control systems biologist' in the study of communication. 'Directed' implies 'with the intention of influencing B' in the sense discussed by MacKay (Chapter 1). 'Adapted' means 'adapted for a signal function' according to evidence of the type reviewed by Cullen (Chapter 4)*

	Socio-biologist	Evolu-tionist	Informa-tion theorist
A affects *B* independently of *A*'s behaviour	√	—	—
A's behaviour affects *B* — *A*'s behaviour not adapted, not directed	√	—	—
A's behaviour adapted, not directed	√	√	—
A's behaviour adapted, directed	√	√	√
A's behaviour not adapted, directed	√	—	√

evidence for evolutionary adaptation for a signal function – this is not the issue with which he is concerned. If the dynamics of social interaction are to be understood, it is essential to regard all patterns of behaviour as potentially significant rather than to exclude some prematurely. Altmann (1965) proceeded by assessing how far the constraints imposed on an individual's behaviour by preceding actions within the group could be accounted for in terms of stochastic models. Some of the recent work discussed in Part C (see especially the chapter by Argyle) involves evidence for the previously unrecognized social significance of certain postures, gestures, expressive movements and positional changes in man.

In so far as a dichotomy based on socially significant behaviour is derived from sequential relations between the behaviour of one individual and that of others, two minor reservations are necessary. First, as MacKay points out, in the study of communication we are usually interested in something more than the merely Newtonian consequences of A on B: to push another individual away involves a quite different mechanism from saying 'Go'. The sociobiologist, however, may be interested in both, since their consequences on the structure of the society may be similar. Second, it must not be forgotten that, at any rate for gregarious species, the mere presence of other individuals has an important effect on behaviour. Thus effects produced by changes in the behaviour of a sender are superimposed on the consequences of his presence.

We see, then, that the dichotomy considered to be important will vary with the interests of the worker concerned. Although agreement about the use of words facilitates progress in any field of knowledge, it is not always possible, especially when words are taken over from common speech. However it is even more important to agree on the distinctions which are useful. To this end, Table 1 summarizes some of the principal points from the preceding discussion, indicating the distinctions which may be useful to a hypothetical socio-biologist, evolutionist and information theorist.

(2) CLASSIFICATION

Given some understanding of the distinctions which may be implied by 'communication', the next question concerns the classification of the subject matter. Lyons, approaching the problem as a linguist, first addressed himself to the problem of defining language. This led him to a series of distinctions which are summarised in Table 2. On this basis, social inter-actions between people involve a number of components. Amongst these it is not profitable to search for an absolute distinction between linguistic and non-linguistic components, but preferable to recognize that there are degrees of 'linguisticness', with the verbal component at the top of the scale. Most of the subsequent chapters in this book are concerned with gestures unrelated to verbal communication or, to a lesser extent, with paralinguistic, prosodic and voice quality components. But Lyons emphasizes that, in the one species which incontrovertibly possesses a verbal language, language and non-language are so closely interwoven that a precise distinction is impracticable. A similar conclusion is reached, though on rather different grounds, by Leach in Chapter 12, and the nature of the relations between human non-verbal communication and language becomes a crucial issue in a number of chapters in Part C.

It is to be noted that the distinctions drawn by Lyons between more and less 'linguistic' do not necessarily correspond with the distinction between communication (in the narrow, goal-directed sense) and non-communication advocated by MacKay: all components indicated in Table

2, except (usually) vocal reflexes, may be used in goal-directed communication. Furthermore Lyons' criteria of language do not necessarily involve all the design features of Hockett (see Chapter 2): in his view vocalization is not an essential feature of language, duality may not be, 'openness' (or creativity) is not a necessary criterion, while semanticity is a feature whose nature depends on the meaning attached to 'meaning' (see below).

Parenthetically, it is proper to say that this classificatory system is only one amongst many in the recent literature (see e.g. Ramsay, 1969). One other, which carries implications which will become important later in this volume, may be mentioned here. Sebeok (1968, 1969), concerned

Table 2. *Relations between verbal and non-verbal communication*

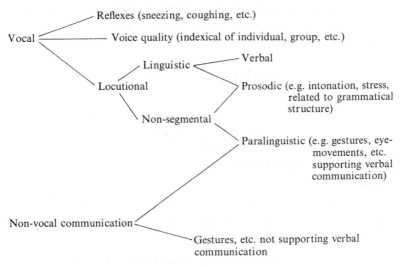

with a much broader range of subject matter – semiotics, 'the general theory of signs' (Sebeok, 1969: 200) – distinguishes between systems which are peculiar to man ('anthroposemiotic') and those found also in other animals ('zoosemiotic'). Within the former he distinguishes, on the one hand, language and those objects and customs which man has elevated to the status of sign systems as he has filtered them through his languages, such as clothing (Barthes, 1967) and cooking (Lévi-Strauss, 1966) in their communicative functions. On the other hand are those semiotic structures in which men are coupled by systems which do not imply any particular linguistic code (e.g. music and painting). At first sight it might be thought that our present concern would be solely with the latter aspect of 'anthroposemiotic' and with 'zoosemiotic'. But, as discussed above, the distinction between those aspects of human communication which do and do not depend on language is not easy to make (see Chapters 11–13). Furthermore, the distinction between anthroposemiotic and zoosemiotic would not be useful to those contributors who are seeking for relationships between animal communication and certain aspects of human communication.

Within zoosemiotics, Sebeok (e.g. 1969) uses the communication engineer's distinctions between six factors: (1) source, (2) channel, (3) destination, (4) code, (5) message and (6) context. These are self-explanatory, and the reader may find that they form a convenient framework for much that is discussed in subsequent chapters. Sebeok relates these to Morris' (1946) three areas of semiotics (see Lyons, Chapter 3) as follows: source, channel and destination belong to zoopragmatics, which deals with the origin, propagation and effects of signs; code and message belong to zoosyntactics; while zoosemantics involves the relations of context with the others.

(3) COMPLEXITY AND THE PROPERTIES OF COMMUNICATION SYSTEMS

Yet another classificatory system, advanced by Tavolga (e.g. 1970), is designed to accentuate the differences in complexity or organization between the communicatory systems of animals at different phyletic levels (cf. Schneirla, 1953). Considering the broad subject of interactions between organisms, he recognizes the following:

1. *Vegetable level*. Growth and tropism, and the effects of physical presence, e.g. effect of plants on each other, and on the animals which feed on them.

2. *Tonic Level*. Based on continuing processes basic to individual development, such as excretion and cellular metabolism. Production of chemical exudates, and some forms of trophallactic and trail-following behaviour.

3. *Phasic level*. Based on discontinuous more or less regular stages in development. Involve broad, multichannel energy outputs with some specialization of emitter and discriminatory responsiveness in receiver, e.g. sex recognition in lower vertebrates, schooling in fishes.

4. *Signal level*. Specialized structures producing specific narrow band signals:

(*a*) Biosocial signal level. Communication within social situations in which organic processes control development of behaviour. e.g. much reproductive behaviour of lower vertebrates and birds.

(*b*) Psychosocial level. Complex patterns of signals, with an increasing role of experimental and social factors. e.g. Lower mammals, bird song.

5. *Symbolic level.* Plastic and variable, dependent on social interactions for its development, with primitive use of symbolic sounds and gestures. Infra-human primates, especially anthropoids.

6. *Language level.* Communication of abstract ideas, speech, meta language. Restricted to man.

Tavolga admits that any one animal may operate at more than one level, and that it is difficult to assign some instances of communication with certainty to one or another level. But in emphasizing phyletic differences he is seeking to avoid 'cross phyletic generalizations with little regard for the differences in phyletic level among the organisms being compared'. In his view a comparative approach must take into account the evolutionary and ontogenetic history of the organisms concerned, and not depend on superficial similarities which conceal qualitatively different mechanisms.

Tavolga's warning against superficial comparisons is timely, but those who are interested in the properties of communication systems as such need features which cut across phyletic levels. Indeed it is only on the basis of particular features that distinctions between levels can be made. It is here that Hockett's 'design features', as discussed by Thorpe in Chapter 2, can be valuable. As Thorpe stresses, and as Hockett and Altmann (1968; Altmann, 1967) were well aware, there are many difficulties in the way of applying the scheme. And, to take Tavolga's point, many of these arise from the fact that each feature covers a range of mechanisms of diverse complexity and sophistication.

Some of the difficulties inherent in the concepts of duality and semanticity are mentioned by Lyons (Chapter 3). Ramsay (1969) points out that significant duality requires that a small number of meaningless units be combined into a large number of meaningful ones: the value of the ratio necessary before one can speak of significant duality is a matter of opinion. Sebeok (1969) takes the extreme view that there are only two semiotic systems which possess duality, human language and the genetic code. With regard to semanticity, some students of animal behaviour would argue that the possession of different calls for different types of predators by many passerine birds (Marler, 1959) and some monkeys (Struhsaker, 1967) is evidence for semanticity. Others hold that man is unique in his ability to communicate information about the environment, as opposed to information about the signaller. But when the matter is pressed it becomes arbitrary to draw a hard line anywhere in the series 'There is a predator behind that tree'; 'I know there is a predator behind that tree'; 'I am afraid because there is a predator behind that tree'; 'I am afraid'. Which, if any, conveys information only about the signaller and which, if any, about the environment?

Again, when is it really justifiable to say that a sub-human communication system, with its limited number of signals and combinations, is open? Is it helpful to imply that the bee dance shares the property of openness with human language, when it can only convey information about the location of a few objects of biological importance?

Many other difficulties and ambiguities in the application of the design features listed in Table 1 or Chapter 2 are discussed in the papers of Hockett and Altmann (see also Bronowski, 1967). They are mentioned here only to emphasize that this scheme, like any other, must not be over-extended or held to carry implications concerning questions to which it is not relevant: the design features provide some information about the capabilities and properties of a system, but not about its mechanism. Especially in its more recent versions, it is fertile in posing new questions about communication systems. But these questions demand a very sophisticated appreciation of the status of 'design features'.

One final point about the complexity of human language. While the complexity must not be underestimated for a moment, it is also important not to get it out of perspective. We must remember that particular efforts have been made to expose its complexity. Furthermore, our estimates of the complexity of phenomena are, in fact, more closely related to the sorts of description we are able to give of them than to the phenomena themselves. Take a much simpler phenomenon – the proven ability of some birds to navigate. One current hypothesis, certainly not adequate but not super-seded, is that, during the day time, a pigeon released in a strange place assesses in a few seconds the altitude and arc of the sun. From this it does the equivalent of computing the altitude of the highest point of the arc and the azimuth of the point of sunrise, using a Nautical Almanac and an accurate chronometer to compute its latitude and longitude, comparing this with the known latitude and longitude of home, using chart and protractor to assess the track to steer, and then sets course by the sun (Matthews, 1968). Other animals navigate, and not all use the sun. Indeed some animals can probably transpose from one set of co-ordinates to another.

(4) ORIGINS OF LANGUAGE

Where does this leave us with respect to the relations between the communication systems of man and those of animals? Some biologists, in part as an article of faith, have argued that human verbal languages must have evolved, and must therefore have evolutionary origins, and that the most probable source lies in the non-verbal systems of animals. Linguists, however, find no signs of evolutionary change within languages themselves, and must thus remain agnostic on the question. Disappointingly, perhaps,

Hockett's design features are of little help here. One might have hoped to find an increasing trend towards the human level in higher sub-human species, but there is little evidence for this. The distribution of design features over species of different phyla (Table 1, Chapter 2) forms a patchwork, with no clear trends towards the higher mammals (see Thorpe, Chapter 6). To some extent, this is a consequence of the fact that communication systems increased in complexity on many separate branches of the evolutionary tree. It is also, as we have seen, a result of the breadth of the design features themselves. On the latter point Chomsky (1968), discussing Thorpe's (1967) view that the characteristic properties of human language are found in the communication systems of lower forms (see also Chapter 6), points out that the sense in which human languages are propositional is entirely different from the sense in which animal communication systems can be said to be so. Chomsky describes animal communication systems as follows: 'Either it consists of a fixed, finite number of signals, each associated with a specific range of behaviour or emotional state, . . .; or it makes use of a fixed finite number of linguistic dimensions, each of which is associated with a particular nonlinguistic dimension in such a way that selection of a point along the linguistic dimension determines and signals a certain point along the associated nonlinguistic dimension.' An arbitrary statement in a human language is entirely different from either of these. Chomsky therefore argues that 'human language appears to be a unique phenomenon, without significant analogue in the animal world'.

But in any case Lyons argues on several grounds that the question of the origin of language is at present unanswerable. First the intimate interpenetration of language and non-language is such that the primacy of one or the other cannot be established, and the difficulty of distinguishing them makes the formulation of precise questions about their relationship difficult. Second, questions about a direct relationship between non-verbal systems as we see them now, whether in man or animals, and verbal language as we see it now, are too crude. Only when the questions are refined as a result of analysis, and we ask about the relations between particular properties of human language and those of non-verbal systems, can we hope for progress. It is here that a sophisticated use of 'design features' could be useful, and the comparisons being made by the Gardners and by Premack between particular aspects of the performance of their chimpanzees and corresponding aspects of the language of human children offer particular hope for progress (see summary by Thorpe, Chapter 2; see also discussion by Bronowski and Bellugi, 1970). But as we have seen, Lyons argues cogently that the lines of evidence used by Lenneberg (1967) to support his view that language depends on properties peculiar to man are not so substantial as appears at first sight – though there is stronger evidence for supposing that particular characteristics of language are peculiar to man.

This leaves open the question of whether any features of language which are peculiar to the human species depend on an 'innate' knowledge of the structural properties of all human languages, or on a more general level of intellectual or cognitive ability. While noting that a human imbecile may show more linguistic ability than an ape which surpasses him in problem-solving ability and other adaptive behaviour, and thus inclining towards the former view, Chomsky (1968) pleads for a non-dogmatic approach to this problem. If we study the development of human compe- tence in a number of different domains, including that of language, and find that the same learning strategies are used in all, then there is no need to postulate a special ability for language, but if different ones are used, then we must postulate separate or partially separate faculties.

Two further points may be made here. The first concerns what is meant by 'innate'. In his earlier writings, Chomsky's formulation was essentially preformationist. For example, in his review of Skinner's *Verbal Behavior*, he wrote (1959): '. . . the capacity of a child to generalize, hypothesize, and "process information" . . . may be largely innate, or may develop through some sort of learning or through maturation of the nervous system'. The double 'or' seems to leave no alternative but that 'innate' means there in the genes. Now without entering again into the old nature/nurture controversy, all characters and abilities are a product of a continuing developmental interaction between the organism as it is at each stage and its environment. If one applies the term 'innate' to characters, one soon gets into the difficulty of what exactly it is that is innate. The genes are related to adult characters through a long process of development which involves more or less continuous interaction with the environment. Empirically the useful distinctions which can be made concern, directly or by implication, differences – differences between two individuals reared in the same environment can be described as genetically determined or in- nate, while differences between genetically similar individuals (or groups of individuals) reared in dissimilar environments can be described as environ- mental (e.g. Hinde, 1968; Lehrman, 1953, 1970). Thus statements that language is innate in man should mean that man differs from other animals in that, if both are reared in the same environment (so far as that is possible), man develops language and animals do not.

Often, however, such statements are intended to imply more than this. On the view that there is a deep structure of language common to all languages, as some linguists have suggested, or on the alternative view, taken by Chomsky, that it is the formal principles which determine both deep and surface structure which are common to all languages, the statement that the ability to acquire language is innate is sometimes taken to mean that structure is present before exposure to language, and that subsequent language development depends only on exposure to language. Such a view begs important questions. The ability to acquire language develops gradually, and may depend on all sorts of non-verbal non-auditory experience. All that can be said is that man differs from non-man in that his interaction with the world is such that after a while language acquisition is possible. And the extent to which the ability to cope with grammatical transformations depends on hearing spoken sentences in relation to events in the outside world is an open question. Just as the organization of visuo-motor co-ordination involves simultaneous experience of self-initiated movement and feed-back from seen movements of the limbs (Held, 1967–8), so also may the organization of language depend on much more than conversation. It may require conversation in relation to events in the physical world, so that some at least of the language universals may depend on universals of cause–effect relations in the physical world. Chomsky's later (1968) use of 'innate', though not explicitly referring to differences, comes nearer to allowing for this. He refers to 'the innate organization that determines what counts as linguistic experience and what knowledge of language arises on the basis of this experience'.* This leaves open the question of the precise role played by the 'very restricted and degenerate evidence', provided by the verbal input to which a child is exposed, in building up the grammar.

The second and related point is that the postulation of a genetically determined language faculty increases vulnerability to the errors which can arise from reification. These can occur at every level of biological enquiry. The psychologist sometimes finds it useful to ascribe certain regularities observable in the behaviour of food-deprived rats to a hunger drive, but it would be a logical bloomer to reify that drive, and to look for it in the brain (e.g. Miller, 1959; Hinde, 1959). To take an example from a quite different field, several decades ago Radcliffe-Brown (1948) pointed out the fallacy of reifying 'culture'. A culture is 'a description of standardized modes of behaviour – of thinking, feeling, and acting'. It is therefore nonsense to say 'The standardization of behaviour in an individual in a

---

* But see Lyons' (1970) and Chomsky's comment in the footnote on pages 113–14 of that book. More recently (1971) Chomsky has indicated awareness of the possibility that language invariants are related to uniformities in the environment, though he considers it implausible. In spite of this, he still refers to 'innate structures of mind' which by implication are distinct from maturational processes and from interaction with the environment (Chomsky, 1971: e.g. 85).

given society is the effect of culture upon him'. Yet this differs hardly at all from saying that the constant basic features of language are due to a reification of the constant basic features of language. However successful the linguists may be in abstracting the principles of a 'universal grammar', assigning those principles to the mind as an 'innate property' does not solve the problem of ontogeny.

This relates back to the question of whether language development depends on a specific or a more general faculty. Language is but one way in which man and ape differ, albeit the most important: are we for instance to postulate also a uniquely human tool-using faculty because chimpanzees seem to possess only the elements of the art of tool-making?

In conclusion, one further point must be made. Even if further evidence does bring us down on the side of the view that human language is a product of uniquely human general intellectual and cognitive abilities, it is still not enough to say that an innate difference between man and animals lies in man's superior abilities, one consequence of which is the development of language. For many of these superior abilities depend on language. Thus an understanding of the intricate interplay involved in both ontogenetic and phylogenetic development demands simultaneous analysis into component qualities of the phenomena not only of language but of all higher functions, and delicate teasing apart of the interactions between them. Whether or not communication with others is *the* function of language, its possession certainly has many other consequences for human intellectual development.

Communication in Animals

INTRODUCTION TO PART B

The discussion in the previous Part has clarified some of the distinctions important for assessing the complexity and properties of the mechanisms by which the members of a species influence each other. We now turn to a consideration of signalling systems found in sub-human forms. It would, of course, be impossible here adequately to survey the full range of signalling systems to be found in the animal kingdom; and it is fortunate that several recent reviews are available (e.g. Busnel, 1963; Frings and Frings, 1964; Sebeok, 1968). It is hoped, however, that the chapters in this section will provide some indication of the diversity of the signalling systems used by animals.

In studying animals we cannot use introspection, and special methods are necessary to show that a given movement or structure is in fact effective in influencing the behaviour of others. Furthermore, full understanding requires knowledge not only of the properties of the signalling system as it is, but also of its development in the individual and the species. It is therefore necessary to consider problems of methodology, ontogeny and evolution. These are introduced in the chapter by Cullen. There follow two chapters by Thorpe which are intended to convey an impression of the diversity of signalling systems to be found in sub-human forms. Birds are selected for special attention because the complexity of their visual and auditory signalling systems rival anything to be found amongst the sub-primate mammals, and also because they have been studied more than any other group. Thorpe devotes special attention to the particular problem of how a signal can be both sufficiently constant to convey species identity and yet have sufficient variation between individuals to convey individual identity. In the last chapter in this section mammalian displays are the subject of a chapter by Andrew: rather than attempting a review, he gives an indication of the sort of signals used by tackling a particular problem – the nature of the information which the signals could convey: this involves discussion of a basic problem in the analysis of display movements, which is introduced on p. 177.

[99]

PART B

Communication in Animals

4. SOME PRINCIPLES OF ANIMAL COMMUNICATION

J. M. CULLEN

Department of Experimental Psychology, Oxford

It is difficult to write a short article on Animal Communication without duplicating and oversimplifying much that is discussed in the monumental work on the subject edited by Thomas Sebeok (1968). However, in preparing this paper I have focussed on some problems related to the theme of the present symposium. Two particular aspects recur in these pages: first the methodological problem of determining what actions actually constitute social signals in animals, and secondly, the evolutionary problems implicit when we assert that these signalling systems have arisen in evolution by natural selection. Where the references to examples used here are not specified, they can be found in the appropriate section in Sebeok (1968), or in Thorpe's articles elsewhere in this book.

All social life in animals depends on the coordination of interactions between them and it is the means by which this coordination is achieved which will concern us here. The term 'animal communication' has often been used to refer to the kinds of signals which pass to and fro between social animals and help to mould each others' behaviour towards some goal which is to their mutual advantage. It should be said at the outset that this goal may be a piece of obvious cooperation like the mutual efforts of members of a bee hive, or equally the mutual advantage in ritualised combats between antelope in settling the winner with a minimum of physical damage to the opponents. Or even the cooperation of two different species such as the 'cleaner fish' which get their food by grazing from the skin and even inside the mouth of their heterospecific 'clients' who come to them to be ridded of their ectoparasites. Most examples of animal communication which we shall consider refer to relations between members of the same species but, as Wickler (1968) and others have argued, the term can quite properly be extended to certain inter-specific signalling systems.

As used here (and indeed widely by zoologists) the term communication is not supposed to imply that the participants are aware of the object of their interactions or that they modify their actions deliberately 'so as to communicate'. Sometimes there may be a degree of this but I know of no instance where it would be possible to satisfy MacKay's formal operational criteria for 'communication' in this full sense (see Chapter 1). Indeed I find it hard to imagine how this could be done. As MacKay makes clear, mere goal-directedness is not the criterion, though if it were, a great part of what we shall be concerned with in the next pages would count as 'communication' in his sense. In fact what has widely been called animal communication would, at best, rate in MacKay's terms as 'signalling'. In this article the terms are used interchangeably, whenever it can be demonstrated that the behaviour of one animal influences that of another.

Various riders need to be added to this bald definition. When the reactor's behaviour is changed by physical force we do not usually speak of a signal: to a man the command 'Go jump in the lake' is a signal, the push which precipitates him is not. Next, the idea that the behaviour of one animal effects a change in another must be taken in a quantitative sense: we mean that over a large number of occurrences of the first behaviour there is a change in probability of what the other animal does. No matter if the signal is sometimes sent out when there is no one there to respond: a bird's song broadcasts ownership of a territory as a lighthouse spreads its warning, and both will here be taken to count as examples of communication on the grounds that *if* perceived by a certain class of receivers, they will evoke a response, a change of behaviour. And among the responses is immobility. Many animals have alarm calls which make their young crouch or 'freeze' until an all-clear is given. Change in orientation is also a response, and so is a change in colour (blushing in man, the banding patterns of many fish in social situations) – anything in fact which involves activity in effectors. Sometimes signal and response are highly specific: the female silkworm moth (*Bombyx mori*) emits a substance which attracts the males to mate. Her scent has no function except to attract the male, and the remarkable sense organs he has evolved to detect it at great distances are equally specialised for just this role. On the other hand the visual and auditory signals of monkeys defy such a simple stimulus–response treatment (Marler, 1965).

While the most conspicuous examples of signals are those where the change which results is rapid, social signals are also known to play a more delayed role in effecting behavioural changes. For instance the sight and sound of courtship induces endocrine changes leading to gonad growth and consequent reproductive behaviour in female ring doves (*Streptopelia risoria*) (references in Lehrman and Friedman, 1969). Possibly other changes in behaviour brought about by signals may be latent, seen only if the recipient comes into a special situation. Any such delayed changes may be of the greatest importance in regulating the social lives of animals but just because they are much more difficult to study than the more rapid effects, we know less of them. For similar reasons there is much of the 'high level' social organisation of animals where we know little of the role of the coordinating signals which weld them together – for instance the cohesion of a troop of howler monkeys or a hunting band of hyenas.

Such examples are mentioned here as a reminder that although they constitute a small part of what we know about animal social signals, they may well be of the greatest importance.

HOW DO WE RECOGNISE ANIMAL SIGNALS?

We stated above that the proper criterion of signal was that it should be demonstrated to evoke a change of behaviour in another individual. Sometimes this is obvious enough: the crying of a kitten makes its mother come hurrying to find it. Often the problem is much more difficult if the signals are embedded in a mass of other behaviour which itself may affect the interactions of the animals. For instance the pre-copulation display of the male eider duck (*Somateria mollissima*) involves as much as $5\frac{1}{2}$ minutes of activity with up to thirteen different actions, including five highly ritualised displays, often occurring many times (McKinney, 1961). As this worker remarks, it is hard to believe that every movement of the male carries a special message to the female, yet it would be equally hard to believe that such a pattern of activity should have no function. Strictly speaking, it has not been established that particular displays of the drake eider have effects on the female, yet in this and many of the other instances of social encounters in animals where the meeting sets off an elaborate song and dance routine between the participants, there is a strong presumption that the performances cannot be there by accident and that they must be signalling something. This argument becomes the more convincing when the displays are actions which are not seen except in such special situations. At its weakest the argument is 'What else can account for these extraordinary actions just during meeting?' But an extensive familiarity with a species can supply considerably better evidence. Tinbergen (1959*a*) has written

> 'the study of unplanned or "natural" experiments has enabled us to make fair guesses as to the function (i.e. the social effects) of most of the gulls' displays and have led to some hypotheses which seem worth checking. While such natural experiments do not provide information about the part of a total display which is effective (e.g. whether sound, movement, posture, or colour is the main agent) they do, if observed with care and with an eye for possible flaws and for "natural controls", and if repeated often enough, tell us a great deal about the functions of the displays as wholes'.

Such a selective method, while it makes for clear cut results of a kind, is not altogether heuristically satisfying since it presupposes that the observer already knows a good deal about the kinds of changes which are likely to be produced. A blinder, blunderbuss technique is to look at *all* sequences of actions between pairs of individuals of some specified categories to see in what ways the observed transition frequencies depart from what would be expected by chance if the animals were behaving independently of each other.

Before considering this approach in more detail it is worth emphasising its limitations, which are also present in the 'selective' method of Tinbergen and explicitly recognised by him. First there is the danger of confounding the event which precedes a change of behaviour with its cause. A display may often induce another animal to run away, but it may have this effect because it is usually accompanied by running towards the opponent. It is easy to forget that it may be the locomotor component which is producing a large part, if not all, of the effect. Secondly there is the danger of the two church clocks fallacy; a time lag of 10 seconds between the striking of the hour by clock A and clock B does not prove that clock A caused B to strike. Even if an event by animal A reliably precedes one by animal B one must examine whether it is possible that both are entrained by some common earlier event. Neither of these difficulties can be wholly satisfied without experimentally manipulating the behaviour of one of the participants and, as will be discussed later, there are often problems about doing this effectively.

For the moment we may consider examples of the blunderbuss approach. Stokes (1962) showed that one can obtain clear quantitative evidence for the signal values of the components of threat postures of the blue tit (*Parus caeruleus*) in spite of the fact that many of the encounters lead to no clear outcome in terms of attacking or fleeing. The results accorded well with what he had found about the motivational bases of these components, i.e., components which were shown by a bird likely to attack tended to have an intimidating effect on opponents. In another study Wortis (1969) used this type of sequence analysis to demonstrate the pattern of interactions which took place and brought about the weaning of ring doves from parental feeding to independent pecking.

The clearest evidence confirming anything which has been derived from such sequences must come from an experiment. Stout and Brass (1969) were able to test hypotheses about the threatening and provocative effects of various gull postures by measuring the responses to wooden dummies with feathered heads and/or wings attached. In spite of the crudeness of these objects they were able to show that the birds discriminated between them, and that the height of the head above the ground was more important in provoking responses than the angle of tilt of head and beak. These results broadly confirmed what the 'natural experiments' of

Tinbergen (e.g. 1959*a*) and his colleagues had indicated, but allowed an analysis of elements which their observations of live birds had not.

Other workers have contrived to alter the appearance of live animals. One of the most extensive series of such experiments was carried out by Immelmann (1959) with zebra finches (*Taeniopygia guttata*), a sexually dimorphic species with juveniles distinguishable from adults. He was able to modify the feather colouring by painting, and the beak colour by lip-stick, and also to use three colour mutant forms of the species (sometimes artificially painted), to show the role of the different parts of the plumage as social signals in sexual, aggressive, parental and filial relationships.

It will be noted that the experiments cited above altered the appearance of the birds or a static posture. It is obviously more difficult to experimentally manipulate an action. But this has been ingeniously contrived by using models. Hunsaker (1962) investigated the nodding displays of seven species of *Scleropus* lizards (Fig. 1). These displays are evoked by intruders on a male's territory, and it is by these actions that the female seems to choose a male of the right species to pair with. Hunsaker was able to show this role of nodding by arranging that it was simulated by a model with a moving head activated by a cam. Different cams produced different nodding patterns and changing the cam altered the relative attractiveness of the model for females of the different species.

In spite of their possibilities, model experiments have often been disappointing for studying the role of social signals because the animals just do not seem to react to their crude resemblance of the real thing. It is probably in their immobility or, at best, crude movements that their shortcomings are most evident.

To use 'natural' variations experimentally gets over some difficulties, but raises others. Lill and Wood-Gush (1965) analysed the preferences shown by male and female domestic fowl between members of the opposite sex, both of their own and of other strains. Considering only their within-strain experiments, two flocks of nine hens each showed a good correlation between their preferences between six individual males. However, Lill and Wood-Gush were not able to find a simple correlation between female solicitation and any of the male's displays, though they suggest that some compound measure of the male's three high intensity sexual displays may raise the female's sexual arousal. These examples serve to show that while natural variation in attractiveness of individuals clearly demonstrates discrimination in this sexual context, it can be difficult to establish the precise signals responsible for it.

Again and again in the investigation of animal signals one meets this problem: the real behaviour involves the perception by a companion of

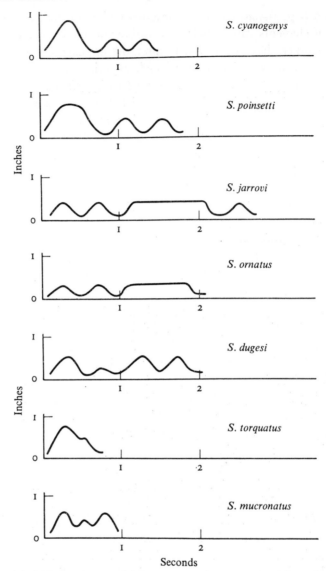

Fig. 1. Head bobbing patterns of *Scleropus* lizards, determined from film analysis. The depth of the bob is represented on the ordinate. When female *torquatus* and *mucronatus* were given a choice of two models differing in bobbing pattern, each chose the pattern of their own species. (From Marler and Hamilton, 1966, *Mechanisms of Animal Behavior*; John Wiley & Sons. After Hunsaker, 1962.)

complex, subtle actions which crude dummies have no hope of simulating, yet without experimentally teasing apart the total matrix of the behaviour in some way one cannot see how the various components contribute to the whole. It has long been clear that one would like to achieve the task of getting a free-living animal to perform a particular action at the command of the experimenter and then to see what effect it had on its companions.

Such methods have become a reality (though still a very expensive one), through the use of electrodes chronically implanted in the brains of monkeys, which can be stimulated telemetrically (Delgado, 1967; Ploog, pers. comm.). Thus for the first time it has become possible for the experimenter to 'switch on' social behaviour of a living animal. This exciting development has yet to be fully explored and only preliminary results are available. Yet this remarkable technical achievement, as it becomes applied to other animals, would seem to provide a most powerful tool for the study of animal communication. It is not to be expected of course, that it is simply a matter of 'switching on' one animal and seeing how the others react; as has been made clear, social signals occur in particular contexts and meaningful results will only come when such experiments are carried out with a thorough knowledge of the sequences of social behaviour in normal animals.

SYNTAX?

Traditionally, animal signals have been supposed to act in a simple way in evoking responses from other individuals. The signals A followed by B would evoke the same responses except perhaps as regards order as B followed by A: 'Man bites dog' is not distinguished from 'Dog bites man'. There appears to be little hard evidence hitherto about the relevance or irrelevance of this kind of ordering, but studies of the pattern of song phrases by robins (*Erithacus rubecula*) and wrens (*Troglodytes troglodytes*) show that the ordering of phrases is of the greatest importance in their recognition as the species song by conspecific males (Bremond, 1967, 1968). At a more complex level Altmann (1967) has claimed that for a Rhesus monkey messages are interpreted differently depending on the sequence. No doubt such order effects may well operate in other cases, particularly perhaps in higher vertebrates, but they are naturally difficult to uncover.

The word syntax has sometimes been used to describe the kind of constraints which appear to operate in such cases. Such a word is attractive in reminding one that animal communication systems may consist of more than elemental signals acting independently of each other; but if it implies a certain equivalence between human words, and for instance, the song phrases of Bremond's robin, the parallel may be merely confusing. What we clearly need at the moment is more information from animals about the importance of order effects. When that is available we may be in a position to consider any parallels with the kind of thing which linguists call syntax (Lyons, Chapter 3).

CONTEXT

The importance of the 'context' of signals has exercised many of those studying animal communication. The term is deliberately vague and may be taken to cover a number of ways in which features other than the ostensible signal (in the narrow sense, of a particular action or structure) affect the way it is responded to by other individuals. It may be that the same action in a different place, for instance in a different part of a territory, evokes a different response. In such a case the spatial context may make the whole difference between threat provoking an attack or provoking flight. Hormonal state can have a similarly drastic effect. An oestrous female may approach a courting male; an anoestrous one will retreat. Equally important can be the personal relationships between two members of a peck-order in how an action by one of them shall be treated. In other cases the response to a signal may depend not on individual so much as group identity: for instance the colony odour which is shared by members of a hive of bees is used in deciding who will be accepted and who attacked as an intruder. Referring to other, less familiar cases, for instance where a very similar display seems to evoke quite different effects in sexual and aggressive situations, Smith (1969) writes, 'There are clearly more functions served by displaying than there are displays . . . Vertebrate communication appears to require extensive use of contextual information by the recipient of the display.' Nor is this restricted to vertebrates. A notable case of the importance of context, is the use by honey bees of their dance during the search for a new home when swarming (von Frisch, 1967). In this situation, as when foraging for food, the same signals are used to convey the distance and direction others must fly out to the new site.

An interesting type of contextual dependency is where signals alter the significance of later signals (Altmann (1967) refers to this as 'metacommunication', i.e. communication about communications). Examples have mostly come from the play of animals (and man), where a characteristic gesture alters, for instance, a threat to a play-threat. Van Hooff (Chapter 8) describes a 'play face' in chimpanzees, a characteristic expression when being tickled or when a group is playing together, and Loizos (1967) gives other examples of such play invitation gestures.

MEASUREMENT OF INFORMATION

Another way of saying that an individual responds to a signal is to say that the arrival of the signal causes the animal to select one response in its repertoire out of a set of possible responses. This resembles the Shannon paradigm (explained in MacKay's article, Chapter 1) and suggests that it might be possible to quantify the 'amount of information' in animal signals in the same way as the signals the communication engineer is concerned with, by the extent to which the arrival of the message narrows down the choice between the responses to be made. Haldane and Spurway (1954) pioneered this approach in animal communication, attempting to estimate the amount of information conveyed in the honey bee's dance,* and the method has been used by several other workers since then. The use of the mathematical tools of communication theory in this biological situation is fraught with problems, some of which are highlighted in MacKay's article. Nevertheless it does seem to have some usefulness in allowing comparisons between communication systems in very different creatures, with very different sense organs and motor systems. Sometimes a direct comparison may be possible without reference to information theory: for instance the chemical trail of the fire ant (*Solenopsis*) specifies the direction in which recruits go out and search for food to a similar extent as the dance of the honey bee (i.e. the variance of the angles of departure about the average is similar in both species (Wilson, 1962)). It adds little to express this in 'bits'. On the other hand when we compare the social interactions of different species of hermit crabs, it is useful to calculate Shannon's 'information statistic', H. This measures the statistical uncertainty of what the animal will do (MacKay, Chapter 1). It is reduced by the reception of a signal, and the amount by which it is reduced tells us how much the recipient's behaviour has been specified by the signal, how far the alternatives have been narrowed down. It is clear that species with signals which have little 'specifying effect' will have more poorly coordinated behaviour than those with a strong specifying effect. It is gratifying therefore that the value of H for the different hermit crabs appear to be related to the social ecology of the species, species with a lower H ignoring each other more than ones with a high H (Hazlett and Bossert, 1965). The same can be said about Dingle's (1969) study of mantis shrimps, where changes in H in time seem to reflect changes in the establishment of a peck order between the individuals. In spite of the assumptions involved in estimating H from the kind of data available in animal encounters, it does seem to draw attention to some average property of an animal's signals in a particular social situation.

SENSORY MODALITIES

The signals which different species of animals use can be of many different physical types and involve any of the sense modalities, either singly or in

---

* The reader should be warned that the importance of the honey bees' dance in transmitting information has been seriously questioned. The matter is discussed in Thorpe's article, Chapter 5.

combination. A large section of Sebeok's book (1968: Parts III and IV) is devoted to illustrating this diversity and Thorpe's articles in this book present a sample of it. For the present purpose it is sufficient to emphasise that the different physical methods of transmitting signals often have intrinsic advantages and disadvantages. Visual signals are bad at night (unless luminous like fireflies or glow worms). Airborne scent is relatively slow acting and is likely to be carried in one direction only from the source; on the other hand for certain social purposes olfactory signals (phero-mones) are particularly suited as long-lasting indicators of, say, the con-tinuing presence of the queen bee in the hive. Water is generally a turbid medium to live in and visual signals are effective only at short distances; on the other hand sound propagates particularly well in water and is much used for signalling by mammals and fish as well as invertebrates. Each of these systems requires specialised sense organs to perceive it and, even when the energy is of a kind we can perceive, the type of sense organ involved may be very different from our mammalian ones and respond to different aspects of the total physical stimulus.

ONTOGENY

Some of the signals of animals have to operate the first time they are used, and must be performed or responded to adequately without the benefit of experience. In other cases an individual may have had prolonged experience through family life or otherwise with older individuals who are already practising the system and may form, or at least perfect, his use of the signals as a result. The extent to which the systems develop independently of experience raises the hoary instinct–learning controversy, and social behaviour has played a considerable part in examples bandied back and forth in such discussions (Lorenz, 1966; Schneirla and Rosenblatt, 1961). The issue is partly a terminological one and for a general discussion the reader may be referred to Hinde (1970).

One way to clarify the problem as it concerns us here is by distinguishing between (1) the actual form of the social signals, (2) the individuals they are addressed to and (3) the responses they call forth in others. In the case of the stereotyped and often species-specific form of displays there are a number of examples where these have been shown to develop normally in animals which have never seen them performed, and with the exception of bird songs and some calls which are of sufficient importance to have a special section in this book, the burden of proof seems to be on anyone wishing to show that such signals are acquired by imitation. On the other hand there is considerable evidence for a number of vertebrates that the class of individuals to whom such signals are given is affected by experience, and particularly early experience (Bateson, 1966). Whether individuals respond appropriately to signals the first time they experience them is the third question and this has been the subject of less research. There are known cases, especially in young animals, where they clearly do: an incubator-hatched young gull will peck at red objects, the colour of its parent's beak. However, such an observation does not necessarily indicate that responsiveness is limited to the naturally appropriate object: in fact the pecking of many young gulls is readily elicited by blue beaks, though not by green or yellow ones (Hailman, 1967), and the demonstration of appropriate responsiveness independent of experience does not mean that responsiveness cannot later be altered, for instance by rewarding pecks directed towards other colours. In any case generalisations about the role of experience on signalling systems are certainly species-bound, and what applies to a stickleback may very well not to a rhesus monkey.

EVOLUTION

While the adaptive 'fits' between signal and response may in some animals be shaped to a considerable extent by their individual life histories, a large part of it is no doubt independent of this kind of experience. Genetic factors can have a considerable effect on the form of signals (Franck, 1969). The effects are usually multifactorial, but in *Drosophila* single genes are known which affect the communication system concerned with mating success in a variety of ways including overall sexual excitability, thresholds of sense organs, and locomotor activity (Manning, 1965). The rapid breeding rate of *Drosophila* has allowed laboratory simulation of microevolutionary changes. For instance Crossley (in Tinbergen, 1965) showed how reproductive isolation could be produced by selecting for forty-five generations for non-interbreeding between populations which initially hybridised freely, and she could then analyse the behavioural differences which were created, showing how the selection had affected the courtship of the males of the two lines and the responsiveness of the females to their own and the other line.

The very existence of complex animal societies is sufficient to show the adequacy, to say the least, of the communication systems which maintain them, and in those cases where they have been studied in detail they often show a high degree of adaptedness to the species requirements. This could not be better illustrated than by the celebrated dance of the honey bee studied in such detail by von Frisch and his students, with the modifications it shows in related species which differ in their foraging ranges (von Frisch, 1967).

One of the problems raised by such specialised and 'improbable' communication systems is how they can have arisen by the processes of natural selection. Leaving aside the special problem, which is not really one at all, of the evolution of such a thing in a species with a sterile worker caste, there remains the question of the evolutionary origin of behaviour such that the signaller should perform an action which corresponds to the location of the food he has just returned from, and that the recipient of the signal should 'read' it appropriately. In more general terms this is the question why a particular act should come to be used as the signal carrying a particular message, and why this act should be correctly interpreted by the recipient. Answers to these two questions can now be suggested for certain types of visual signals concerned with the threat and courtship displays of vertebrates, and it is quite possible that answers on similar lines might apply in other cases.

If we turn from complicated displays to the simplest signals, we find on the one hand that the signal action may be identical in form with, or in fact be, an action with a secular function. The male *Aedes* mosquito responds to the normal wing beat of the female and approaches her for mating. In certain Braconid wasps the wing beat frequency alters when the animal is ready to mate, while many *Drosophila* species, though courting on the ground, still use their wings to create species-specific courtship songs. In such cases the normal locomotory pattern has become specialised as a signal or 'ritualised'. In a similar way many of the simple displays of birds and fish can be seen as being modified locomotor acts relevant to the situation at the time (Bastock, 1967; Hinde, 1970). Threat displays look like 'intention movements'* of attack, or sometimes a mosaic or alternation of attacking and fleeing, both things which are likely to be associated with threat: indeed the behaviour of a threatening animal becomes much more comprehensible when it is seen as subject to a 'conflict' between tendencies to attack and to flee. In a similar way, courting animals often seem to be in conflict between tendencies to attack, to flee, or to behave sexually and copulate, and many courtship displays depend on a three-point interaction between them.

However, from Tinbergen's first attempts to systematise the sources of 'derived activities' (1952) it was clear that besides intention movements there was another large class of actions which seemed to provide potential material for the evolution of displays. While they occurred in the same situations as the intention movements they seemed irrelevant and quite unadapted to those situations, unlike the attacking, fleeing or mating actions. In the middle of a fight the animal might suddenly start to pick up nest material, or groom itself. These irrelevant actions were called

* Not 'intention' in the sense of MacKay. The word is merely used to describe preparatory movements.

'displacement activities' and a mechanism was proposed to explain why they should occur in threat and courtship. (See also Andrew's chapter, Chapter 7.)

Since it was first formulated, the theory was found to make sense of a lot of data (Hinde, 1970). A number of other sources of display were subsequently recognised: balancing movements, protective movements of structures like the ear pinna or eye, alerting responses (the so-called orienting reflex), autonomic responses preparing the animal for muscular activity, e.g. flushing of the skin, changes in erection of the hair or feathers, increased or decreased body tonus associated with arousal. This large class of actions, though not locomotory intention movements in the original sense, are clearly all preparatory for, if not a direct response to, the situation in which the animal finds itself. So is the 'redirection' of behaviour, e.g. attack, from an opponent the animal dares not strike to some other object. The displacement activities have proved less satisfactory and the category is now recognised as being causally heterogeneous (Hinde, 1970). Nevertheless, whatever the reason for the occurrence of one of the so-called displacement activities, they provide raw material for signals.

The conflict theory (see above) also offers an explanation for why there might be more than one display in certain situations. The different displays reflect a difference in strength or relative strength of, say, locomotion towards and locomotion away. Andrew's (1961) review of passerine displays showed how many species have a forwardly directed threat, sometimes with beak open, representing a display likely to lead to attack, as well as an upwardly directed, more nervous type of threat less likely to lead to attack and more likely to lead to escape.

Given the regular association of certain actions with subsequent actual attack, it is not difficult to imagine that they would come to be learned by an opponent as an indicator of impending attack, just as now domestic animals can learn to react to another species' signals. If modifications of one action were associated with different probabilities of attack or flight, these too would no doubt come to be discriminated appropriately and one can see that a more and more pronounced demonstration of the proto-signal would benefit the signaller, and an increased sensitivity to such signs would benefit the 'signalee' – in other words how the signal came to carry the message it does, and how it came to carry the meaning it does. Appeasement actions, those which reduce an opponent's hostility, might seem to be something of a problem to understand in this way, but the kinds of action from which appeasement displays have been evolved (crouching low, rolling over on the back, concealing the weapons, avoiding a direct gaze, inviting grooming) would seem to have a good chance of working just

8

because they are the opposite of threatening, attack-provoking gestures (Tinbergen, 1959*b*). Why the performer should be motivated to do them may be less obvious, but sometimes they represent ritualised withdrawal or retreat. If, at what stage, and how, these newly acquired response systems would become genetically encoded is even more speculative but this problem is the same as for any 'acquired character' (Waddington, 1957).

The displays of many animals are very far removed from the form of secular movements from which they might be supposed to have originated. Yet by judicious use of the 'comparative method' one can sometimes show links between the highly specialised display and its ancestral form. The method hinges on the existence across a number of closely related species of actions which can be regarded as 'homologous' with the specialised display, meaning that they are supposed to have evolved from a common form in some ancestor. Now there is no fossil record of the past history of display actions, yet following the well-worked system of biology for establishing lineages when the fossil record is absent or incomplete, one can apply the same kind of principles to behaviour. The establishment of such homologies has often been done intuitively, but Baerends (1950) and Wickler (1968) set out the kind of criteria which need to be met, though Atz (1970) has recently criticised the whole procedure. It may be stressed that the method depends on a good knowledge of the behaviour of a number of species whose taxonomic relationship is independently established.

The displays of the peacock (*Pavo cristatus*) may be taken as an illustration of the way the comparative method has suggested the evolutionary history of a display (Schenkel, 1956) (Fig. 2). The male uses his fan, which is not strictly speaking his tail, as part of his courtship of the female. When she appears he erects it and the true tail behind, he turns *away* from her somewhat, rattles the tail feathers against the fan and begins to make little backward mincing steps until she comes into view around the side of the fan. He then turns sharply towards her, arching his fan further over his head and assumes a rigid 'ecstatic' posture with body slightly tilted forward and head stiffly bent. If the female now crouches in front of him, he mounts and copulates. Schenkel examined the displays of a number of pheasants, which are related to the peacock, and showed that one can recognise a homologous display in many of them, which he calls a frontal display, as opposed to a lateral display, which many of them also have. But the frontal display differs somewhat from species to species. The forward tilted body is common to them all, while the ecstatic phase may be present but less rigid. In some species the male pecks the ground when in the position and the female may run to see what he has. The opposite end of the series from the peacock is represented by the domestic cock which prior to copulation calls the female to food in a way comparable to the courtship feeding of many other birds. The cock's display is clearly a signal, but it is one in which the feeding significance of the act is evident. In the series represented by *Gallus–Phasianus–Polyplectron–Lophophorus–Pavo* there is a progressive overlaying of the courtship feeding act to such an extent that the feeding origin has quite disappeared in the peacock – except in the female, which has been observed to peck at food while being courted by the male, but this was earlier dismissed as a displacement activity. In Chapter 8 another example of the use of the comparative method illustrates the phylogeny of human laughing and smiling.

An interesting but difficult case concerns the honey bee's dance. (The reader who wishes for a brief description of the dance may turn to Thorpe's chapter, Chapter 5, in this book.) There are regrettably only four species in the same family as the honey bee and their dances are so similar that they tell one little of the probable evolution of the dance (von Frisch, 1967), except that it must have come from a performance in which the dancer ran horizontally (and not on a vertical comb) and, using the sun, orientated her waggle dance directly towards the food source she was indicating. In tracing evolutionary trends one is always looking for gradual transitions, so that apparently novel features such as this faculty to transpose an angle relative to the sun to one relative to gravity require special attention. However, it has turned out on further investigation that this surprising 'abstraction' can be performed by a number of other insects with no social pretensions, nor indeed, so far as one can see, any benefit from being able to make the transposition. Other special features of the honey bee's dance cannot be precisely matched with performances in social bees from other families, but evidence from insects as different as flies and moths throws some light on the way the distance of food may be coded in the form of the dance (references in von Frisch, 1967).

The honey bee example illustrates well one of the problems commonly encountered in seeking the phylogeny of displays, namely that none of the species still alive today may correspond to stages in the putative ancestry. This is hardly surprising as each species has no doubt acquired its own specialisations. However, provided that the departures are not too great, the information provided by contemporary species may suggest a plausible phylogenetic history for the display. The results can seldom be regarded as certain, but many of them are certainly better than plausible guesses.

RITUALISATION

The kind of change which displays have undergone during evolution in the service of their function as signals has been called 'ritualisation' and has been reviewed a number of times (see references in Huxley, 1966; Wickler, 1968). Since differences in the signals of related species are generally supposed to be due to adaptive pressure of some kind – in some cases merely to be different from a sympatric species – differences between existing species provide clear evidence for the kinds of alterations to the signal which ritualisation entails. Hinde (1959) has listed:

Differences in frequency of occurrence or intensity of a display.

Differences in coordination of components or complete loss of some components.

Differences in speed of performance.

Development of rhythmic repetition.

Differences in orientation.

Such species differences in movements are often correlated with the development of specialised structures – tufts, plumes, bare skin, bright colours – which considerably enhance the conspicuousness and often the distinctiveness of the displays (Fig. 2).

Another type of change which certain signals have undergone is a shift in their functional context. Nelson (1969) describes a highly specialised (and therefore unlikely to be convergent) display, skypointing in the Sulidae, which is used in several species for sexual advertisement, though there is little doubt that the situation where it occurs in *Sula bassana* as announcement of a pre-flight intention is the evolutionary origin. Another conspicuous example is the genital presentation which, though undoubtedly derived from the sexual act, has come to have a largely non-sexual, social significance in many higher primates. The term 'emancipation' is sometimes used to describe the supposed underlying motivational changes which must be responsible in these cases, but it seems better to regard them as not different in kind from the threshold shifts which are needed to account for many of the other kinds of ritualisation, and indeed specialisations of non-signal behaviour too (Blest, 1961).

A feature of many ritualised signals which has not been sufficiently brought out in this account is their stereotypy. Some signals are amazingly fixed: thus the durations of various displays of the goldeneye (*Bucephala clangula*) have coefficients of variation as low as 6 per cent, though other displays are more variable. Morris (1957) has written of the way displays may evolve 'typical intensity', in the sense that the variations in the unritualised signal corresponding to motivational shifts are ironed out to produce a relatively invariant signal corresponding to a range of motivational levels. While actual measurements of the constancy of signals is only beginning (except in the case of sound signals) it is clear that some signals are essentially digital – constant and distinct from others – others are variable, while yet others are hybrid systems with some features constant, others variable. Such differences in form take on significance when considered in relation to the meaning of the signal. As Morris and others have recognised, there is an essential contradiction between the ambiguity of signals and the precision with which they can coordinate the behaviours of

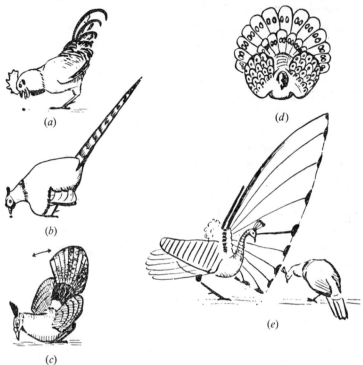

Fig. 2. Series to illustrate the ritualisation of the peacock's fan display from the calling-to-food action of other species of the pheasant family. (From Eibl-Eibesfeldt, 1970, after Schenkel, 1956.) (*a*) *Gallus domesticus*, (*b*) *Phasianus colchicus*, (*c*) *Lophophorus impejanus*, (*d*) *Polyplectron bicalcaratus*, (*e*) *Pavo cristatus*. The food, normally present to elicit the action in species (*a*) and (*b*), becomes more optional in later members of the series. Note the 'eyes' on the wings and tail of *Polyplectron*. These look like the 'eyes' on the peacock's fan but are constructed on the feathers in a completely different way (i.e. convergence). Such series are not intended to illustrate the actual evolutionary stages of the peacock's display in detail, but to show the broad features of the process which can be inferred by the 'comparative method'. (From *Ethology: the Biology of Behaviour*, by I. Eibl-Eibesfeldt. Copyright © 1970 by Holt, Rinehart & Winston, Inc. Reproduced by permission.)

two individuals. The more distinctive and clearcut, the less adjustment there can be for fine differences in behaviour. The bee dance illustrates the problem (von Frisch, 1967). Food located near the hive causes the finder to perform the 'round dance' which has an alerting function for potential foragers but gives no directional information. It is a relatively unspecialised performance not too different from the active circling of a fly searching for food. Such a signal would be highly inefficient for distant food as few individuals would be likely to find it. Thus the alerting signal for distant food, the waggle dance, is analogue rather than digital, doubly analogue in fact since the approximate distance and direction are both coded.

After surveying the acoustic signals of primates, Marler (1965) concluded that whereas in *Cercopithecus* and *Colubus* monkeys each species tends to have a number of distinct calls, in the baboons, macaques and the chimpanzee the calls intergrade much more. Such differences might be related to differences in the type of social organisation of the species.

Besides the issue of analogue or digital signals, there are many other features of communication where the functional requirements have no doubt done a great deal to shape the communication system in general, and sometimes even particular signals. One beautiful example was recognised by Marler (1957) in the resemblance between the hawk-alarm calls of a number of small passerine birds which was seen as a true convergence towards the most suitable sound to announce the danger with the minimum risk. (This example is detailed in Thorpe's chapter, Chapter 6.) The contrary principle was seen to operate for those calls where it is necessary to tell the location of the caller: songs, owl-mobbing calls, flight calls. Here the physical characteristics of the calls are such as to make them easily locatable.

The shaping of signals by the requirements of natural selection is also clear in some cases of reproductive isolation. If there were nothing to prevent it, closely related species whose distributions overlap (i.e. which are sympatric) would be likely to interbreed and produce less viable or less fertile offspring. In general this does not happen (Mayr, 1963), and it is through behavioural mechanisms, such as courtship, that species are often kept apart. There are two complementary views about the evolutionary origins of these reproductive isolation mechanisms and the evidence suggests that sometimes one, sometimes the other has been more important. One of them supposes that the courtship differences arose during the period of geographical isolation which is generally supposed to accompany species formation (Mayr, 1963), as a secondary consequence of other adaptations. The other supposes that the differences arose, or at least were greatly exaggerated, when the species came to overlap again following their formation and hybrids were penalised.

Closely related species often show differences in courtship signals, and it is of interest that in those vertebrates which form a pair in advance of actual copulation the reproductive isolation operates in the early, rather than the late, phase (Hinde, 1959). For example the behaviour differences described earlier for the lizard species studied by Hunsaker (1962) were present in the nodding seen in early encounters with the female. The pre-copulatory nodding display was similar across all the species. Alexander and Otte have emphasised the same point for crickets. The genitalia provide a possible reproductive barrier, yet between closely related sympatric species genitalic differences are slight and it is the sound signals which keep them apart. This type of evidence supports the second view of the origins of reproductive isolation: otherwise one would expect that precopulatory or genitalic differences should be just as great as the earlier ones.

Additional evidence comes from cases where it can be shown that species are more different in respect of sexual behaviour in areas where they over-lap. Thielcke (1969) has recently reviewed the alleged evidence for this in bird songs and found no convincing examples, but examples in other vertebrates are known (e.g. frogs, Blair, 1964). Crickets once again provide a number of instances (Alexander, 1968). Other examples, for instance Blair (1964), show equally convincingly that reproductive isolation may exist between populations which are not and probably never have been in contact with each other.

The link between the biological requirements of signals and the general habits and ecology of the species can operate at a higher level too. Types of social organisation are broadly correlated with differences of environment and with the dispersion of food which the species lives on (Crook, 1965; and references in Lack, 1968, for birds; Eisenberg, 1966, for mammals). Such differences involve things like the degree of monogamy or polygamy, which in turn will affect the degree of sexual dimorphism since in polygamous species stimulation of the mate – consequently bright colour, conspicuous signals – will be at a greater selective premium (Darwinian sexual selection). Crook (1964) investigating one family of birds (Ploceidae) in some detail, showed how the food dispersion pattern was correlated with the extent of colonial nesting and polygamy (and also other features like the shape of the nest), and how the polygamous species tend to have more ritualised threat and courtship displays. In fish a similar adaptive radiation may centre on the requirements for protecting the young. Within the large cichlid family those species spawning and tending their eggs in relatively open sites lay numerous, small eggs. Other species keep their eggs in more protected sites, either in the mouth of the male and/or female, or in crevices or holes. Wickler (1966) showed that, amongst the forty-five species studied, the open nesters are monogamous and the sexes look much alike: both parents are necessary to protect the young and a strong pair bond is formed. For those with concealed broods the majority are dimorphic but may be polygamous or monogamous: in such sites the young can be protected by a single individual. Correlated with these differences in pair bonding are others concerned with relations between the pair and with the brood. Simple dichotomous classifications of this kind are, of course, oversimplifications, but the general trends are illuminating. So are some of the exceptions, such as the species *Tropheus moorei* which is a monomorphic mouth brooder which forms closed groups of individuals which recognise each other and drive off strangers (Wickler, 1969). The species is also unusual in possessing a special yellow band around the dark body which is used not only in courtship but by juveniles as well to allay hostility in moments of tension within the group (cf. the sexual presenting of monkeys). Even within a genus there may be correlations between signals and the style of brooding. For the *Tilapia* cichlids substrate brooders go in for mouth fighting while those which carry their eggs in the mouth have more ritualised threat displays. Investigations of this kind showing correlations between types of signals and social organisations, and offering plausible explanations for such relations, are still rare. But they are increasing fast and should go far to account for the tremendous diversity in the communication systems which we find today, even amongst closely related species.

INTER-SPECIFIC SIGNALS

Social communication is normally taken to comprise reactions to fellow members of the species, but under certain circumstances it may be considered more widely. Baerends (1950) argued that there are a number of the inter-specific relationships, including symbioses, which can properly be considered together with intra-specific ones because they involve mutual advantage and in many cases the evolution of a signalling system between species not less complex than that operating within a species. Even plant–animal relationships can be considered social on this basis, such as that between bees and the flowers they pollinate. In practice the criterion of mutual advantage has to be applied with discretion since sometimes that advantage may seem distinctly one-sided.

Wickler (1968) has drawn particular attention to the way in which the study of inter-specific signals can throw light on the evolution of intra-specific communication. The occurrence of mimicry illustrates that certain features of the model are being recognised by what Wickler calls the 'signal-receiver' (the animal on whose responses the whole mimetic system depends). A predator who fails to distinguish between an unpalatable butterfly and the palatable species which mimics it is taken as the signal receiver. Viduine weaver birds parasitise the nests of other small finches, the Estrildines, each species of parasite being fairly specific to one species of host. The parasite's young are raised in the nest with the host by the host parents and the number of ways in which the parasite young have come to resemble the host, for instance in the highly distinctive mouth pattern and the very peculiar begging posture which young Estrildines have (but no other species of birds except for the parasitic Viduines), show that the host species must be discriminating against parasites which fail to look and behave like their own young (Nicolai, 1964). In other words that this kind of stimuli play a part in releasing the parental responses of the host species.

Wickler has also shown how the evolution of intra-specific signals may depend on taking advantage of a pre-existing response in the signal-receiver in what he calls intra-specific mimicry. The male mouth-brooding cichlid *Haplochromis burtoni* has evolved bright orange spots on its anal fin to which the female responds as she would to her eggs after spawning, which she sucks into her mouth for incubation. By reacting in the same way to the male's 'egg dummies' the female draws sperm into her mouth at the same time so as to ensure fertilisation of the clutch.

From a consideration of interspecies responses one can see how proto-signals can come to work in intra-specific signalling, by 'parasitising' existing response systems. For instance the predatory angler fish *Phrynelox scaber* has evolved a pink worm-like tip to its first dorsal fin ray which it moves so that this lure is dangled close in front of its mouth as it lies still and camouflaged on the bottom. Prey fish approach the lure and are engulfed by the predator. The ruse works because the angler fish parasitises the normal response to the normal feeding stimuli of its prey. Similarly Blest (1957*a*) showed that small ocelli on the wings of butterflies and moths can benefit their owners by deflecting a bird's predatory attack from some more vulnerable spot such as the body.

Inter-species signals may not only originate but also develop through ritualisation in the same way. In this context too Blest (1957*a*, *b*) was able to show how the small ocelli which distract a bird's attack have sometimes evolved to large, solid looking eye-spots which are flashed at the predator at the moment of attack and may scare it at least momentarily, giving time for escape by flight, or by falling to the ground. Different species of moths under attack display their eyespots in different ways, some flashing them rhythmically, others having a more sustained static posture. The kinds of differences between the species parallel closely the types of differences seen in the ritualised *intra*-specific displays listed earlier, such as change of coordination of components, altering amplitudes and thresholds of response, changes in rhythm and of course the enhancement of the locomotory effect of the movement by conspicuous markings. For the moths, evolution in the direction of increased conspicuousness is clearly advantageous when the startle response is the system being parasitised. As to why one particular type of display was ritualised rather than another there was a suggestion, for instance, from the nine species of Automerine studied, that rhythmic displays were more likely in moths which were more ready to take flight subsequently, while static displays were found in more sluggish species. In such cases the stereotyping of a response cannot depend on an aim for the kind of unambiguity which would be relevant in a conspecific situation.

COMMENTS

(1) *Scope and definition*

It is immediately apparent that Cullen's chapter, like those which follow, is concerned with a much wider range of phenomena than those comprehended by MacKay's narrow definition of communication. Lyons pointed out another difference between the linguist and the student of animal behaviour. The latter is forced to define a signal by whether it does or does not affect the behaviour of another individual. Lyons writes:

> 'This criterion, strictly applied to language, would imply that many utterances are not signals, since they "evoke" no discernible change in behaviour. It may, nonetheless, be quite easy to demonstrate that they bring about a change in the receiver's beliefs or knowledge. Behaviourist psychologists have, of course, tried to escape from this difficulty by reducing "knowledge" and "belief" to "dispositions to respond". But this is not very helpful operationally, unless they can specify the manner in which the behaviour would change and under what conditions. One might postulate, *à la* MacKay, that every signal that is informative for a given organism brings about a change in the internal state of that organism (whether this be a permanent or a transient change) and that this change may have, consequentially or concomitantly, an overt behavioural correlate. In the study of animal signalling one is perhaps obliged to restrict oneself to an "external" account (in terms of the behavioural correlate); in the study of human signalling one is not.'

And, some would add, such a restriction in either animal or human case may focus attention away from underlying principles (cf. Chomsky, 1968: 58).

(2) *Context*

(*a*) Many studies of animal signalling systems have attempted to catalogue the signals in the species repertoire, and then to identify for each on the one hand the particular circumstance and motivational condition under which it is emitted, and on the other the particular response it elicits in another individual. Subsequent research has revealed that such an approach is often over-simple, for some signals are emitted in more than one set of circumstances, and the response elicited may depend in part on additional contextual factors (see also Thorpe, Chapter 6). This point is so important, and comes up again so often in subsequent chapters, that an additional example from Smith's (1966) study of Tyrannid flycatchers is worth while. One of the vocalisations, described by Smith as the Locomotory Hesitance Vocalisation, is given in a wide variety of motivational situations involving an internal conflict with a tendency to locomote. Some examples are: a male patrolling his territory gives the call whenever he ceases flying; a member of a newly formed pair gives the call whenever approaching its mate; and begging young may give the call when approaching an adult who may respond aggressively. The message can be regarded as the common factor in all such cases. The possible meanings are numerous. For instance if the caller is a patrolling male, the call may stimulate an unpaired female to approach and initiate pair formation activities; but a territorially-inclined male may fly away or else prepare to fight. The meaning thus varies with the nature of the caller, the attendant circumstances, and the status and state of the recipient.

(*b*) Cullen uses the term 'context' in an intentionally vague way to cover all these other factors which determine the response to a signal. Firth (in correspondence), while agreeing with the importance of contextual factors, pointed out the difficulties in defining context, and suggested that a satisfactory theory of non-verbal communication might depend on a more satisfactory classification of context. MacKay wrote as follows:

'When talking about "context" the presuppositions appropriate to serial communication (e.g. verbal) may have to be questioned. If a system works on the basis of parallel logic, what you mean by "context" may equally be regarded as the remainder of the parallel input. "Ordering" in a series of signals may be functionally equivalent to permutation among parallel inputs. Whether this is so will depend on the details of the engineering of the nervous system concerned, and no sharp methodological distinction between signal and context can really be made in advance of empirical knowledge of this kind. In other words, I wonder whether it might not be useful to translate talk about context into talk about the combined selective function of multiple signals, whether these are in series or in parallel. This would then be distinguishable from situations in which the overall state of the animal was different, so that its response to identical configurations of signals was correspondingly different.'

(*c*) In his discussion of 'context', Cullen referred to the special type of contextual dependency where one signal alters the significance of a later signal. Altmann and others have referred to this as 'metacommunication', but Lyons points out that this use of the term differs from the philosophical distinction between 'metalanguage' and 'object language'. Here the concern is merely with one signal 'modulating' the function of a subsequent signal: there seems no difference in kind from the function of paralinguistic features.

(3) *Message and meaning*

Cullen does not use these terms, and thus avoids the difficulties inherent in their diverse meanings (see MacKay, Chapter 1, and Leach, Chapter 12).

(4) *'Idealisation' and the study of animal communication systems*

In Chapter 3, Lyons described the stages of 'idealisation' in the linguist's description of the raw data of human speech. The parallels in the study of animal behaviour are interesting. The first stage involves the discounting of performance phenomena and, in so far as they occur, the student of animal behaviour does likewise. The second, the discounting of variations attributable to personal and sociocultural factors, is implicit in many studies of animal communication, but in others, it is just these factors which are the centre of attention. Indeed one of the most important properties of the signal may be its ability to convey information about the identity of the signaller. Some cases of such 'indexical' properties (cf. Lyons, Chapter 3) are discussed by Thorpe in Chapter 6.

Lyons' third stage of idealisation involves the distinction between actual utterances and the system-sentences on which the linguist's analysis can proceed. In the main, of course, this is not applicable to the study of animal communication, but it is noteworthy that one difference is that system-sentences must contain elements which are provided by the context in utterances. Looked at from this point of view, students of animal communication have been hindered by attempting 'idealisation' and neglecting the contextual elements which, as the work of Smith and others show, are essential parts of the system.

(5) *Information theory*

In Chapter 1 MacKay underlined some of the difficulties inherent in the direct application of the mathematical theory of communication to biological communication systems, while arguing that the concepts derived from it were valuable for biologists. Cullen, however, believes the quantitative application to be useful in some contexts. It is important here to distinguish between Shannon's measure as a measure of correlations of behaviour patterns, and as a measure of information received by an organism. On general grounds one would not expect the two to be the same, particularly where there might be high degrees of redundancy (for the organism) that might not be apparent to the observer.

5. THE LOWER VERTEBRATES AND THE INVERTEBRATES

W. H. THORPE

The purpose of this chapter is to survey very briefly the communication systems of the vast assemblage of animals included amongst the lower vertebrates and the invertebrates. As there are, in all probability, more than a million species of insects in the world, it will be realized how tentative and cursory this chapter must be. In producing it I have been much indebted to Part IV of Sebeok's (1968) valuable survey.

It is not proposed to discuss in any detail the means whereby information is communicated by visual signals – there are excellent books which cover this aspect of the subject. I need mention only three – apart from the 1966 volume dealing with ritualization which was edited by Julian Huxley and which gave rise to the meetings which have produced the present volume. These are Hinde (1970), Marler and Hamilton (1966) and Eibl-Eibesfeldt (1967). Nor do I propose to pay much attention to the new field of the study of communication by secretion and dissemination of chemical substances (now called pheromones); a field which is now being explored with greater energy and expertise than ever before and which includes topics of far-ranging and basic significance.

Since the functions of, and the methods employed in, the communicative systems of the amphibia and reptiles are in some ways so similar to those of birds and other higher animals, albeit incomparably less well developed, it is natural to start with those animals more familiar to the general reader and work our way downwards towards the invertebrates, which are, except to the zoologist, so much less known.

LOWER VERTEBRATES

Amphibians

The first group of the amphibia is the Apoda which comprises the Cae-cilians, a group of worm-like creatures occurring throughout the tropical regions of the world and living mainly in burrows in the soil. Since the eyes are vestigial and there is no evidence of vocalization, it can be assumed that tactile and chemical signals are the most likely means of communication. The Urodela or tailed amphibia comprise for our purposes chiefly the salamanders and the newts. Here again, chemical signals appear to be the most important. In the newts (Tinbergen, 1953) sex recognition and court-ship are mediated by a series of signals, involving:

(1) *Visual signals*. Most species are markedly sexually dimorphic (Smith, 1954), and the male assumes specific attitudes during courtship. Lack of movement by the female intimates to the male that he is to continue courtship.

(2) *Tactile signals*. In many species a sudden leap by the male sends a strong water current to the female, sometimes even pushing her aside.

(3) *Chemical signals*. Scent glands are distributed over most of the body in both sexes, and appear to produce sexually stimulating secretions. The tail-waving of the male creates water currents carrying a chemical stimulant to the female. There is no evidence that sound-signals of any kind play a part in communication amongst the tailed amphibia. The salamanders also communicate the identity of species and sex and the physiological readiness, or otherwise, to mate, mainly through chemical, tactile and visual signals, probably in this descending order of importance. But their communication signals appear more complex and varied than in the other tailed amphibia.

When we come to the frogs and toads (Anura), the situation is strikingly different in that vocal communication is of major importance. But as Marler and Hamilton (1966) point out, whereas a number of non-human primates have a vocal repertoire of up to about twenty sound types and the birds a similar number, various frogs and toads are known to produce only up to four recognizably distinct sounds.

In general, the sounds produced by the Anura can be classified under five headings, of which the first three are the only really important ones for the purposes of the present discussion. These are (1) *the mating call*. This covers the sounds produced by most adult frogs and toads which are commonly heard as choruses from breeding aggregations, or sometimes from isolated individuals. These mating calls are produced only by the adults, in most species only by the male, and are thought to be clear evidence of sexual maturity. When a frog or a toad approaches a member of its species in a breeding pool, one can fairly safely assume that the vocal signals have already effectively communicated species identification. There remains the question of identification of sex and of the reproductive state of the female that is clasped by a male. Hence we have (2) *the male release call*, normally uttered at a breeding site. He clasps males or females indiscriminately, but differential behaviour in the sexes determines whether he retains his grip. If the individual seized is a male he struggles, accompanied by sounds variously described as chirps, croaks, grunts or clucks, usually elicit his release. By contrast a ripe female seized by a male of her own species emits no sound, remains passive and the male retains his grasp. Bogert (1960) describes the male release call as a short explosive sound repeated at irregular intervals. This is in contrast to the often much more complex mating calls of some of the frogs, which may be repeated at regular intervals with a characteristic rhythm, are often trill-like, and sometimes have a relatively complex accoustical structure. The third anuran type of vocalization is *the female release call*. In some species unreceptive females utter sounds similar to the release call of the male. Thus Noble and Aronson (1942) mention warning croaks in leopard frogs (*Rana pipiens*) of both sexes. If clasped by a male the female utters the croak, except during the ovulatory period when she becomes receptive as well as silent. She is mute during sexual congress and the male's clasp does not elicit the croak until several hours after oviposition is completed. This female release call, while certainly widespread, does not appear to be of anything like the biological significance of the two types of call previously discussed. Still less important seems to be (4) *the post-oviposition male release call*. According to Lutz (1947, quoted by Bogert, 1968) the Brazilian tree frog (*Phyllomedusa guttata*), after spawning is completed, begins to cluck softly (quite a different call from that of the mating call) and then leaves the female, who remains in the same position for another 30 minutes. Finally there are (5) *ambisexual release vibrations* which are not vocal in the sense that they do not depend upon the vibration of the vocal cords, but are probably produced by accentuated respiratory movements that cause a cup-shaped cartilage (the arytenoid) to vibrate – the vibrations being transmitted to the body musculature. It is supposed that the inflated lungs may then serve as resonating chambers and hence render a vibration audible to man.

Reptiles

The reptiles present a much less elaborate picture of vocalizations than do the amphibia. The roaring vocalizations of the American alligator (*Alligator mississippiensis*) (Evans, 1961) and the Nile crocodile (*Crocodilus niloticus*) (Cott, 1961) are, however, of much interest. In both animals it seems that calling by the males is answered by other males and is a territorial defence device; and that females are attracted by the roar of the males during the breeding season and, in addition, by the secretions of musk glands which are situated near the jaws and are primarily functional during courtship. The so far but little-studied vocalizations of turtles and tortoises seem to be employed in a similar way and are also supplemented by the use of chemical signals. Snakes are notoriously silent but the rattling of the rattlesnake (*Crotalus* spp.) provides positive identification for an animal that is well protected by its venom apparatus. Lizards have elaborate visual displays but are, on the whole, silent creatures. However, a striking exception to this is provided by the geckos which, being nocturnal, have developed extensive vocalizations. The leopard lizard (*Crotaphytus wislizenii*) is particularly remarkable in this respect (see Wever, Hepp-Raymond and Vernon, 1966).

Fishes

Here again we have to deal with a large group of species in which form and colour are often highly distinctive both of the species themselves and, during the breeding season, of sex within the species. It is consequently not surprising that a great deal of communication is by recognition, sometimes minutely precise, of patterns of form and colour and subtle combinations of these. That many species are territorial and many others show the elementary social behaviour of schooling further emphasizes the communication problems that must arise in this group. Moreover the fact that fishes are aquatic animals raises still further points of special interest for the biologist studying communication systems.

Although a great many fishes which live in shallow water, in clear streams and in such situations as coral reefs, are highly visual animals, there are equally an enormous number which live at great depths where light is sparse, or in water which is highly charged with concentrations of microplankton which severely limit the penetration of light and the efficacy of visual organs. The same difficulty arises with fish living in muddy fresh and coastal waters. Indeed Dietrich (1963) expresses the opinion that the range for effective vision in the marine environment is generally less than one metre, and in areas of high turbidity the effective range may be reduced to only a few centimetres. It is thus not surprising that the olfactory organs of many fishes are extraordinarily sensitive. Froloff (1925, 1928) and Bull (1928–39) were the pioneers here. These workers found that two species of *Blennius*, when offered sea-water extracts of natural food substances, were able to differentiate between concentrations of 0.000375 per cent and 0.00075 per cent weight of living food substance in sea-water. Later, the work of Hasler and Wisby (1951) and also of Kleerekoper and Mogensen (1959, 1963) has further shown the extraordinary powers of fishes such as trout and lampreys in detecting specific amines secreted by particular species. Von Frisch (1941) demonstrated that a minnow (*Phoxinus laevis*) that has been frightened or is in some way under stress produces substances in its body mucus that stimulate fright behaviour in other animals of the same species. There is little doubt that the study of fish pheromones has only just begun and is of great promise.

An outstanding landmark in the knowledge of the sensory powers of fish was the discovery by Lissmann (1951, 1958) that some fishes of tropical rivers, notably the Mormyridae and Gymnotidae of the great turbid river systems of Africa and South America respectively, use electric pulses for orientation. They detect objects by perceiving the electrical field distortion of the discharges which they produce. It is known that each species of Gymnotid fish has a characteristic resting pulse pattern of electric discharge which may be modified by particular circumstances and activities. Lissmann (1963) has recently discovered electrical receptors in other species, the function of which has not yet been fully investigated. Among fish species producing both weak and strong electrical fields, casual observations suggest that the electrical discharge may have a social function, though it remains to be seen whether such signals as have yet been described permit individual and species recognition: on general grounds, it seems highly probable.

It has been shown by Harris and van Bergeijk (1962) that the swim bladder acts as a transducer for pressure waves; transforming them into local 'near-field' effects. Pure pressure waves are rapidly and effectively propagated in water but are essentially non-directional except at close range and high intensity. As Tavolga (1956) points out, there is, as yet, no evidence that fish can assess direction in a far field. He says 'In effect, fishes have but a single ear since the two inner ears are not only close together but coupled to a single swim bladder'. However, whether or not fishes can localize sounds with any high degree of efficiency (and the evidence is still contradictory), much further evidence of widespread sound production amongst fishes has come to hand as a result of the development of underwater microphones. The significance of this is, as yet, not understood; but playback experiments have demonstrated that responses can be elicited by recorded sound. The grunts of male gobies (*Bathygobius soporator*) in collaboration with visual stimuli encourage female courtship, and they also elicit the approach of males.

In the satin-fin shiner (*Notropus analostraneus*) Stout (1963) has shown that territorial males produce single knocks when chasing and during courtship, rapid series of knocks during display fighting, and purring during the courtship activities of approach, circling and 'male passing over nest-site'. Playback of the knocks tends to elicit aggressive behaviour from other males while purring sounds favour courtship activities. There is evidence (Delco, 1960) that sounds of other species fail to elicit responses in some cases but are effective in others (Tavolga, 1958). For a valuable summary of the present state of knowledge concerning communication in fishes, see Tavolga (1960).

INVERTEBRATES

Honey bees

Among the invertebrate animals the Insecta, which contain more species than all the rest put together, are of particular interest from our point of view. Among these, by far the best known and most important instance is provided by the dance communication of the honey bee (*Apis mellifica*) which was worked out by Karl von Frisch and his pupils, and first announced in detail in his classic paper of 1946. The first full account in English was provided in von Frisch's book *The Dancing Bees* (1954) and further details are to be found in his great book *The Dance Language and Orientation of Bees* (1967*a*). The main outlines are so well known that they can be summarized briefly here.

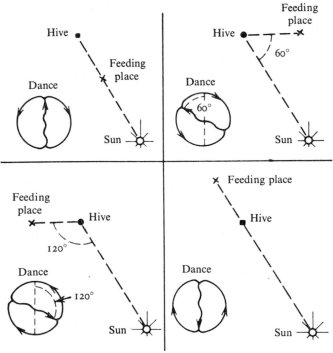

Fig. 1. The waggle dance of the honey bee performed on the vertical face of the combs inside the dark hive. The direction of the waggle run across the diameter of the circle conveys the direction of the food discovery. In the hive the direction is relative to the vertical, in flight relative to the sun. Thus a following bee must transpose the angle of the dance to the vertical to an angle relative to the sun when it sets out to locate the discovery that the scout has announced. (After von Frisch, 1954.)

Von Frisch found that scout bees, when they have found a rich source of nectar, return with a full load of pollen or nectar to the hive and communicate accurately the location of their find by dancing in the dark hive on the vertical combs. This communication is effected by the famous waggle dance or figure-of-eight dance (see Fig. 1). The angle between the waggle run and the vertical is determined by the position of the sun relative to the food, and this angle changes as the sun travels across the sky. It was then shown that bees can compensate for sun movement. Then Lindauer (1961) discovered that, when swarming is imminent, scout bees communicate the location of possible new sites for the colony by dances that are the same as those used to announce food, except that they last much longer, sometimes several days. During the course of these dances the relative direction of the sun changes and the orientation of the dance does so too; only when very few, or no, dissenting directional votes are cast does the swarm move to the new hive site that they have thus agreed upon.

Persistent dances can also be induced by placing food inside the hives. When the scout bees dance under these circumstances they indicate the location of the most recent discovery outside the hive. Even if the sun and sky are not visible to them before they enter the darkened hive to dance, the waggle runs continue to indicate the correct direction of the previous site. Thus the orientation of the waggle run across the comb changes through the day, shifting counter-clockwise relative to the vertical (Marler and Hamilton, 1966).

Distance communication in the bee dance is based on the energy expended in the outward flight to the nectar source. It can thus be rendered to some extent inaccurate by variations in the strength and direction of the wind encountered. With this limitation the rate of the waggle run is correlated with the distance to the nectar source, and this is in turn correlated with the time of each circling movement of the whole dance. At very short distances the method of signalling changes to a round dance, which communicates distance but not direction.

The complete waggle dance thus indicates both the direction and the distance of a food source, and constitutes for the colony of bees an extraordinarily efficient method of harvesting as rapidly as possible a new and abundant source of food. The dance is only performed when a worker has discovered, or is foraging at, a particularly rich source of supply. Each performance of the dance on the comb results in a number of other bees setting out and finding that particular food source. If the supply is still rich when the newcomers reach it, they also dance on the combs on their return, and so a numerous band of foraging workers is quickly recruited for work at a rich source of supply. As soon as the source begins to fail the returning workers cease to dance although they themselves continue exploiting that source as long as an appreciable yield is obtained. So the whole mechanism serves not only to direct new foragers to food supplies that have been discovered by fellow members of the colony but also to get them there in approximately the right numbers to exploit the food supply properly. Hockett (1960, see Table 1, Ch. 2 above) has discussed these results, evaluating the performance in the light of his series of design features. He shows that this communicative device is unique amongst the invertebrates in showing semanticity (Item 7), productivity (Item 11), and also, in his view, unique in the animal kingdom in always showing displacement (Item 10). Whether it shows a tradition (Item 13) is doubtful.

Wenner and Johnson (1967) and Wells and Rohlf (1967) have recently questioned the theory of dance communication as set out by von Frisch and his co-workers, arguing that the dances lack communicative importance. According to them the bees find their way to food sources by means of other cues such as odour. These investigators take the view that those experiments of von Frisch that apparently demonstrated directional and distance communication lacked the controls needed to eliminate the possibility that a used food source (the experimental dish) would accumulate attractive odours lacking at the control dishes. They claimed that an experimental design eliminating this factor failed to demonstrate either directional or distance communication from foragers to inexperienced bees. Michener (1969) also suggested that inadequacy of controls in earlier experiments could be a justifiable criticism. But when we consider the sum total of experiments over the years, the adequacy of the controls is impressive.

Von Frisch claims, with I think, strong evidence, that his experiments with scented feeding places and scented places with no food show conclusively that the bees' behaviour was not only a question of guidance by scent. This theory of guidance by scent was indeed von Frisch's first theory as to the methods of communication employed by the bees. But after a long series of exhaustive experiments he found that this could not be the sole explanation and that an entirely different system of signalling must be operating. In his experiments few newly-alerted bees came to the scented places but very many to the one with food. He points out that Johnson's work (1967) extended only over a small distance whereas in the Austrian experiments the distance from the hive to feeding place has often been much greater: 750 metres, 1050 metres, 2000 metres and 4000 metres. He says, 'Again I must state that local scent marks cannot possibly be held responsible for the arrival of recruits at the proper place over such great distances'. Von Frisch (1967b, 1968), in strongly defending his conclusions, ascribes the results of Wenner and Johnson to the use of strong sugar solutions which excite the bees and cause them to fly about erratically.

Wenner and Johnson also argue that sound is another communication mechanism; for, during the straight part of the dance, the bee makes a sound the pulse rate of which is approximately 2.5 times as fast as the rate of waggling. Another such mechanism may be what has been called the 'rumbling dance', a zig-zag performance recalling the behaviour of the tropical 'stingless' bees (*Trigona*) when they return to the nest from foraging expeditions. While Wenner and his associates have thus made an important contribution to the understanding of the communicative behaviour of the honey bee worker, the general consensus of present opinion must I think be that von Frisch and his colleagues have answered their objections cogently and satisfactorily. The work of von Frisch, his pupils and his associates over twenty-five years is so overwhelming in detail that it seems most unlikely that the dance is simply a non-functional performance on the part of the bees. As I myself found on a memorable visit to von Frisch at his experimental station at Brunnwinkl, Austria, in 1949 (Thorpe, 1949) even a human observer can predict from the dance both the distance and direction of the food source being used. However, Wenner and Johnson's experiments do demonstrate that there are alternative mechanisms for inciting bees to forage – mechanisms which can function under certain circumstances. Sound is one of these. Now, at the time of writing, von Frisch's work seems to have been finally substantiated by the paper of Gould, Henerey and MacLeod (1970).

Other Hymenoptera

Although the dance language of the honey bee has, rightly, taken pride of place in this section of our discussion, nevertheless it must not be supposed that it is entirely unique amongst the insects or even amongst the Hymenoptera. There are, in fact, a number of less well developed, but not dissimilar, methods of communication found in the bees. Those who wish to follow this line further must consult the writings of Lindauer (1961, 1967).

Now leaving aside gestural–visual communication and audio-communication in the Hymenoptera, we must make brief mention of the extraordinary development of chemical communication systems. There is a great and rapidly accumulating literature on this subject and those interested are specially recommended to the works of new pioneers: C. G. Butler (1967) and E. O. Wilson (1968, Wilson and Bossert, 1963). Figure 2 displays diagrammatically the elaboration and diversity of communicative substances secreted by a number of those few ant species which have been investigated. These substances function either by evoking immediate behavioural responses with what may be called releaser effects, or by means of endocrine-mediated primer effects that alter the physiology and subsequent behavioural repertoire of the receptor animal (Wilson, 1968). The sensitivity to these substances is astonishingly high. It has been estimated that in many cases no more than a few molecules strike the chemosensitive sense cells. In fact, Boeckh, Kaissling and Schneider (1965, quoted by Wilson, 1968) have shown by means of the air-stream dilution technique that a single molecule can, in certain circumstances, be sufficient to fire one of the chemical sensory hairs. Pheromones, as in the bumble bees (*Bombus*) and innumerable ant species, can be left behind as a continuous signal after the animal has departed: so they are thus used as territorial and trail markers as well as to provide individual nest odour. Wilson estimated that the transfer of information can be surprisingly rapid and that the potential transmission rate for spatial information in the fire ant's

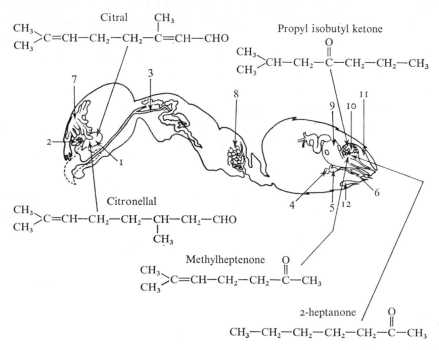

Fig. 2. Some ant alarm substances which have been identified and the location of the glands which secrete them. Citral identified from *Atta rubropilosa*, citronellal from *Acanthomyops clariger*, propyl isobutyl ketone from *Tapinoma nigerrimum*, methylheptenone in *Tapinoma* and other species, and 2-heptanone from *Iridomyrmex pruino.us*. The exocrine gland system represents *Iridomyrmex humilis*. (After Wilson and Bossert, 1963; Wilson, in Marler and Hamilton, 1966.)

(*Solenopsis saevissima*) odour trail is about one bit every eight seconds: a rate comparable, as he says, with that which has been computed for the waggle dance of the honey bee with respect to purely spatial information. Bossert (1968) has argued that patterned transmission of pheromones consisting of amplitude and frequency modulation (regular patterns in frequency of release over a given time period) is technically feasible over very short distances or in a low steady wind. If such transmission actually occurs, Wilson estimates it could be used to process 100 or more bits of information per second – though there is, of course, no evidence that the receiving animal absorbs it at anything like this rate. The number of possible organic odorants is astronomically large and there are, no doubt, innumerable instances yet to be discovered and studied. In ants the pheromones used for trail-forming are relatively volatile and a natural single trail would evaporate within a minute or two to below threshold density, depending, of course, on how absorptive is the substratum. This means that a single trail-laying worker can pinpoint a food source only within a few feet of the nest, for the trail fades before other workers can follow it for greater distances.

The idea that chemical 'shapes' can be perceived by insects has often been discussed but good evidence seems yet to be lacking. Trails left by ants in the form of a drop of fluid from the tip of the abdomen often have a characteristic shape in that the fluid drop points against the ant's direction of travel; but I know of no evidence that ants which are lost refer to the shape of the marks to tell direction.

The trail substances used by ants tend to be species-specific in action affording, so to speak, a private line for communication amongst species where the foraging territories overlap. It is also evident that in many species of ants, as also in bees, individual colonies may have their own characteristic odour within their species odour. A most striking case of recruitment by insect pheromones (a phenomenon very widely known in the insects) is provided by a certain bark beetle (*Ips confusus*). When males of this species work their way into the phloem-cambial tissue of a host tree they release a volatile substance from the hind gut that is attractive to both males and females. Other individuals, exploring in the vicinity, are drawn to the penetration gallery, producing a mass attack which is only too familiar to students of forest entomology.

Besides pheromones which carry orientational or directional information, there are many examples of substances which, just as in the 'Schreckstoff' described in fish by von Frisch, act as warning of danger to other members of the species. It seems that in the ants such alarm substances are not by any means always characteristic of the species – not even always characteristic of the sub-family. Wilson and Bossert (1963) have shown that in two genera of ants, *Lasius* and *Acanthomyops*, the terpenoid alcohols produced (if they are as they appear to be, sex substances) are remarkable in that they occur in medleys – each species manufacturing a blend in which the proportions of the constituents are peculiar to the species. How far this may turn out to be a general feature is anybody's guess, but if applicable it increases the potentialities of pheromones for information purposes very greatly indeed.

Wilson (1968) has listed the various ways in which chemical systems can be adjusted to enhance the specificity of signals or to increase the rate of information transfer. His classification is as follows:

(1) Adjustment of fading time.

(2) Expansion of the active space; thus if the pheromone is expelled down wind only a relatively small amount is required since orientation can be achieved by the following insect zig-zagging up wind and so keeping in the zone of stimulation – as do many male moths flying up wind from truly astonishing distances to find a single female ready for mating.

(3) Temporal patterning of single pheromones.

(4) Use of multiple exocrine glands (see Fig. 2 above).

(5) Medleys of pheromones (the phenomenon just discussed above).

(6) Change of meaning through context.

(7) Variation in concentration and duration.

(8) New meanings from combinations.

It is still largely guesswork which of these various categories are the more important, but it is certainly true that all are possible means of information transfer, not only in the Hymenoptera (ants, bees, wasps, etc.), but widely throughout the insects and other invertebrates.

It remains to say a few words about auditory communication in the insects. Here again, the use of the method is widespread, although the Orthoptera (including the grasshoppers and crickets) and the Hemiptera (including the cicadas) provide the major examples. In this field, the work and the summaries of R. D. Alexander (1960, 1967, 1968) provide the main door of easy access to the subject. Alexander (1967) arranges the acoustic signals of arthropods under nine functional headings: omitting those which we have already dealt with under Hymenoptera, namely food and nest site directives which are limited to social species, these are:

(1) Disturbance and alarm signals (predator repelling and conspecific alarming).

(2) Calling signals (pair forming and aggregating).

(3) Aggressive signals (rival separating and dominance establishing).

(4) Courtship signals (insemination timing and insemination facilitating).

(5) Courtship interruption signals.

(6) Copulatory signals (insemination facilitating and pair maintaining).

(7) Post-copulatory or inter-copulatory signals (pair maintaining).

(8) Recognition signals (limited to sub-social and social species and functioning as pair and family maintaining stimuli).

(9) Aggressive mimicry signals (prey attraction by production of pair-forming signals of prey species).

In considering the sound communication of Orthoptera, we must remember that the insect auditory organ, usually and perhaps nearly always, consists of a system which is highly sensitive to amplitude modulation and to temporal spacing but normally insensitive to frequency modulation. (Michelsen has described an exception to this in the locust

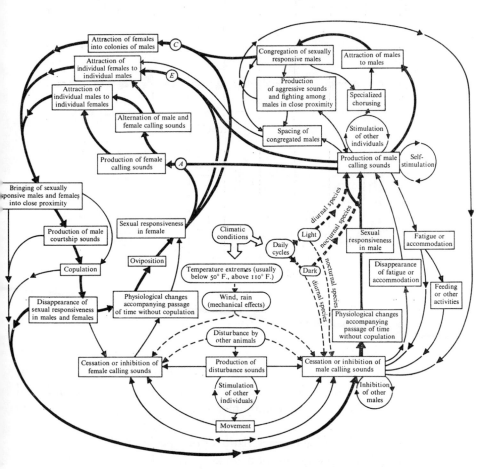

Fig. 3. A diagrammatic representation of the behavioural sequences and cycles associated with sound communication in adult Orthoptera and Cicadidae. Heavy lines indicate more important sequences, and the symbols (C), (E), and (A) designate sequences most characteristic of the Cicadidae, Ensifera, and Acridinae, respectively. (From Alexander, R. D., in *Animal Sounds and Communication*, W. E. Lanyon and W. N. Tavolga (eds.), 1960, A.I.B.S. Washington, D.C.)

(Michelsen, 1968).) Figure 3 gives an indication of the behavioural sequences and cycles associated with sound communication in adult grasshoppers, crickets and cicadas. Figures 4 and 5 show the importance of rhythmic organization.

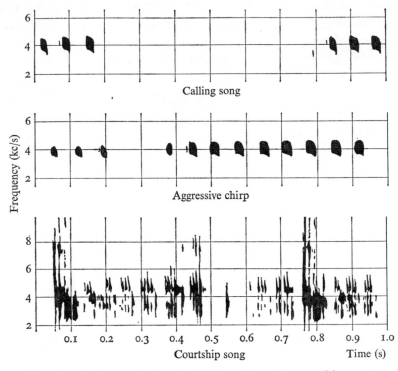

Fig. 4. The three basic sound signals of field crickets, illustrated by spectrograms of tape recordings of *Acheta firmus* (Scudder). (2nd generation reared adults from Grand Isle, Louisiana.) (From Alexander, R. D., in *Animal Sounds and Communication*, W. E. Lanyon and W. N. Tavolga (eds.), 1960, A.I.B.S., Washington, D.C.)

The special instance of visual communication provided by the fireflies, beetles of the family Lampyridae, is of peculiar interest (Fig. 6). Lloyd (1966) describes the flash communication of fireflies in the following remarkable passage:

'At the time of evening characteristic for the species, males arise from the grass and fly and flash, most of them keeping within an ecologically well-defined area such as a lawn, forest edge, stream bed or wet corner of a pasture. Male flight paths during moments of light emission are characteristic for the species; some species can be identified by this behaviour alone. Females are found on the ground and on grass or other low vegetation. When a male receives a flashed answer from a female after the species-characteristic interval following his own flash, he turns and flies towards her. After a few seconds he repeats his flash pattern; if he again receives the correct flash response he continues his approach. Flight terminates a few centimetres from the female after from 1 to 10 flash exchanges. After landing the male usually completes the approach

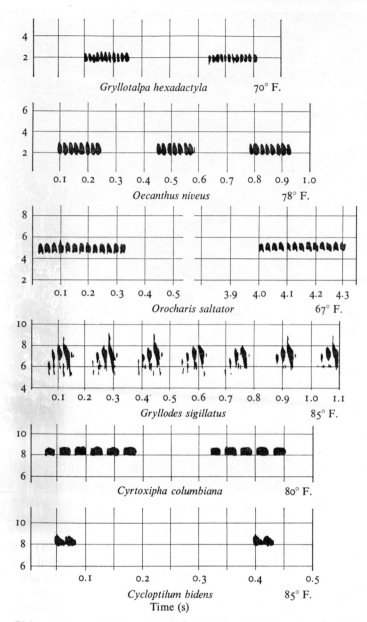

Fig. 5. Chirp patterns of calling songs in six different subfamilies of crickets. Top to bottom: *Gryllotalpinae*, Champaign Co. O., 24 Aug. 1954; *Oceanthinae*, Erie Co. O., 26 July 1955; *Eneopterinae*, Dyar Co. Tenn., 24 Sept. 1955; *Gryllinae*, Florida laboratory culture; *Trigonidiinae*, Lenoir, N.C., 2 Aug. 1955; *Mogoplistinae*, Raleigh, N.C., 8 Aug. 1955. Ordinate, kc/sec. (From Alexander, R. D., in *Animal Sounds and Communication*, W. E. Lanyon and W. N. Tavolga (eds.), 1960, A.I.B.S., Washington, D.C.)

Fig. 6. Pulse (flash) patterns and flight during flashing in male fireflies of the genus *Photinus* as they would appear in a time-lapse photograph (modified from Lloyd, 1966). The species illustrated are not all sympatric. Small triangles near numbers designating species indicate direction of flight: (1) *consimilis* (slow pulse), (2) *brimleyi*, (3) *consimilis* (fast pulse) and *carolinus*, (4) *collustrans*, (5) *marginellus*, (6) *consanguineus*, (7) *ignitus*, (8) *pyralis*, and (9) *granulatus*. (After Alexander, in Sebeok, 1968.)

by walking and exchanging flashes with the female. Males mount females immediately upon contact. For complete attraction, only flash signals are necessary, and in all species tested, males were attracted to females caged in air-tight glass containers. During approaches, females frequently fail to answer some of the male flashes, and when they resume answering, males continue their approaches. Males remain in the area of a previous response, emitting their flash pattern for several minutes after females are removed.

In most species, activity lasts for about one half-hour and then decreases slowly over the next 30 to 40 minutes, until eventually only an occasional flash can be seen.'

Figure 7 shows how the temporal patterning of the firefly flashes resembles the temporal patterning of the stridulatory calls produced by the Orthoptera and by the very special sound-producing mechanism of the cicadas.

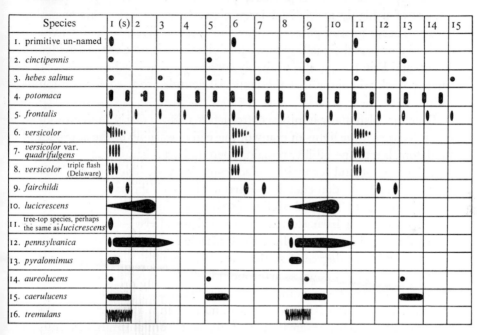

Species	I (s)	2	3	4	5	6	7	8	9	10	11	12	13	14	15
1. primitive un-named															
2. *cinctipennis*															
3. *hebes salinus*															
4. *potomaca*															
5. *frontalis*															
6. *versicolor*															
7. *versicolor* var. *quadrifulgens*															
8. *versicolor* triple flash (Delaware)															
9. *fairchildi*															
10. *lucicrescens*															
11. tree-top species, perhaps the same as *lucicrescens*															
12. *pennsylvanica*															
13. *pyralomimus*															
14. *aureolucens*															
15. *caerulucens*															
16. *tremulans*															

Fig. 7. Pulse patterns during flashing in male fireflies of the genus *Photinus*. Species 2 and 14, with the same flash pattern, are quite different morphologically, were taken in Maryland and Minnesota, respectively, and may be allopatric. Delay times in female answers may be different between species. (After Alexander, in Sebeok, 1968.)

Crustacea

This is another arthropod group containing a very large number of species the vast majority of which are marine: a much smaller number live in fresh water and a relatively few are terrestrial. Only a very small proportion of this great total of species have been studied sufficiently thoroughly to allow much to be said about the methods of communication. However, enough is known to indicate that the general situation is broadly similar to that in insects. A valuable general survey is given by Frings and Frings (1968). Visual signalling by means of special gestures has been particularly thoroughly worked out in the Fiddler crabs of the genus *Uca* by Altevogt

(1957, 1959) and Crane (1943, 1957, 1958). In this genus the males have elaborate and highly specific gestures involving rhythmic waving and other movements of the specially developed large claw on one side of the body. In recent years it has been discovered that a great many crustacea make noises of one kind or another. These involve tapping, bubbling, stridulation, etc., specially developed structures for the purpose often being involved. Thus the snapping shrimps, Alpheidae, produce loud cracks by the sudden closure of a highly specialized large claw. In some cases this snapping seems to serve to stun potential prey and it seems likely to be involved in territorial or sexual signalling as well. While there is no clear-cut evidence, Frings and Frings suggest – probably correctly – that it is highly likely that further work will show that these special sounds are used for communication.

There are also many luminescent crustacea, particularly in deep waters, and it looks as if the lights of many of these give warning of the approach of predators, but the possibility remains open that aggregational attraction and sexual attraction and recognition also result from the light signals. Finally, there is little doubt that chemical signalling is also important in this group. Crisp (1961) and Crisp and Mellers (1962) have found that extracts of the bodies or shells of barnacles, when applied to otherwise clean slates and then submerged in waters where the barnacle larvae (*Cypris*) are present, cause a significant degree of settlement. It is thought that the factor here may be a substance known as arthropodin which is a regular constituent of the exoskeleton of these animals and not a specifically produced signalling material. Carlisle and Knowles (1959) find that in many marine crustacea the moult in which the female is transformed from an immature into an adult ready for reproduction, is of a special type which they call the copulatory moult. Females just before or at the time of this moult seem to be recognized by males which often seize them and hold them for mating. In some cases the male is attracted from a distance to the pre-moult female, while in others it is necessary for him to make antennal contact before recognition can occur.

Arachnida

Table 1 shows the generalized course of display in spiders and summarizes the results of a number of papers by Crane published in *Zoologica* from 1943 to 1957. Other important books and papers include those of Bristowe (1939–41) and Cloudesley-Thompson (1958). In the case of the web spiders the male moves about until he accidentally touches a web of the female: from then on he is capable of detecting the species and sex but generally not the sexual readiness of the female. In other cases the male finds the

Table 1. *Generalized course of display in salticid spiders*

Male	Female
Becomes aware of female; starts display. Stage 1. (*Minimal releaser:* several sight factors; airborne chemical stimuli also involved.)	
	Retreats, or watches male, usually in braced, high position, often vibrating palps. Rarely attacks. (*Minimal releaser and director:* several sight factors.)
Approaches, in zigzags, or follows (if female retreats), continuing or resuming display. (*Minimal releaser and director:* above sight factors, plus type of female motion or lack of it.) Special female signs, such as vibrating palps and light abdominal spots probably have directive value.	
	Becomes completely attentive; sometimes gives weak reciprocal display. (*Minimal releaser:* summative effect of display motions.)
Speeds up display tempo. (*Releasers and directors:* reduced motion of female, plus chemical stimuli. Self-stimulation is doubtless also a factor.)	
	Ceases motion and, usually, crouches low, legs drawn in.
Enters stage 2. (*Releasers:* primarily, proximity of female; also involved, usually, her lack of motion, low position, and doubtless, reinforced chemical stimuli.) Copulation follows unless female withdraws. (*Director:* sometimes a pale abdominal crossbar.)	

From Crane, 1949; in Sebeok (ed.) 1968.

female by the scent of her dragline which she lays down as she walks. Full pre-copulatory behaviour depends upon whether the female is in a web or not. If she is, the male generally uses vibrations of the web as signals. These may be taps in the lair in trapdoor spiders or plucks of the thread of the web in orb-spinning spiders. If the female responds she may give aggressive signals, in which case the male departs, or if she is mature and sexually ready, she may give signals of a different type or remain passive, whereupon the male enters the web and is able to copulate. In view of the predatory habits of the female when in the web, it is obviously important that means of identification of species and reproductive state should be rapid and unequivocal (see also Parry, 1965). In some of the parasitic water mites (e.g. *Unionicola ypsilophorus*) (Welsh, 1931) a chemical is given off by the host – a fresh-water clam – which when responded to brings the parasite into the shell of the host. In other species of water mite males can apparently recognize females and *vice versa* by signals which are both chemical and tactile. In one case, *Eylais infundibulifera*, recognition is said to be entirely tactile, no evidence for chemical signals having been detected (Böttger, 1962, quoted by Frings and Frings, 1968). To complete our brief survey of the Arthropoda it is worth mentioning that some centipedes and millipedes are known to produce sounds by stridulation and it seems likely that further work will show that these are communicative in function.

Mollusca

There is little doubt that many land snails and slugs communicate both by chemical secretions and mechanical stimulation; though the details of the former process have not yet been worked out experimentally. In the oyster, *Crassostrea virginica*, Galtsoff (1938) has shown that females release eggs when stimulated by a chemical in the sperms, whereas the male is unspecific and will release sperms as a result of temperature changes and the chemical substances provided by a number of species of invertebrates not even closely related. With the Cephalopoda (octopuses and squids) – a group that is outstanding for mobility, highly developed eyes similar in structure and in action to those of the vertebrates, and high brain development (all, of course, features which are linked with one another) – no definite evidence of chemical communication seems to have been produced, whereas there is every reason to suppose that recognition of sex and readiness to mate is signalled by visual displays. Many octopods are remarkable for their ability to effect rapid changes of colour and pattern; but whether these are of significance in communication is not known.

Other Groups

When we come to the group Annelida, which includes a vast number of marine worms and also the earthworms and leeches, there is little doubt that the precise synchronization of spawning in many marine forms must be controlled by chemical stimulation. In the land leeches there are some casual observations which suggest mechanical signalling and it has been noticed that a male may approach a tapping female and set up a duet with her. 'The two alternatively tap and curl the front ends of their bodies together, this ultimately leading to copulation' (Leslie, 1951, quoted by Frings and Frings, 1968). There are similar suggestive observations with regard to earthworms but, as far as I am aware, no experimental confirmation. In marine annelids there is one case on record of a signalling system using luminescence. In a Bermudan polychaete, *Odontosyllis enopla*, females rise to the surface of the sea at the swarming season where they form a glowing mass and the males are attracted to this. If the females stop glowing the males signal by flashing and the females resume their light (Galloway, 1908; Galloway and Welsh, 1911). Even animals as structurally simple as the coelenterates and the echinoderms may show associative learning and patterns of complex behaviour for which precise signals may be necessary. (For a general discussion see Thorpe and Davenport, 1965.)

Bullock and Horridge (1965) have shown that the nerve nets of simple animals such as sea anemones and jelly fish are by no means simple in function. The more complex behaviour in these organisms appears in the symbiotic sea anemones, which have associations with fishes and with hermit crabs. Much material of interest will be found in Davenport (1966). From this work it seems that precise recognition of specific chemical agents may be part of the normal behavioural biology of these animals. It seems not impossible that in the widely known partnerships between giant anemones and their inquiline fishes, the activity of the fish may in some way be directed to effecting recognition of the fish by the anemone. It even seems likely that an anemone may recognize its own individual fish, for it does not sting it when reintroduced to it, whereas another fish of the same species is stung on first contact.

The following extracts from Davenport's work, quoting in the first place Ross (1960), and secondly Davenport and Norris (1958), give a varied and intriguing picture of the complexities of behaviour and communication which may be taking place in these 'lowly' animals.

The crucial experiment, which has such a direct bearing on coelenterate physiology and behaviour, in which a hermit crab *Eupagarus* takes little or no part in the activity which results in the change of position of the sea anemone *Calliactis* to the shell of its host, was conducted by Ross at Plymouth. The behaviour of the anemone is so precise and so interesting that it would be best to quote Ross' words directly in a description of the process:

'The attachment of *Calliactis*, which have been removed from shells, to stones, glass plates and other objects, is in no way remarkable. An animal lying on its side, or supported in some way, merely secures a foothold by the edge of the pedal disc, and from this foothold the attachment spreads until the whole surface is adherent. The entire process may take many hours and is accomplished by the combined effects of muscular suction and cementing secretions. Essentially this process is the same as the method employed by *Actinia equina* and *Anemonia sulcata*. Compared to these, however, the pedal disc of *Calliactis* attaches much more slowly, but once the attachment is made it is firmer, and because the attached foot is virtually immobile, it is more permanent.

The attachment of an unattached *Calliactis* to a shell occupied by *E. bernhardus* is very different indeed. Faurot (1910, 1932) described this briefly and the following expanded account, based on my own observations, agrees with his in the essential points. First, the tentacles explore the surface of the shell very actively and many of them adhere to the shell, perhaps by glutinant nematocysts, forming an attachment firm enough to hold the anemone on the shell even when the crab moves about. This tentacular attachment develops within a few minutes into a firmer attachment in which the whole oral disc is involved as well. Apparently, the radial musculature contracts and pulls against the expanded margin and its adhering tentacles and so produces an immense suctorial disc. At no other time does the animal seem so much to deserve its specific name *"parasitica"*. Once this is achieved, the anemone is virtually safe from being dislodged, even by the most active movements of the crab.

All this may take 5 to 10 minutes and is followed by a slow bending of the column which, in another few minutes, brings the pedal disc up to the shell so that the animal is bent double. The pedal disc, meanwhile, becomes greatly distended and begins to adhere to the shell immediately contact is established. Within another few minutes the whole pedal disc has spread out, its swollen surface and edge becoming bedded down firmly to fit the grooved, often encrusted, surface of the shell. Finally, when the pedal disc is firmly attached the tentacles and oral disc let go, and in the space of another 2 to 3 minutes, the column straightens out and the animal assumes its normal extended posture.'

The behaviour of the fish *Amphiprion percula* to its host anemone has been described during what is called the 'process of acclimation'. Very little indeed is known about how frequently this process occurs in nature, how much of the adult life of these inquiline fishes is spent in a free state or how much they move from host to host. But when the individual 'unacclimated' *Amphiprion percula* was introduced into the observation tank with the anemone, a fairly stereotyped series of events occurred which terminated in acclimation. To quote from Davenport and Norris (1958):

'An unacclimated fish introduced into the tank a foot or so away from the anemone usually approached the anemone within a few minutes and began to swim under the disk around the column, and occasionally over the top of the disk a centimeter or more away from the tentacles. Such fish spent most of their time under the disk at this stage and sometimes were seen nibbling at the column of the anemone. Most fish seemed to "recognise" the anemone within a few minutes and swam towards it . . . As the process proceeded, passage over the disk became more and more frequent and the "acclimating" fish moved closer and closer to the tentacles. Swimming was accomplished by a distinctive series of slow vertical undulations, in which the tail was usually held a little lower than the rest of the body. Eventually, on one of these trips over the disk, the fish would touch a tentacle or two, usually with the ventral edge of its anal fin or the lower margin of its caudal fin. Commonly this resulted in a moderate adherence of the tentacle to the fin and contraction of the tentacle. The fish then jerked itself free with a violent flexure of its body and usually raced off the disk. Not all newly introduced *Amphiprion* caused clinging upon their first contact with tentacles, but it was the general rule. However, this adherence failed to deter the fish, which nearly always returned immediately to the anemone, either under the disk or over the tentacles . . . .

After this initial contact the fish typically came closer and closer to the tentacles, touching them with increasing regularity. The reaction to the clinging of tentacles became less and less violent until a sudden flexure of the animal's body was the only reaction given by the fish. Mouthing or nipping of tentacles was often observed in this and later stages.

The clinging and contraction of tentacles upon contact with the fish gradually became less until it ceased altogether. At the same time the fish began to swim deeper among the tentacles using the same slow undulating movements as when it had cruised above the disk.

Once the fish was swimming in fairly constant contact with the tentacles of the anemone, a very striking change in its behaviour occurred. The general speed of swimming suddenly increased until the *Amphiprion* was dashing back and forth over the disk of the anemone, flailing un-reactive tentacles aside with violent movements of its body. Often the fish raced beneath the anemone and appeared in one of the folds of the disk margin, its head completely ringed in tentacles. The fish frequently maintained this vantage point for a few seconds, holding position with rapid alternate fanning movements of its pectoral fins, after which it might dash onto the disk again for another foray among the tentacles. The powerful swimming typical of this stage of the acclimation process was accomplished by rapid and strong lateral body flexures. The impression given by the swimming behaviour of the fish after final acclimation was that the fish was "bathing" its entire surface among the tentacles.'

This final activity was so marked, so sudden, and so dramatic that upon observing it one was obliged to recall the rejoinder of Chuang-tzŭ to Hui-tzŭ quoted below,* for once an *Amphiprion* reached this level in its behaviour, it exhibited every indication of 'joyful' activity in its mad dash in and out of the tentacles of its host.

Finally, it seems appropriate to mention that even in the Protozoa and in creatures such as the slime moulds (*Acrasiales*) cellular aggregation has to take place before the reproductive phase, which leads to the production of spores, can occur. Here communication takes place by means of the discharge of a chemical called acrasin (Bonner, 1959). Acrasin, which may be more than one compound, initiates release of further acrasin by other amoebae nearby, which sets up centrifugal wavelike pulses of the material. Acrasin not only induces acrasin discharge but also induces amoebae to stream towards the central producer (Frings and Frings). Shaffer (1953–7) has studied the properties of extracted acrasin in a cell-free medium. The substance seems to be rapidly destroyed by enzymes produced by the amoebae. In this way a gradient is produced. He has shown, however, that it is not the gradient itself that orients the cells; rather it is the time sequence in which the relaying amoebae produce acrasin which may be responsible. This being so, it is possible that a specific pulse structure is the signal to which the amoebae are responding.

COMMENTS

(1) *Specificity of responsiveness.*

Throughout its life an animal's sense organs are being bombarded by changes in physical stimuli of many sorts. Behaviour of any sort, but most especially intraspecific social signalling, demands selectivity in responsiveness. In most higher vertebrates specificity in responsiveness to signals is largely a function of central mechanisms, but in some of the forms discussed in this chapter it is peripheral. Thus the great sensitivity of some

---

* 'Chuang-tzŭ and Hui-tzŭ were strolling on the bridge over the Hao. Chuang-tzŭ said: "The white fish stroll out freely; this is fishes' delight". Hui-tzŭ said: "You are not a fish. How can you know fishes' delight?" Chang-tzŭ said: "You are not I; how can you know I do not know fishes' delight?" Hui-tzŭ said: "If, because I am not you, I do not know you, then since you are not a fish, your ignorance of fishes' delight is total". Chuang-tzŭ said: "Please go back to the beginning of your discussion. When you said: 'How do you know fishes' delight?' you asked me knowing already how I knew it. I knew it from my pleasure in strolling over the Hao while the fishes strolled under the Hao".' (From the book known as *Chuang-tzŭ*, compiled about 290 B.C. Translation kindly supplied by Dr L. E. R. Picken.)

moths to female pheromones depends on specificity in the sensory cells (Schneider, 1963). Recent work by Capranica and others (Capranica, 1965, 1966, 1968; Frishkopf and Goldstein, 1963; Capranica and Frishkopf, 1966) is of peculiar interest in showing that specificity in responsiveness to the species-characteristic mating call may be mediated by the peripheral auditory system in frogs. Among the more interesting points are:

(a) The mating call of the bullfrog (*Rana catesbiana*) has a low frequency peak at 200 Hz and a high frequency peak at 1400–1500 Hz, with a noticeable dip at 500 Hz.

(b) Calling can be elicited by synthetic signals, but responses are suppressed if the amplitude of the 500 Hz component is too high relative to the 200 Hz component.

(c) The suppression of vocal response by a frequency component peaking at 500 Hz seems to have the functional significance of preventing males from joining in chorus with immature males: the calls of the latter have a low frequency spectral peak at 600–700 Hz and not at 200 Hz.

(d) Single fibre recording from the 8th nerve revealed two types of unit. 'Single' units were responsive to frequencies in the region from 1000 to 2000 Hz; 'complex' units to frequencies below 1000 Hz. The response of each complex unit to an excitatory tone could be inhibited by a second high frequency tone: for those complex units most sensitive to tones around 200 Hz, the best inhibitory frequency was 500–600 Hz. This inhibition is peripheral as it is not affected by nerve section. Simple and complex units derive from different parts of the inner ear.

(e) Cricket frogs (*Acris crepitans*) show considerable geographic variation in their mating calls, and respond selectively to calls from frogs of their own area. Recording in the medulla yielded two types of units, one sensitive to frequencies from 200 to 1000 Hz and the other tuned to a higher frequency which varied with the geographical origin of the animal. In each case the maximal sensitivity of the high frequency units corresponded to the spectral peak in the mating calls of frogs from the same area. The frequency discrimination probably resides in the basillar papilla of the inner ear.

No doubt such peripheral mechanisms are less versatile than more central ones.

(2) *Fishes*.

A further recent review is given by Moulton (1969).

6. VOCAL COMMUNICATION IN BIRDS

W. H. THORPE

Birds are the most vocal animals apart from man. They use sounds to convey information between the members of a pair or a potential pair, between parent and young, between siblings, between conspecifics in a feeding or migrating or roosting flock, and often between different species within such a flock. Moreover, their vocalizations are incomparably more complex and more precisely modulated and controlled than are those of other animals; and the apparatus for achieving these results is, as far as is known, unique both in principle and in structural design. It follows that the group is as important as any in the animal kingdom in the comparative study of non-verbal communication.

Before proceeding it may be as well to remind the reader of the distinction between 'message' and 'meaning' (Smith, 1968). We often find that a given utterance by a bird contains much potential information, but that only part of this is used in a given set of circumstances. Let us take the example of the song of the robin (*Erithacus rubecula*). This is normally a proclamation announcing occupation of a territory and readiness to defend that territory against all rivals. The song itself is an elaborate succession of sounds, some of them individually quite complex. And this sequence of sounds may be in some respects distinct from one robin to another. If, when a robin has first established a territory, one plays back to it the song of another robin, or of that robin himself, he gives a violent aggressive reaction. But just as violent a reaction can be produced by playing back a very simple schema of a robin's song consisting of the alternating high and low notes at the correct time intervals but omitting all else; so in this experimental situation all the other 'information' in the song is disregarded (Bremond, 1967). But we know, from work on this and many other species, that in natural circumstances a great deal of the detailed structure of the song, which in this experimental situation may be ignored, can be perceived and acted upon.

Thus it has long been realized (e.g. Thorpe, 1958; Hinde, 1958) that a rival chaffinch (*Fringilla coelebs*) may recognize a given male not merely by the position of his territory, but by the individuality of his voice. Since then many new instances have come to light (e.g. Falls, 1969), and as Hinde (1958; see Thorpe, 1961) proved, the territorial singer can sometimes quickly adapt to a rival by matching its song pattern to that of the intruder. As Bremond (1967) has recently described, the European robin's instantaneous imitation of an invader's signal amounts to saying 'I am talking to you, invader of the moment' (see Busnel, 1968). All these examples, and there are many more, indicate a syntactic element, a real combination of signals. Moreover, the tropical shrikes *Laniarius aethiopicus* and *L. erythrogaster* (Thorpe, 1963, 1966, 1972) can modify the messages encoded in their duets by changing the acoustic and temporal patterning of their antiphonal contributions.

In fact, a robin, in its response to intruders, responds selectively to auditory information in just the same way as it does to visual information. Many years ago Lack (1939) showed that a mere bunch of red feathers stuck on a wire in a robin's territory was sufficient to release a violent and continuing attack. Nevertheless, a robin in a different situation, for instance when it is choosing (or responding to) a potential mate, can react to very subtle differences in the colour and attitudes of the approaching robin. In this case the context of the message determines to a considerable extent what part of the message is responded to, and thus its meaning. If a recipient uses the context in responding to the message, one message can carry several meanings. The constant message, yielding different meanings in different contexts, is a simpler device than one which varies the message together with the meaning. In practice both devices are used in different species of birds. In general, in the case of simple calls which for instance give warning of predators, inform the other members of the flock that food has been found, or tell the parents that a chick is hungry or lost, discrete meanings cannot well be transmitted by means of a single message, for the context is not sufficiently well perceived by the recipient. But the examples of song already discussed show how important it is to distinguish between message and meaning in every instance, and how dangerous to assume that all potential information in a message is perceived or understood by the hearers and affects their behaviour appropriately.

Having referred to 'call-notes' it is now convenient to say something about the commonly made distinction between call-notes and song. Call-notes are usually simple in structure, consisting of one or a few bursts of sound, in contrast to the longer and more complicated sequences of song. Call-notes in the main convey information which may warn of danger, help to control the movement of a flock, indicate the whereabouts of food, and so forth. Song, on the other hand, is a type of vocalization appropriate to, and often confined to, the breeding season; it is given primarily by the male under the general physiological control of the sex hormones and, as we have seen, is often capable of a high degree of modification by imitative learning.

Call-notes are often adapted to the particular function that they serve, and we may consider one example, the call-notes used to warn conspecifics of the presence of predators. Such notes are specialized in their acoustic structure for the particular type of predator encountered. Fig. 1 shows the type of call given by many small birds in response to some enemy either on the ground or perched in a tree; a predator such as a hawk, an owl or a weasel. It is clear that these sounds are all loud, repetitive and cover a great

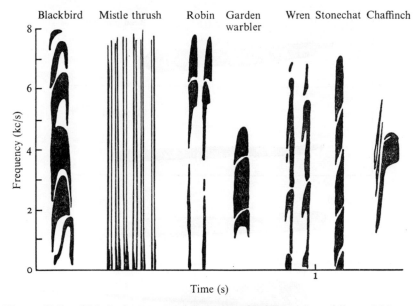

Fig. 1. Calls of birds from several Passerine families given while mobbing owls. (From Thorpe, 1961, after Marler, 1959.)

frequency range. They are often given in a situation in which it is presumably advantageous for the bird concerned to draw attention to this particular danger and warn others of its exact location. An example is the mobbing response of many small birds to a perched owl in which they stand around at a fairly safe distance and chatter at the enemy. The result is to attract neighbours to join in the mob; and it is to be noted that the call structure is such that it is very easily located – by a feature which our ears experience too. For loud, click-like sounds are very easily and accurately referred to the right direction and roughly to the right distance by both birds and men.

Fig. 2 shows an entirely different kind of call which is given when a flying predator is in the air above. If a hawk or an owl, then the threat is immediate and the danger great. So instead of mobbing, the bird dives into the nearest shelter and gives a type of call which starts and stops gradually and maintains a fairly constant pitch which is somewhere around 7 kHz. Other birds, hearing these calls, likewise fly to shelter and repeat the same call from their hiding places. The immediate point that strikes one on hearing these calls is that they are extremely difficult to locate. This is due, in the first place, to the fact that they lack the characteristics previously mentioned which make localization easy. But besides this there is another reason. A bird employs the same methods as we do to locate the origin of a sound. These consist in comparisons between the sound as received by the

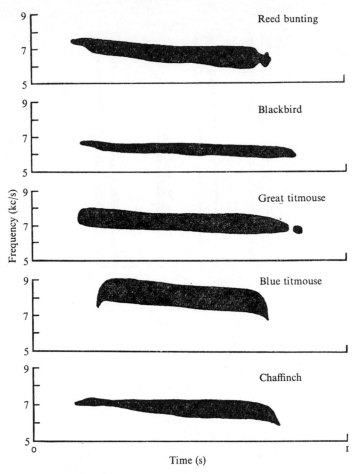

Fig. 2. Calls of five different passerine species given when a hawk flies over. (From Thorpe, 1961, after Marler, 1959.)

two ears involving three different types of data: (1) phase difference, (2) intensity difference and (3) time of arrival. Phase difference is most likely to be valuable at low frequencies: for instance, the data become ambiguous when the wavelength is much less than twice the distance apart of the ears. Intensity differences are most valuable at fairly high frequencies and the differential effect is due to the 'sound shadow' cast by the head of the listening bird: sound shadows only become important when the wavelength approximates to the diameter of the obstruction. Differences between the time of arrival at the two ears will, of course, be more obvious the larger the distance apart of the ears. Calculation suggests that sounds of about the frequency of 7 kHz are likely to be maximally difficult to locate for the ears of a bird of medium size – say a hawk or an owl (Marler, 1955). This frequency is too high for effective binaural comparisons of phase difference, and too low for there to be enough sound shadow perceptible by the ears. Thus the 'seet' call of the male chaffinch probably gives no clues as to its location, either by phase or intensity differences. This leaves only the third method, the appreciation of binaural time differences; and just because the sounds begin and end gradually or imperceptibly, no clues for time differences are supplied. This may, in fact, be the most important point of all, and Dr Andrew has pointed out to me that the same effect may be achieved by a long, very regular trill, as used by the domestic fowl in response to a hawk, even though at a lower pitch. Finally, the wavelength is small enough to result in fairly free reflection from the trunks of small trees.

Now let us consider song. This must serve both as a species and as an individual signal, and we may first consider the question of securing and maintaining species distinctiveness. Here we have to establish the mechanism for the maintenance of the overall specific pattern of the song from generation to generation. How much of this can be put down purely to genetic programming? If we examine young birds kept in complete isolation from conspecifics – isolated, that is, either from the egg or from the early fledgling life – we find a number of instances of birds so treated in which the vocalizations emerge little or no different from those of individuals reared normally. This certainly seems to be true of many of the call-notes above referred to. It also seems to be true of some songs. Thus the song sparrow (*Melospiza melodia*) will produce an almost completely normal song even when it has been fostered from the egg by canaries and allowed to hear only canary song, which is entirely different in pattern and structure (Marler, 1967a).

The same situation seems to obtain with many species, perhaps all species, of the pigeons and doves (Columbidae). The acoustic structure, which provides the basis of the tonal quality of the notes of birds of this family, is, on the whole, remarkably similar as we pass from species to species. To put it very crudely: almost all doves 'coo'. But though all the notes sound similar, probably no two species the world over 'coo' in exactly the same rhythmical pattern or with the same accent. So it should, in principle, be possible to construct a table for identification of the doves of the world based solely upon the numbers of the notes in the song, the overall duration, the length of the individual notes, the times, the rhythm and the accent. And, in fact, one can often recognize a given species throughout an enormous geographical range solely by the temporal organization of the song. Experiments on some of these species (Lade and Thorpe, 1964) and evidence from earlier work (Thorpe, 1968) indicates that the vocal pattern is completely resistant to cross fostering or other means of exposure to alien song patterns during early life. Another example of this resistance to early experience is provided by the European cuckoo (*Cuculus canorus*), where the song produced by the male is entirely independent of the vocalizations of its foster parents.

As a result of such examples one is forced to the conclusion that there must develop, independently of experience of the species-characteristic song, a system which is fully competent of itself to activate and control the motor processes of the vocalization. This has been shown to be the case in doves by Nottebohm (1970) and in the domestic fowl by Konishi (1963), who have shown that these birds can develop and maintain the normal repertoire and forms of their vocal signals even though complete deafening has been carried out operatively in the first days of life. In other words, in the call-notes of the domestic fowl, just as, in all probability, with those of the doves and the song of the song sparrow, the mechanism is fully competent without the necessity for experience of the species pattern.

In many birds, however, we find something less than this full competence in that some relatively specific auditory stimulation is required to render the mechanism fully effective and produce the specific vocal signals. In the chaffinch, this stimulation can come from the hearing of chaffinch-like songs during the bird's first autumn, some months before it starts to sing itself. As one might expect from this, the normal song structure cannot be produced if the chaffinch has been operatively deafened as an early juvenile. The overall song structure – what one might call the 'sound skeleton' of the song – which constitutes the specific vocal signal, is not produced by birds so treated. (Though again, some call-notes seem unaffected by this procedure of deafening (Nottebohm, 1968, 1970).) However, once the normal song structure and patterning has been allowed to establish itself in the natural manner, operative deafening at a later stage in the life does not affect its performance and maintenance. Thus the stereotyped movements of singing are independent of at least that source of feedback which, from comparisons with the difficulties which some deaf human beings experience in maintaining normal speech, we should have expected to be the most important. Nottebohm (1970), in discussing the establishment of a song memory in *Fringilla coelebs*, points out that the imitation of a song model consists of two processes: (1) establishment of the memory as an auditory template; (2) its conversion to a motor pattern. If the auditory template is to be matched by song development, hearing must remain intact. Adult male chaffinches, deafened after one full season of singing experience, retain their song pattern with great fidelity. By contrast, the song patterns of a year-old male, deafened when it had almost but not quite perfected its two song themes, were not retained but regressed in quality. (This suggests that there are two stages in the development and stabilization of song – a transient memory, and later, a permanent memory retained even in the absence of auditory feedback. Whether these memories are just an efferent programme or an integration of output and proprioceptive feedback is not known.) Clearly then, the songs of such species are at least very different from speech in that the song template requires development by auditory experience which must be, in at least some respects, specific if normal bird song is to be produced; whereas a child can learn the phonetic character-istics of any language to which it is regularly and fully exposed.

But even the stimulus of a very poorly organized or almost completely disorganized vocalization from another chaffinch can canalize song develop-ment. Hand-raised chaffinches reared in groups acoustically isolated from other members of the species produce songs more complex than those of hand-raised individuals kept alone. However, the songs are not necessarily more like the normal species song, for each such group produces its own idiosyncratic pattern. But autumn-caught birds, which have had some previous experience of the species song, do produce more normal songs if they are subsequently kept in acoustically isolated groups than if they are kept alone: apparently in both cases the experience of counter-singing facilitates song development (Thorpe, 1958, 1961).

In the chaffinch, learning of refined details of song is at its maximum early in the first spring – namely when the bird is about eight months old – and increases up to a peak at about fourteen months, after which it ceases entirely. Singing behaviour becomes fixed and unalterable after the first breeding season. By contrast, in other species such as the red-backed shrike (*Lanius collurio*) and the canary (*Serinus canarius*), a period of ability to learn new song elements may recur annually.

Thus it appears that, in the chaffinch, an innate template does exist though by itself, without any auditory experience, this template cannot express itself in full control of the motor pattern of vocalization. Indeed, if there was no such template and it were all a matter of imitation, then the song of the chaffinch could hardly be as stable as it is. A good instance of this is given by the introduction of chaffinches to New Zealand in 1862 and to South Africa in 1900. After this long period of isolation, during which it can be reasonably assumed that no further introductions were carried out, the basic song structure has not drifted substantially away from that characteristic of the ancestral population in Europe. If everything were imitatively learned, it is almost inconceivable that this would be the case. But the bullfinch (*Pyrrhula pyrrhula*) is very different: a young bullfinch seems to learn songs only from its own father (Nicolai, 1959). If fostered by canaries it will adopt canary song and ignore the songs of bullfinches even if kept in the same room.

Other cases of song learning have been analysed in various species of finches and in some of the buntings (Marler, 1967*b*), in addition to the chaffinch. One of the species discussed by Marler, the Arizona junco (*Junco phaeonotus*), is peculiar in that the situation found in many other species is in some degree reversed. That is to say, the birds seem to acquire the overall song pattern by imitation, but the syllabic structure of the songs seems to depend upon birds stimulating each other to invent or improvise diverse syllable types. The diversity of songs of the European blackbird (*Turdus merula*) provide another remarkable instance of vocal inventiveness (Hall-Craggs, 1962, 1969). As our knowledge extends there will probably be found an almost infinite gradation of combinations of these various factors in the song development of different species.

In some species the imitative contact necessary for effecting maintenance of the species distinctiveness also produces or maintains group distinctiveness of vocal signals. The chaffinch, by means of the control exercised over its imitative ability, maintains the normal outline and main features of the song, but fills in the structure by taking, perhaps, most of the finer details from its singing neighbours. The white-crowned sparrow (*Zonotrichia leucophrys*) provides another instance. In this way a group dialect is produced such that the birds from one locality tend to approximate to one another in the details of their song; and those from a locality say fifty or a hundred miles away tend to have a slightly different average pattern of fine detail. Sometimes such dialects are restricted by quite small barriers such as river systems. Such local dialects often seem to serve no particular function, being no more than an accidental outcome of the way in which the song matures. Perhaps (Thielcke, 1969) they help to make the song signal more effective by reducing local variability. But in certain cases they must surely help in the early stages of genetic divergence. Be this as it may, within the local dialect the individual birds can develop their own peculiarities of song so that this can now serve the additional function of individual distinctiveness. An interesting case of this is shown by songs of the bou-bou shrike (*Laniarius aethiopicus major*) depicted in Fig. 3 (Thorpe and North, 1965, 1966; Thorpe, 1966, 1972). Here, the members of a pair learn to duet with one another and, while adopting certain phrases and

rhythms which are characteristic of the locality, work out between themselves duets which are sufficiently individual to enable the bird to distinguish and keep contact with its mate by singing duets with it (or, to be
more exact, singing antiphonally with it) in the dense vegetation in which
they usually live (see also Hooker and Hooker, 1969).

Fig. 3. Antiphonal songs of *Laniarius aethiopicus major* from various areas in
East Africa. *x* and *y* indicate the contributions of the two members of the pair,
but *x* may sometimes refer to the male and sometimes to the female, and similarly
with *y*. (The figure '8' above the treble clef indicates that all that follows should
be read as 1 octave higher than would otherwise be the case, e.g., middle 'C'
becomes 512 Hz instead of 256 Hz scientific pitch.) (From Thorpe, 1966.)

In some birds, no doubt, vocal quality alone may suffice to distinguish a
species, and in that case the imitative and inventive powers of the bird can
be used if necessary to produce a completely individual vocalization. In the
wild, most birds with imitative powers restrict these to the task of copying
members of their own species and do not, even in captivity, mimic other
bird species – still less man himself. But, as everyone knows, there are
other species which mimic widely: the mocking bird (*Mimus polyglottos*) in
the United States, the starling (*Sturnus vulgaris*) in Europe, and the
racquet-tailed drongo (*Dicrurus paradiseus*) in India are good examples.
These birds are something of a puzzle since it is not at all clear what
biological advantage is achieved by imitating other species. The Indian
mynah (*Gracula religiosa*), however (Thorpe, 1967), shares with the parrots
the peculiarity that, while supreme imitators in captivity, they have never
been heard to imitate any other species in the wild. Thanks to the work of
Bertram (1970) we are now beginning to understand the situation in this

very puzzling case. It appears that these mynahs do, in fact, imitate, but only other mynahs: since mynah calls are of such a wide range of patterns and of such acoustic complexity, this imitation is undetectable by man's ear without long training and close study of neighbouring individuals. When, however, these birds are hand-reared, they seem to imprint very easily and thus they imitate all sorts of noises in their own environment, particularly the noises most closely associated with human beings. Hence their fame as mimics.

We now come to a problem where the emphasis is somewhat different. Successful social organization in animals often requires a capacity for mutual recognition, at some distance, between mates and also between parents and offspring. That is to say, it is often necessary not only for adults to recognize specific characters of their young and *vice versa*, but also for them to recognize one another individually, so that parental care and familial organization and cohesion can operate. So far I have been discussing birds in which the young must be cared for or maintained in a nest by the parents, where individual recognition of parents by young, or *vice versa*, is usually unimportant. With nidifugous species, where the young move about and may scatter soon after hatching, individual recognition is often crucial. This implies both individual distinctiveness, and the ability to learn and respond appropriately to individual differences. Indeed, when we consider the problem of reproduction and survival faced by birds such as many species of gulls, terns, gannets, penguins and so on – birds which nest in very dense colonies and obtain their food during the nesting period in rather restricted areas of sea or coast line near their colony – we can at once see a number of ways in which such abilities for individual recognition could be advantageous. Without individual recognition, the feeding of the young, at least as soon as they become mobile, could be a very wasteful process. Hordes of young would be competing for food from each individual adult as it returned to the colony; with the result that the strongest, the most fortunate, the most mature or the quickest, would obtain ample food and many others would starve. And those which did survive would be no better fitted to take their place as adults as members of a colony-nesting species. In the circumstances of colony-nesting, the eugenic need is for selection to operate so that adults which ensure that their own young are fed are at a selective advantage without, thereby, decreasing the chances of other young in the assemblage. This is the *sine qua non* of social life.

Again, it is important that the parent should bring food – for example, fish – of the right size for young of a given age: a fish three inches long may be too large for a small chick to swallow; whereas occasional meals of fish 0.25 inches long would result in under-nourishment of the larger chicks. Similarly, it is often apparent that the smaller young receive a different kind of food from that brought to the larger ones. So both size and quality may vary with age. The same problem arises in the brooding of the young birds, especially during bad weather. If all were haphazard, some young would get too little brooding and some too much, which would presumably result in disastrous mortality.

Later still in the life-cycle, when the young are nearly ready to fly or are first flying, it may be necessary for them to follow a particular parent rather than to accompany anonymously a flock of adults. Following an individual parent will give the young bird a much better opportunity to learn by example and experience how to find places where the right food can be obtained in quantity, and how to learn the best way of catching and eating it. Furthermore, as is usual in mammals, parental care is of immense importance in protecting against predator attack and learning how to avoid it.

That fidelity to nest-site and mate is probably widespread in birds and are, indeed, of adaptive value in enhancing breeding success has been demonstrated by the work of Coulson (1966 *a*, *b*) on the kittiwake (*Rissa tridactyla*). Coulson found that a female which retained her mate from the previous breeding season bred earlier, laid more eggs and had a greater breeding success than one which had paired with a new male. More recently it has been shown (Ashmole and Humbertotova, 1968) that parental care in sea-birds may be much more prolonged than has hitherto been thought; and, of course, parental care at sea can only be achieved if there is mutual recognition.

Now if the visual powers are sufficiently well developed, vision will probably provide a larger number and greater variety of cues for recognition than any other sense. But it also seems certain that in birds, just as in primates, because of the greatly varying conditions of visibility, lighting, apparent size and angle of approach, the difficulty of instantly recognizing a complex visual pattern must be very great indeed. In fact, the visual responses of sea-birds must constantly be at the mercy of sudden, unpredictable changes – dazzle, fog, cloud and, of course, darkness (Thorpe, 1968).

Consequently, if a sense that does not involve such great difficulties over perceptual constancy could be used for recognition, either in addition to or in place of vision, there would be great selective advantage in doing so. So the questions arise: how far can the auditory sense serve this purpose? Is the acuity and complexity of auditory perception great enough? And is the sound production of each bird sufficiently constant to that individual and sufficiently distinct from other individuals to allow for such a system to operate?

Now it is likely that, if sounds can be used, the message they carry is far less liable to distortion and interference than the visual patterns perceived by the eye from a distance. The only important advantage which vision seems to possess is the precise directional adjustment of the focussing power of the eye whereby the gaze can locate and track the object and so maintain fixation. But studies of the so-called 'cocktail party' phenomenon indicate that, with the human ears at least, the capacity of hearing is able to overcome these obstacles created by loud and continuous random noise and so maintain contact with a preferred sound signal in very much the same way as the eye can locate and track an object. So it is not too much to expect that the roar of wind and sea and the hubbub of the colony may be similarly overcome by the hearing of colonially nesting sea-birds.

As to the parameters which are available for recognition of a complex sound, they are much greater than might at first be expected. Meaningfully patterned sounds vary in more elaborate ways than simply pitch, loudness and duration. To quote Gibson (1968) such sounds

'. . . instead of being of simple duration vary in abruptness of beginning and ending, repetitiveness, in rate, in rhythm, and in other subtleties of sequence. Instead of simple pitch, they vary in timbre or tone quality, in vowel quality, in approximation to noise, in noise quality and in changes of all these in time. Instead of simple loudness they vary in the dimensions of loudness, the rate of change of loudness and the rate of change of change of loudness. In meaningful sounds these factors can be combined to yield higher order variables of staggering complexity. But these mathematical complexities seem, nevertheless, to be the simplicities of all auditory information, and it is just these variables that are distinguished naturally by an auditory system.'

To come now to colony-nesting birds, Tschanz (1968, summarizing nearly ten years of work) was a pioneer in his demonstration that young guillemots (*Uria aalge*) learn to react selectively to the calls of their parents and that, during the first few days of life, the parents similarly recognize their own young. Indeed, there is some evidence that the young, while still within the egg, may learn to recognize some aspects of the sounds produced by the adults.

There has long been considerable reason for suspecting that in some terns and gulls also the adults can recognize their mates and young, and the young their parents, by call alone (Thorpe, 1968). Recently some further progress in this field with relation to the terns and, rather surprisingly, the gannets has been made. Hutchison, Stevenson and Thorpe (1968) found that, with the Sandwich tern (*Sterna sandvicensis*), the so-called 'fish call', uttered by the parent when returning to its young with food, has just the kind of structure required to provide auditory data for individual recognition. In the 40 different individuals with which it was possible to obtain a series of samples of the 'fish call', each bird had a call measurably distinct from all the others, and the successive calls given by any one bird were extraordinarily similar. Fig. 4 shows the sound spectrograms of typical recordings of two consecutive calls of three different individuals of the Sandwich tern. Table 1 gives correlations between certain measures of successive calls for 20 individuals.

Table 1. *Correlations between two successive calls of the same individuals of the Sandwich tern. Total sample 20 individuals*

Measure	Correlation coefficient
Total duration	0.98
Duration of segment *a*	0.92
No. of vertical bars in segment *a*	0.82
Duration of segment *b*	0.94
Lowest frequency in segment *b*	0.95
Duration of segment *c*	0.98
No. of vertical bars in segment *c*	0.98*

* These are all significant at less than the 0.01 level.

Having provided this evidence that each Sandwich tern has its own individual 'fish call', we may ask the question: If other birds of the same species were to identify an individual's call, which features might they use? If the variation of a particular characteristic were large it would presumably be easier to discriminate differences between the birds than if the variation were very small. Furthermore, if two measures of a particular characteristic have about equal variation, and the mean of one measure is smaller than the mean of the other, the data from human subjects suggest that it might be easier to discriminate differences over the low absolute values. For example, if the standard deviation were 1s., it would be easier to discriminate difference from a mean of only 1s. than from a mean of 100 ss. Thus, the standard deviation divided by the mean value of a series of measures should give a ratio for which low values indicate small variation and/or a large mean, or a difficult series of measures to discriminate; and high values indicate an easier one. If we assume that a similar rule can be applied to the Sandwich tern, then those measures which have the largest values of the ratio *standard deviation/mean* with respect to a particular characteristic (for example, duration) would be most likely to yield sources of discriminable, inter-individual differences. It could be argued also that the measure having the smallest values of this ratio might indicate

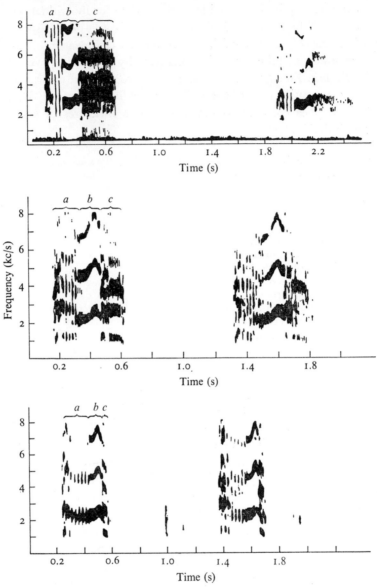

Fig. 4. Sound spectrograms of typical recordings of two consecutive calls of individual Sandwich terns. Note that each call is divided into three segments, *a*, *b* and *c*. Segment *a* contains anything from 7 to 15 vertical bars which indicate brief pulses of sound of great frequency range. Segment *b* shows the fundamental and usually two harmonics rising and falling during the duration of the phrase of 0.07 to 0.15 s to give an inverted 'v' or inverted 'u' pattern. Segment *c*, of duration between 0.01 and 0.28 s, again shows vertical bars varying in number from 3 to 28 and showing peaks of intensity at particular frequencies rather than the very wide and uniformly distributed frequency range displayed by the bars in segment *a*. (From Hutchison *et al.*, 1968.)

the features which identify a call as that of *Sterna sandvicensis*. This ratio
was obtained for each measure (Table 2). Because the duration of segment
c has a higher ratio than the other three duration measures, and because the
number of vertical bars in segment *c* has a higher ratio than the number of
bars in segment *a*, it is more likely that both duration and number of bars

Table 2. *Ratios of standard deviation over mean for 40
individuals of the Sandwich tern*

Measure	Ratio
Duration of segment *c*	5.85
No. of vertical bars in segment *c*	3.62
Lowest frequency of fundamental of segment *b*	3.09
Duration of segment *b*	1.81
Total duration	1.74
Duration of segment *a*	1.55
No. of bars of segment *a*	0.52

in segment *c* are used to identify the individual's call than are those
measures in other segments. Finally, the lowest frequency of the fundamental of segment *b* has a high ratio, and if terns are more or less equally
sensitive to changes in the frequency as they are to changes in duration,
then this, too, might contribute to the discrimination of calls of different
individuals. All this seems to amount to an interesting confirmation of
Marler's (1960) suggestion (also supported by Falls, 1969) that the two
functions of species recognition and individual recognition will often be
relegated to different parameters of a song.

In summary we can say that, while the extent to which these separate
individual characteristics in the Sandwich tern are recognized as distinctive
by the birds themselves has not been investigated, it is clear that if the 'fish
call' is used, as it seems to be, as an effective means for individual recognition in a large colony of two thousand or more pairs, patterning of the call
could play an important part.

With the common tern (*Sterna hirundo*), Stevenson *et al.* (1970) have
demonstrated that the young bird, at four days of age, while quite unresponsive to the playback of calls of other members of the colony, responds
immediately on being played the returning call of one of its own parents,
its response being a sudden alert, 'cheeping', turning and walking towards
the loudspeaker. It seems then that the returning call of the parent
common tern carrying food is quickly learnt and responded to by the
individual young concerned. To put this in anthropocentric terms, with
both the Sandwich tern and the common tern it appears that this call is in
effect saying, 'Here is Mum (or Dad) with food'.

White, White and Thorpe (1970) have investigated the individuality of the calls uttered by returning adults of the gannet (*Sula bassana*) on the Bass Rock, Scotland. Here the problem is somewhat more difficult in that the calls do not have the clear division into three sharply defined sections that is found in the two species of tern studied; and it has not yet been possible to establish clear differences in frequency/time pattern in the gannet like those found in the Sandwich tern. Nevertheless, the striking

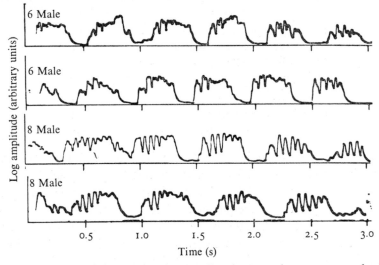

Fig. 5. Amplitude displays of landing calls of two male gannets; each call was uttered on a separate occasion. Normal speed (7.5 in/s). (From White *et al.*, 1970.)

fact has emerged that, if the variation of amplitude with time is studied, rather than the normal features of the sound spectrogram (frequency, time), then consistent individual differences are displayed just as in the Sandwich tern. Fig. 5 shows, for the landing calls of gannets, the amplitude of the sound envelope as a function of time. Each full call is made up of a series of components which are fairly similar. First inspection shows that individual consistency resides in the first few peaks and troughs. That this conclusion is justified was shown when, for each component, ordinates were measured at intervals of 0.01 second; corresponding ordinates were averaged to give an overall 'profile' of a full call. One call profile obtained from each bird was designated as a 'reference' (see Fig. 6) and the remainder were called 'samples'. Each reference was compared with all the samples using product–moment correlation. Samples of calls given by the same male on different occasions were more highly correlated with their own reference call than with references from other males (White *et al.*, 1970). Figs. 6 and 7 show the results of matching gannet calls with the sample profiles using progressively smaller amounts of each profile. Not only do these results confirm the earlier tests that gannets have individual calls; they further show that there is enough information in the initial part of each call (i.e. approximately the first 0.15 second of sound) to distinguish one individual from another.

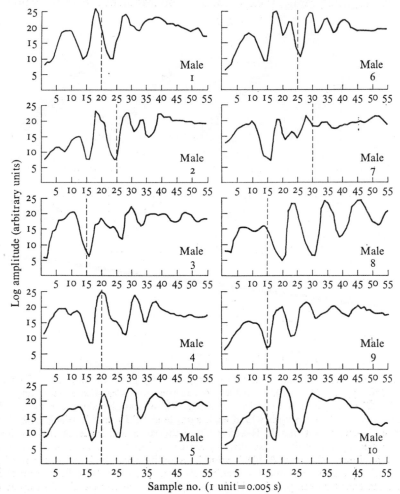

Fig. 6. The reference profiles with which the sample profiles of the gannets were correlated. Dotted lines show the smallest amount of that sound which was necessary to match with one of its own sample sounds. (From White *et al.*, 1970.)

If there is enough information in the first part of the call to identify the caller, what is the function of the end part? It is perhaps worth mentioning that by looking merely for individuality in a call we may be preventing ourselves from detecting much information other, than the identity of the bird, which the call may be carrying. But it seems reasonable that the identifying part of a call should be at the beginning (after comparatively long periods of silence) so that sitting individuals, who probably pay low level attention to all sounds coming from the proper direction, can receive individual information efficiently. But here we must remember the peculiar environment to which the sitting gannet is exposed. This environment is extremely noisy

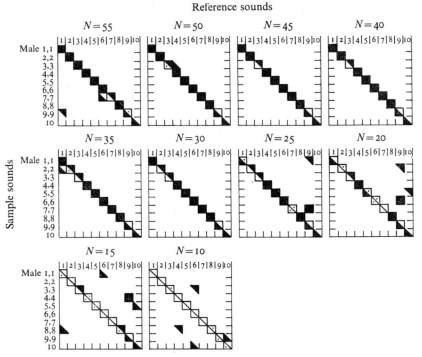

Fig. 7. The results of matching using progressively smaller amounts of each profile ($N = 55$ to $N = 10$). N refers to sample number, where each unit is approximately 0.005 s of the call in real time. White triangles show the expected best match (the reference with which each sample should match). Black triangles show the observed best match. Reading along the rows will indicate with which reference a sample identified. Reading down the columns shows which samples matched a given reference. A black spot in a white triangle is an error mark, showing when a sample stopped matching its own reference. The figures on the left indicate that two samples in addition to the reference were available for each bird except number 10. (From White et al., 1970.)

and the birds usually call while in flight. In conditions of movement and noise, what would be the most effective signal? The possibility is, as has been suggested for other vocalisations in this article and in White et al. (1970), that intermittent signals such as those given by the gannets might be very easy to locate. Sitting birds do not appear to be searching visually for their absent mates. Birds tend to approach their colony upwind, diverging towards their particular nests only when they are some yards away. White *et al.* argue that the acoustic structure at the beginning of each call probably facilitates accurate localization, and by tracking approaching sources of sound, a sitting bird might quickly detect one which was approaching its nest and at the same time gain further information from the individual features; whether amplitude fluctuations or something else. Birds certainly have an acute temporal sense (Schwartzkopff, 1960) and it is possible that individuality may reside less in the amplitude *per se* than in its temporal patterning as a series of impulses.

While those few birds which have been investigated certainly have a finer temporal sense than do human beings, their frequency (pitch) perception is of the same order as that of the human ear. How well they perceive amplitude (loudness) is not known. It is perhaps worth speculating that perception of amplitude fluctuations could be a way of accommodating for changes resulting from movement (White *et al.*, 1970). There are changes in the temporal structure of a sound in proportion to rate of movement of its source (the Doppler effect). It follows that all sound frequencies received are also changed. This would mean that the frequency and time structure of an approaching bird's call would be heard with a more or less constant shift from their true values, the shift depending on the caller's speed of approach. Amplitude would, however, change only gradually as the bird came nearer. We do not know whether birds are capable of discriminating these changes, though it should be possible to test this.

Until recently, the most extensive work on the individuality of sounds has dealt with humans. A great many experiments suggest that amplitude fluctuations of speech are relatively unimportant and, indeed, may be removed electronically without seriously affecting intelligibility. Computer recognition of individual speech shows that frequency/time parameters are more useful than amplitude characteristics (for reference see White *et al.*, 1970). In the gannet, amplitude changes are highly constant to the individual though whether the birds themselves use this characteristic, or some other feature not yet detected, has not yet been unequivocally proved. It has, however, been shown (White, 1971) that when landing calls of known individual gannets were recorded and subsequently played back to their sitting mates, the filmed responses of the mate and the rest of the colony gave strong evidence that the calls of both males and females were recognized individually by their respective mates. There was, however, no evidence for similar recognition of neighbours.

But let us return to the question of imitative ability. It is clear that a capacity for vocal imitation must be widespread in birds to account for the way in which they use their vocalizations for communication and for the production of local dialects. In some species, such as the Indian hill mynah, this ability is very large indeed, in others it is very slight or perhaps absent. Such ability as there may be in the terns discussed above (and it must be very small indeed) must be under strict control, as otherwise the individual terns in the colony would not have the idiosyncrasy of voice that we find. In other words, the problem here is to inhibit or control the imitative powers in certain circumstances in order to maximize the tendency to produce individual peculiarities of voice. This suggests that the motor learning involved in the production of sound signals is a different process from the perceptual learning involved in recognizing those signals; and, in certain species, the two must be kept very strictly apart. Moreover, this work on the vocal communication system of sea birds emphasizes the importance of auditory powers as the perceptual basis for group organization. Clearly, the young must have an extremely high ability for learning the individuality of their parents' calls, and the adults the same ability for learning the individuality of their mate's calls. But they *must not* imitate them slavishly. If they imitate at all they must imitate the component parts and then randomize them. But perhaps for this task pure inventiveness would be better!

To summarize, the variability of the 'songs' of birds is adapted, primarily by genetic programming and secondarily by imitative adjustment, to distinguish the species from every other species, and to distinguish one individual more or less certainly from all others in a population.

These ends may be accomplished in different species by differences in the degree of competence of the innate neural template for song structure. Thus the degree of individual experience necessary to develop the full song by addition to and adjustment of the innate template, may range from nil or minimal on the one hand to something near totality on the other. The song of some species (for example, the chaffinch) can be so precisely controlled and adjusted by limited use of the imitative faculty that the song can serve both functions at the same time, and local dialects may arise as a by-product of the mechanism of development; though it is easy to think of ways in which dialects could be adaptive. In species, such as the sea birds which nest in dense colonies which we have been discussing, individual recognition is combined with specific recognition, but there seems to be little group distinctiveness. The considerable advantage which recognition based on acoustic signals offers as against visual recognition, especially under the difficult conditions imposed by life in dense colonies on the seashore, seems clear. But more interesting still, to my mind, is the possibility hereby raised (it is as yet no more than a possibility) that the great breeding assemblages may be social units to a degree hitherto unrealized.

Moreover, I think it can be said that as this kind of work proceeds, we find ourselves viewing with an ever increasing respect the auditory capabilities of birds; for they possess ears which are in some respects inferior to ours and in others markedly superior, but in either case showing a capacity for organizing their auditory perception and their vocalizations with a precision exceeded only by man and seldom even approached by other mammals.

In man language is, of course, the prime vehicle for the transfer of social information. There has, perhaps, in the past been too much readiness to use the term 'language' for the vocal–auditory transmission of information amongst animals. Nevertheless, in spite of the dangers of this nativistic view, dangers which are well exemplified by Table 1 in Chapter 2, such usage has recently been supported by well-known students of human language (Chomsky, 1967; and Teuber, 1967: 75 and 84), where language is defined as 'specific sound-meaning correspondence'. Chomsky (1968) considered the matter further and retreated somewhat from his first position. But his difficulties have in fact now been dissolved by the work on Primates described in Chapter 2. As a result of studies by many workers, Teuber (1967) said, 'It has become clear . . . that linguists are ethologists, working with man as their species for study, and ethologists linguists, working with non-verbalizing species'.

I sometimes amuse myself by imagining an intelligent visitor from another planet arriving on this earth just before the differentiation of the human stock – say somewhere about one million years ago. If such a visitor had been asked by an all-seeing Creator which group of animals he supposed would the most easily be able to achieve a true language, I feel little doubt that he would have said unhesitatingly, 'Why, of course, the birds'. And indeed, if we now look at the birds together with all the mammals other than man, we have little hesitation in saying that the birds are by far the more advanced, both in their control of their vocalizations and by the way in which they can adapt them collectively and individually to function as a most powerful communication system. But having said this, I am sure that the recent studies on the ability of chimpanzees to learn sign language (referred to in Chapter 2 above) must give us cause for hesitation. Perhaps it is the human capacity for conceptualization that is the key to the riddle although, even here, it is not certain that the conceptual powers of birds are less sophisticated and effective than any of the sub-primate mammals. But if powers of conceptualization are the answer, it is clear that they must have come *before* the evolution of language as we now know it in the human stock. On the other hand, it is also clear that we must not run the risk of being too simple-minded about human language. It is unique, as we have seen in Part A, and it is unique particularly in the possession of numbers 14–16 of the design features listed in Table 1 of Chapter 2. Yet human language is made up of many features which are found in the various communicative systems of animals – fish, insects, mammals and birds. Birds certainly have more of these features together than does any other group. But perhaps it will turn out that these three key abilities – numbers 14–16 – can only be fully possessed by animals with a large mammalian type brain – although there is as far as I know no neurological evidence to explain why this should be so. Perhaps birds, with the brain they have, cannot get any further on the road to true language, and indeed, perhaps none of the existing mammalian stock, not even the Primates, can do so either. This in any case brings us to the question of there being some special type of *mental* organization necessary for the production of human language, as Chomsky has maintained. If we grant this for human language, why should we not admit the possibility also for certain systems of animal communication? If we look closely at the more advanced vocalizations of birds, and if we consider the recent experiments on gestural language in chimpanzees, one begins to wonder about this. We certainly cannot rule out this possibility on neurological grounds; and it is part of the basic assumptions of the physiologist that there must be in the brain neural structures or adaptations appropriate to the exercise of all the mental capacities possessed by the organism in question.

We may never come to know how this great gap between the highest animal 'language' and the language of man was bridged by the human ancestral stock. No anthropologist nowadays ever hopes to find a language which is primitive in the sense of forming a link between the language of man and the communication systems of animals. The search for a primitive language was abandoned by sociologists and anthropologists fifty years or more ago. And this recognition of the uniqueness of the human language and still more, the understanding of those respects in which it is unique, constitute a great step forward in comprehension of the relationship or lack of relationship between man and the animals. But if we are ever to find out how such a gap might possibly have been bridged it seems, to some of us at least, that the fullest possible study of bird language is an essential preliminary.

There is one final point of great interest – namely the problem of the 'musicality' of bird songs and its relation to biological function. Do these two aspects reinforce, or run counter to, each other? I would like to conclude by a summary of the views of my colleague Mrs J. Hall-Craggs who has kindly provided me with the following brief summary of her conclusions.

'Charles Hartshorne (1958) suggested that when a song is highly functional it deteriorates musically. For example: if a blackbird is singing well and a neighbouring blackbird approaches its territory, the singer doesn't sing more vigorously or more musically in order to intimidate the intruder; on the contrary, the song becomes loose and disjointed; phrases are left unfinished and the pauses between the phrases become longer. It seems as if the bird has to attend to the form of its song in order to be able to 'sing well'; 'well', that is, by our standards. This principle may also be applied in a broader context. If one records the song of an individual blackbird daily throughout its entire song season it is always found that initially the song is fairly primitive or musically unsophisticated. Most of the brief song motifs are there but there is no evidence of organization; the longer phrases have yet to be formed from the motifs, and the compound phrases or sentences from the phrases. At this time the song is highly functional; the bird is in competition with others, securing a territory and attracting a mate. It is later in the season, when these immediate and pressing needs have been fulfilled, that the song becomes organized in a manner so closely resembling our own ideas about musical form. It would appear, then, to be moving towards the realm which we call art music – where our experience of musical form enables us to predict what is likely to happen next. This is discussed further by Hall-Craggs and by Thorpe (1972) in a monograph.'

COMMENTS

(1) *Song-learning and human language*

Marler (1969, 1970) has recently emphasized the many parallels that are to be found between song-learning and the development of human speech. In both, acoustical stimulation plays a double role, both allowing external sounds to be heard and remembered, and also allowing the organism to hear its own voice and match it to the memory of what has been learned previously. In both auditory feedback is critical for development, but becomes more or less redundant later. The learning involves in part a transition from an amorphous subsong or babbling, to the adult type of articulation. In both vocal learning is to some extent independent of extrinsic reinforcement. Learning is most likely to occur at certain periods of life, and there are predispositions to learn some sounds rather than others. But Marler stresses that, while the parallels may be manifestations of a basic set of rules that any organism engaging in vocal learning is likely to evolve, 'in no sense does song-learning generate a language'. Indeed, although in song-learning birds show a design feature which Hockett regards as critical for language development (see Table 1 in Chapter 2, by Thorpe), the capacity to learn new sounds has quite different consequences in birds and man. Marler argues that birds employ vocal learning 'in a very specific way, actually reducing the complexity of signalling behaviour within a population. These birds demonstrate traditional learning, but lack some of the other key characteristics which, coupled with the capacity to learn traditions, then made for the great expansion of our own linguistic system' (Marler, 1969: 51; see also Table 1 in Chapter 2).

Since Marler wrote this paper, Nottebohm (1970) has demonstrated yet another resemblance between the control of bird song and human speech: in both cases there is lateralization of control from the central nervous system. Furthermore in birds, control can be shifted from what is usually the dominant side to the other side before song-learning occurs, but not subsequently, just as in man a unilateral lesion to the speech area affects speech if it occurs in adulthood, but usually does not if it occurs early in life. Nottebohm is careful to point out that lateralization in other organisms occurs with respect to quite different kinds of behaviour (e.g. Kandel, Castellucci, Pinsker and Kupperman, 1970) and, rather than being a characteristic of vocal performance, may be characteristic of many kinds of integrated behaviour.

(2) *Individual recognition of voice*

Further extensive data have recently been reviewed by Beer (1970).

PREFACE TO CHAPTER 7

The next chapter provides an introduction to the signalling movements used by mammals. Rather than reviewing the nature of the systems to be found in this group, Andrew centres his discussions around an issue which has been controversial amongst students of signalling systems in vertebrates. This has already been referred to briefly by Cullen but some additional description of the issue is necessary.

In the breeding season many species of birds defend territories. Territorial defence often involves prolonged skirmishes along the boundaries in which the rivals in turn attack, flee and show a bewildering sequence of postures and movements. To the naive observer the birds' behaviour often seems incomprehensible: the bizarre postures adopted, the unexpected sequence of displays, the alternation of attack and fleeing, seem without rhyme or reason. But with the additional knowledge that each individual attacks in encounters on its own territory and flees when on its neighbour's, the boundary being an intermediate zone where tendencies to attack and to flee are more or less in balance, understanding begins to come. The display movements begin to be comprehensible when they are seen as associated with even finer and more temporary balances, each display being related to different absolute and relative strengths of the two tendencies.

In a similar way, much apparently bizarre courtship behaviour can be understood as involving conflicting tendencies to attack, flee from and behave sexually towards the mate. For example, early in the breeding season pair formation in many small birds is initiated by threat postures directed at the female by the male. If she is receptive, she adopts a submissive posture instead of fleeing or attacking back, and the male's behaviour gradually changes from aggressive to courtship. The latter can be understood as involving all three tendencies, but gradually the male's tendency to attack the female decreases and the primary tendencies are to flee from her and behave sexually towards her. Only when the latter predominates sufficiently can mating be accomplished (Tinbergen, 1952, 1959; Hinde, 1970).

Of course the detailed course of events varies greatly between species, and many displays must be understood in terms of other tendencies in addition to the three mentioned above (Andrew, 1961a). Furthermore, ritualization of the displays for their signal function has often obscured the conflict with which they were primitively associated (see Chapter 4). But over a wide range of species the idea that many displays are based on conflict has considerably facilitated their understanding. In the present context it is of particular interest that, to put it colloquially, when fighting, animals signal 'I may attack or I may flee' more clearly than 'I am about to attack' or 'I am about to flee'. Uninhibited attack or fleeing is usually swift and silent – and there are obvious functional reasons for this. Similar considerations apply to courtship and various other contexts, but of course it does not follow that all signalling movements are a consequence of conflict – they clearly are not.

Criticism of this conflict view of display has been of two main sorts. First, conflicts were originally described in terms of 'drives'. The term 'drive' had previously been used in diverse ways by students of animal behaviour holding widely differing theoretical views, and had in particular been associated with 'energy' models of motivation which have severe limitations in the explanation of behaviour. Furthermore, it was often used to imply that diverse patterns of behaviour (e.g. all aggressive responses) could be understood in terms of a single factor. In an only partially successful attempt to escape these associations, the term 'tendency' came to be used instead (Hinde, 1955–6).

The second point, which is more a refinement than a criticism, concerns the level at which the conflict takes place. Is the ambivalence best described as involving tendencies each of which is associated with a number of discrete responses (e.g. aggressive tendency associated with flying towards the adversary, pecking him, striking him with the wing, etc.), or as involving conflicts between those responses themselves?

7. THE INFORMATION POTENTIALLY AVAILABLE IN MAMMAL DISPLAYS

R. J. ANDREW

School of Biological Sciences, University of Sussex

Direct studies of communication between animals must rely on either the experimental presentation of a signal, preferably as an interjection in a normal exchange, or studies of the effects of the behaviour of one animal on another, based on correlations between the acts of the first and those of the second. Both approaches have inherent difficulties. Experimentally simulated signals may have misleading effects because they are not accompanied by an appropriate set of other signals. Although this need not be an issue in the interpretation of correlations between the behaviour of naturally interacting animals, systematic relations between the two behaviour sequences may well depend on an earlier synchronizing exchange (e.g. both might be recovering from a previous bout of threatening or from a sudden meeting or a shared input from a third animal: see also Cullen, Chapter 4).

Such studies are in any case little advanced in mammals, and have been recently extensively reviewed (Sebeok, 1968). It therefore seemed appropriate here to examine mammalian signals from a quite different point of view, namely that of the information which, potentially, they could convey to another animal. In order to do this it will be necessary to examine the causation of displays,* which, on other grounds also, is a topic which should be carefully considered in any discussion of communication. If, for example, displays are held to be caused by conflicts between aggression, fear and sex (which I believe to be misleading), then it may well be assumed that they could convey information only about the likelihood of attack, fleeing and copulation.

Any discussion of causation must begin with the definition of units of study which, in the case of behaviour, are small groups of responses which vary together very closely in their distribution over time, and are relatively constant in form. 'Fixed action pattern'† is often used to describe such

---

* 'Display' will be used throughout this chapter to mean a signal or pattern of signals, whose exaggerated or stereotyped form suggests that it has been affected during evolution by selection for use in communication. There is often no clinching evidence for this in the displays discussed here.

† The term has not been used here, since in its original strict definition by Tinbergen it is that part of a species-specific pattern whose form is not controlled by external stimuli.

a group if the constancy is sufficient. These patterns should properly be defined only by descriptions of form, without reference to data or hypotheses about their causation. It is usual to proceed next to describe the interrelations of such patterns in terms of a few hypothetical intervening variables such as aggression and fear. Originally, each of these had all the properties of a drive, in that it in some way caused the responses associated with it, and that it was reduced by their performance (or might increase in the absence of performance). More recently, it has become common to use the term 'tendency' rather than drive, and to use it to stand for all the causal factors of a group of apparently related behaviours (Hinde, 1970: 360). It will be argued later that even this usage is not helpful. However, most recent studies of the causation of displays still use aggression, fear and sex, or equivalent terms. It will be best therefore to consider first the reasons for this particular choice of intervening variables, and its consequences. It may be illustrated in the more recent literature from Moynihan (1958) who states that the hostile displays of a number of gulls (e.g. the ringbill, *Larus delawarensis*) are 'produced by simultaneously activated attack and escape drives (plus other factors in some cases)'. Different displays are said to be given at different combinations of intensities of these drives. These hostile displays are, in fact, very diverse, including calls, which are sometimes given in the absence of any opponent, and patterns like 'choking' (bows and regurgitation movements) given at the nest site, as well as threat displays. A similar position to that of Moynihan has been restated for male–male hostile displays more recently (Tinbergen, 1966), and is being used currently, with the addition of other drives in special instances, to explain an equally wide range of displays in primates (van Hooff, 1967; Moynihan, 1964; Blurton Jones, 1968). It should be stressed that these authors do not state that aggression and fear are the only major variables to be considered in the causation of hostile displays, though in practice this has commonly been implied.

The activation of aggression, fear or sex during a display is inferred by methods set out very clearly by Tinbergen (1959a: 50). First, display components may be recognizable by form as elements of one of those response systems. Secondly, overt attack or fleeing may occur before, after or during the display, and thirdly, stimuli may be present which are known at other times to evoke overt attack or fleeing. Clearly such methods do not allow one to conclude that any particular major tendency is essential to the causation of a display. One way of testing this for current explanations of hostile displays is to examine situations in which aggression and fear are unlikely. This has, of course, been considered before; thus the fact that the mew call may precede courtship feeding and feeding of the young gull is held to suggest that it has a more complex causation than other hostile displays (Tinbergen, 1959 a). However, it is always possible to argue that a conspecific of whatever kind may evoke attack or fleeing responses, even though at low intensity. It is much more difficult to argue in this way when display components are evoked in the absence of any social companion.

Situations in which an expected reward is not forthcoming are particularly interesting. In primates (Andrew, 1962), the same main series of vocalizations as is produced by social encounters of progressively increasing hostility, can also be evoked by non-reinforcement in a situation where a social partner is absent or of minimal importance. Most striking of all is the demonstration in my laboratory (Pearson, 1970), that isolated male guineapigs will show a wide range of courtship and threat displays on finding empty a food dish from which they have been trained to feed. The displays include direction of urine sprays at the food dish just as at the female in courtship sequences, and tooth chattering, which is common in hostile interactions.

Findings of this kind are not clinching. It would still be quite possible to argue that aggression is activated when an animal is frustrated; indeed attack is known to be facilitated in pigeons by non-reinforcement (Azrin, Hutchinson and Hake, 1966). Alternatively, new intervening variables might be used. In the present case, some displays could be said to be caused by frustration, both when they occurred in the absence of expected reward and in their normal social contexts. What is made clear is that the choice of intervening variable is a relatively arbitrary one.

Some workers (e.g. Wiepkema, 1961) have avoided this by turning to factor analysis as a means of objectively extracting a small number of variables, which are sufficient to explain most of the correlations between the different behaviours studied. The parallels between this method of statistical analysis and the use of major intervening variables to explain the causation of displays deserve careful examination. Firstly, although factor analysis does not specify in advance the number and nature of the variables to be extracted, it does assume that they provide a useful way of describing causation. In both cases new data can be accommodated by adding new factors (e.g. frustration), or altering the quite elastic properties of existing ones. Most importantly, any particular behaviour pattern must nearly always be described as dependent on ('loaded on') more than one factor. This may be compared with the statement that a display depends on a conflict between fear and aggression. Both statements are essentially condensed descriptions of the situations in which the display occurs rather than accounts of causation; the difficulties encountered when they are treated as the latter are considered later.

A case has been made so far that the choice of major intervening variables is commonly influenced by tradition; to this should be added that it is difficult to avoid an implicit assumption of various drive properties when terms like aggression (or even 'tendency to attack') are used. However, I would argue that, even if these pitfalls are avoided, the use of major intervening variables is always likely to be misleading. In general, attempts to measure major tendencies (or drives) have been discouraging. It is clearly illogical to hope to represent the functioning of a complex network with a variety of inputs and outputs by the value of a single variable. This is now generally agreed (e.g. Hinde, 1970), and yet there is probably equally general agreement that major tendencies are useful (or indeed essential, since no other alternative has been found) at the kind of preliminary stage of analysis with which, for example, the present account of mammalian displays will be concerned. I believe such a position to be mistaken. Preliminary analysis of the organization and causation of behaviour is quite possible without employing major tendencies (or comparable variables like emotionality or frustration). One can ask how much of x an animal will eat after manipulation y, just as early in experiment or observation, as one can ask how hungry it is after y. The question has the advantage that it can receive a clear and unambiguous answer, and such an approach can be elaborated into a quantitative control theory treatment, with no change in the logical status of the concepts involved.

The approach suggested here for preliminary analysis is to concentrate on small coherent groupings of behaviour patterns, defined by description, and to consider their immediate causation. In the case of running, for example, the immediate causation is often that the animal finds itself too far from a particular stimulus. The reasons for responding to the stimulus will be various, and often a hierarchy of causation will exist: thus a dog might run a certain distance in order to be near enough to bite a man. Quite reasonably, immediate causation of such a type is often taken to be too obvious to mention, and the dog is said to run in order to bite. The same assumption is contained in the statement 'aggression (or tendency to attack) makes the dog run', which, however, goes further, and implies that the direction and timing of locomotion can be explained in some simple way by variations in a single causal factor. In fact, it is far from safe to assume this even for behaviour as simple as locomotion. Mobbing displays evoked by a predator provide an excellent example. In a bird like a chaffinch (*Fringilla coelebs*) (Hinde, 1954) or blackbird (*Turdus merula*) (Andrew, 1961b), short flights are often made towards or away from the owl which provokes the display. Hinde (1954) concludes that avoidance in mobbing is dependent on the fleeing drive, but that approach, although primarily dependent on the attacking drive, may also involve curiosity. In other language, this means that locomotion may be partly or entirely controlled by central instructions to see the stimulus object more closely or from a different angle: the direction of such flights need not always be towards the object. Indeed, direction might sometimes be irrelevant as long as the animal did not come within a previously calculated distance around the predator. The timing of flight also may be independent of likelihood of approach or avoidance. In blackbird mobbing, flight (and associated behaviour such as calls) occurs periodically, independently of distance from the owl, suggesting that the bird may move when a threshold state has been reached in the system controlling locomotion (Andrew, 1961*b*). Clearly an explanation in terms of an aggression/fear conflict loses nearly all its attractiveness if it is impossible to predict from it either direction or timing of locomotion.

One consequence of ignoring immediate causation is that behaviour patterns tend to be subdivided into a number of different patterns which, although identical in form, are regarded as quite different because they are caused by different major intervening variables. This has been justified most clearly by Blurton Jones (1968) who showed that, in the great tit (*Parus major*), hammering at food with the bill is promoted by food deprivation, but hammering at the ground in response to a stimulus which evokes attack, is not affected by such deprivation. However, despite his arguments, it is not clearly essential even in such a case, to consider the two situations separately: the same type of visual stimuli may well evoke hammering in both cases, for example. Further, when the two situations are examined separately, it is difficult to see any resulting justification for the use of terms such as hunger and aggression.

Here, as in all cases where they are used, such variables impose a fixed hierarchy of causation which can obscure quite obvious relations. When, for example, a protective reflex like ear flattening is said to be caused by tendencies to flee and attack, it is genuinely difficult to see how it can also occur in boisterous play with an infant. The difficulty disappears if ear flattening is recognized as evoked by possible danger to the face. Obviously any full analysis of the causation of a behaviour pattern will reveal that it is not fixed, but that it varies greatly from situation to situation, particularly once we pass beyond its immediate causation. Furthermore, many causal relations will not be hierarchical: the most extreme possibility is parallel causation, where a single stimulus object (like a mother or a mate) is simultaneously competent to evoke a wide range of responses.

The assumption that displays are caused by conflicts between tendencies or drives has already been mentioned. There are considerable advantages

(Andrew, 1956 *a*, *b*) in replacing such a description by a consideration of incompatibilities between specific responses. Thus, tail-flicking is common in small birds in a variety of situations, some of which might be described as conflicts between tendencies to attack and flee, or to copulate and flee. In fact, the tail movements are essentially those of take-off for flight, given when the animal is likely to fly, but inhibited from moving far; that is, they are intention movements of flight. Some hostile displays are probably caused similarly. In the great tit, for example, the forward threat occurs whenever full attack is prevented, whether because close approach is inhibited (e.g. by the evocation of fleeing) or because of a physical barrier (Blurton Jones, 1968). It is also clear from the form of the display that it is a checked attack.

It is, in fact, far from certain whether any display can be said to depend on two major intervening variables in such a way that, for the display to occur, both must be present, and cannot be replaced by any other tendency. Whether this is true or not probably depends very much on the way in which displays are defined. If a large number of loosely associated components are listed as making up a threat display, then some might well be attack, and others fleeing responses. However, it is difficult to find a single stereotyped pattern which definitely has such causation (although there are a great many which might be so caused). Wiepkema (1961) reached a very similar conclusion as a result of factor analysis of courtship sequences in the bitterling (*Rhodeus amarus*).

Perhaps the most instructive examples of a rigid application of the conflict theory to mammalian displays are the classical analyses of cat threat displays and dog facial expressions in terms of aggression and fear. As a result, complex and wide changes of form are explained as due to varying superimposition of two extreme and clearly specified patterns, one characteristic of pure aggression, and the other of pure fear (Eibl-Eibesfeldt, 1970). The arbitrary nature of this hypothesis is clear from the controversy initiated by Brown and Hunsperger (1963), who argued, on the basis of brain stimulation experiments, that the cat defensive threat display is not caused by an aggression/fear conflict but is a unitary pattern in its own right. Some of their original argument was dubious: thus, stimulation at a single site could produce a conflict since two or more systems could well be activated at a single electrode. However, the pattern as evoked by central stimulation does appear to vary only within clear limits, with no sign of alternation of attack or fleeing: this is even clearer for defensive threat in the opossum (*Didelphis virginiana*) (Roberts, Steinburg and Means, 1967). Now that the two-pattern explanation has been questioned, further investigation of the organization of the responses involved in these displays will certainly follow. Later in this chapter the immediate causation of some of them will be discussed. It should be emphasized that none of these arguments make it any less necessary to consider the relation of displays like defensive threat to the other behaviour evoked by an approaching con-specific. Both facilitative and mutually exclusive relations with attack are possible, and the likelihood of approach or retreat (which could be des-cribed in terms of an approach/retreat conflict) is likely to be important, since defensive threat typically involves remaining still.

A final rather separate consideration is that the conditioning of a display has been demonstrated in the Siamese fighting fish (*Betta splendens*) (Thompson, 1963), and it may well be common. The changes leading to the evocation of a display by a new stimulus can only be described, when major intervening variables are used, by postulating a complete change of causation; they are much more easily accommodated by an approach which concentrates on the immediate causation of the display.

In the account which follows the immediate causation of a number of important groupings of display components is examined (cf. Andrew, 1963 a): it is assumed that this is likely (but not certain) to be similar in all displays in which the components occur. It is also argued that the full extent of potential information available in displays is made much more apparent by this approach than by an account which effectively concen-trates on information about the likelihood of a few admittedly important groupings such as attack or copulation.

RESPONSE GROUPINGS IMPORTANT IN MAMMALIAN DISPLAYS

1. *Responses elicited by novel and significant stimuli*

There has always been particular difficulty in accounting for the causation of the responses which are sometimes lumped together in the Pavlovian 'orientation reflex'. These all have in common that they can be evoked by stimuli which are novel (i.e. differ from preceding patterns of stimulation), significant (e.g. conditioned stimuli for feeding presented to a hungry animal) or startling. There is actually little advantage in treating all these very varied responses as a single coherent group. They are here divided into a number of separate groups, the most important of which are alert responses, which are reflexes evoked when major sense organs are actively collecting information, protective responses, which serve to protect sense organs and other vulnerable structures, and a range of responses involved in exertion. These groupings, and the responses which make them up, differ in causation in a variety of ways, including differences in the stimuli which specifically evoke them. However, they, and to a lesser extent some other responses like vocalizations, do appear to be facilitated simultaneously.

An influential attempt to provide a mechanism to explain this facilitation was that of Sokolov (1960), who showed that the orientation reflex was produced when a repeatedly presented pattern of stimulation was slightly changed. He suggested that this was because a central model of the pattern, which had been established during previous presentation, now no longer matched the perceptual input. A very marked mismatch, such as might result from completely unexpected (startling) stimulation, might produce protective and exertion responses as well. Horn (1967) has argued cogently for alternatives to the physiological mechanism suggested by Sokolov, which need not concern us here. However, the neural circuits which he proposed in its place retain the essential features of a memory and a device to compare the information stored in it with current input.

The most difficult thing to explain is that many of the responses which occur as a result of a mismatch following a comparison between past and present perception, also appear when a sought-after (or significant) stimulus is perceived: in man, accelerated heart beat and other symptoms of a 'shock of recognition' are familiar examples. Here an exact match, rather than a mismatch, must be assumed. Andrew (1964, 1969) described calls of the domestic chick which are evoked both by a wide range of changes in stimulation, and by familiar stimuli that are significant because they are conditioned stimuli (e.g. for feeding). Clearly such calls have rather similar causation to alert responses; a term, 'stimulus contrast', to describe stimuli competent to evoke alert responses was introduced, and it was pointed out that the chick calls and some further responses could be said to depend on stimulus contrast, as so defined. It is likely that a good many display components are facilitated in a somewhat similar way, particularly those best developed in greeting situations following a period of separation.

It is possible to explain all these findings by an elaboration of Sokolov's original hypothesis: following initial perceptual analysis, a stimulus is normally matched against already stored central models for recognition. During such comparison a mismatch causes the response groupings under discussion to be facilitated. It is assumed that such facilitation begins during prolonged comparison, and continues until all parameters specified by the central model have been found to match perceptual input. Facilitation is therefore to be predicted whenever recognition follows a period when a central model has already been retrieved and is, for example, being used as a 'search image' (e.g. of a particular type of prey). Similar extended comparison could also occur during recognition, if preliminary perceptual analysis were to cause the retrieval of a detailed stable model, requiring comparison in many parameters. It will be convenient to speak hereafter of periods of 'recognition comparison', whether a mismatch is found or not.

The duration of recognition comparison, and therefore of alert responses and similar behaviour, will be affected by at least four factors: (*a*) the accuracy with which the model specifies parameters, (*b*) the criterion of acceptable match, (*c*) the stability of the model during comparison and (*d*) the persistence of the model (e.g. does it disappear after comparison or is it re-activated). Two extreme cases are a perfunctory identification of a familiar but indifferent stimulus, and the marked responses induced by an equally familiar, but sought-after or significant stimulus. In the first case parameters are probably broadly specified (perhaps because the model represents averaging of many rather varied past experiences), criteria of match may also be broad, and the model disappears as soon as match has been demonstrated. In the second, parameters are specified accurately, and criteria may be strict; the model is likely to be persistent, just as it must be postulated search images are persistent.

A stable and persistent model, which specifies some parameters clearly but leaves others vague, is likely to produce unusually prolonged recognition comparison. An interesting special example is that of the calls and other responses given by passerine birds in mobbing, which paradoxically are evoked both by strange and novel stimuli, and also at uniquely high intensity by owls (Hinde, 1954; Andrew, 1961*b*). Passerines clearly have (innately) considerable information about owl characteristics, and yet they treat an owl as a uniquely strange object. On the present explanation, recognition comparison is extended during owl mobbing, because a central model already exists which is stable but incomplete. Comparison therefore involves a large number of clearly specified parameters, and will continue until a detailed central model is completed. It is suggested that, on the other hand, novel objects in general (which elicit mild mobbing) are compared against much vaguer and less detailed central models, which can rapidly be changed to match on the few parameters (e.g. position and size) which they specify. The obvious signs of recognition comparison during greeting behaviour may be caused by short term degradation and loss of some of the details of an elaborate and stable model of a conspecific. Another interesting type of situation is one which provides inherently vague perceptual inputs (e.g. objects seen at dusk). The prolonged recognition comparison which results presumably occurs because no model can match perfectly an input which gives inadequate information about some of its parameters. Fleeing may also develop: it is worth emphasizing that persistent mismatch may also induce responses other than those discussed here.

Not all examination of a stimulus will involve recognition comparison. No persistent stable central model will be involved when a new perceptual pattern (e.g. a new landscape) is being learned; it could be argued that a labile central representation, changing to match newly perceived features, is necessary to allow storage.

The specific information potentially available in various groups of responses facilitated during recognition comparison is considered later. In general, the intensity and duration of signs that a stimulus is found significant or novel will be relevant to the interpretation of almost any group of other signals.

2. *Alert responses*

These reflect at least two strategies of collection of information: scanning the environment or concentrating on a particular stimulus. Either strategy could under some circumstances be appropriate to recognition comparison, and it would not be helpful therefore, even as a first approximation, to consider alert responses as a single grouping facilitated during recognition comparison.

The differences between strategies are most obvious in visual reflexes. Sustained examination of a particular stimulus is accompanied by a frown (eyebrow approximation and constriction of the circular muscle about the eye) in many mammals with mobile faces. These movements may serve to exclude distractions in peripheral vision, whilst pupil constriction at the same time may increase the depth of sharp focus. In man, frowns of puzzlement or close attention occur during detailed and prolonged examination; this need not involve vision, suggesting that such patterns of reflexes are associated with general strategies of attention rather than being specific to a particular sense. Animals with mobile ears show similar phases of attention by ear movements as well, and they can also be recognized for olfaction and whisker movement.

Scanning eye movements, on the other hand, are often accompanied in many mammals, including man, by raised eyebrows and wide open eyes, both of which give a wider field of unimpeded vision. (Eyebrow raising is reflexly part of glancing upwards). Pupil dilation also probably occurs at such times; certainly it is present when raised eyebrows and widened eyes accompany a fixed stare. Its effect is probably somewhat to increase the likelihood of response to stimuli in the peripheral field of vision by increasing the amount of light entering from the periphery relative to that from the centre of the field, which will be little affected. (Note that pupillary changes of this kind are superimposed on the basic pupillary responses to level of illumination.) A wide range of stimulus objects are effective. Thus, cats show pupil dilation in defensive threat and when stalking a rat, whilst sustained eyebrow raising occurs in man, both when staring at a terrifying object, and when awaiting some move by an opponent in a physical contest. Such combinations of alert responses represent to some extent a mixture of the two main strategies. Functionally, it may be useful to increase responsiveness to unexpected changes in the peripheral field when a central model appropriate to the object being examined in the centre of the field has already been retrieved. Patterns of alert responses of this kind indicate protracted recognition comparison even in the absence of other signals; when combined with other signals, they potentially communicate such diverse information as that an object is regarded as frightening or prey, or (in man) an infant or a lover.

Other patterns of alert responses are potentially equally revealing. Thus focussed attention, with the impeded peripheral responsiveness which results, is usually a sign of confidence during a social interaction; one special instance of this is the sustained stares and frowns of confident threat in for example, Capuchin monkeys, apes, men and dogs.

3. *Protective responses*

Although emphasizing the importance of specific causal factors, Andrew (1963 *b*) treated the whole group of protective responses as if they were facilitated in the same way as alert responses; a rather similar position is taken by Sokolov (1960). This is, in fact, misleading. Protective responses are evoked by startling and painful stimuli, and so often occur with alert responses. They are also evoked by stimuli potentially dangerous to sense organs and other vulnerable areas; clearly a mismatch during recognition might be taken to indicate such danger, and so facilitate protective as well as alert responses, but appropriate specific stimuli will produce protective responses after recognition is complete. It is equally misleading to state that protective responses are caused by a tendency to flee. Protective responses and fleeing are sometimes simultaneously evoked; at other times (e.g. friendly play, marked exertion, pushing through obstacles) protective responses occur in the complete absence of fleeing.

Two phases of the full pattern of facial protective responses may conveniently be distinguished. The first involves eye closure, orbicularis oculis contraction (causing brow lowering), ear withdrawal, and two higher threshold components: a check in respiration with glottis narrowing or closure (which, if followed by a sudden expiration, results in a cough) and mouth corner retraction. If the second phase develops, movements serving to free mouth, head and body of substances or objects may be added: repeated tongue protrusion, and lateral shaking of head and body.

Two of these reflexes have apparently changed in form during the evolution of primate facial displays: ear withdrawal has remained conspicuous, despite reduced ear mobility (particularly in the Old World monkeys), because of the involvement of scalp retraction in ear withdrawal (Andrew, 1963*b*). Changes have also occurred in the evolution of mouth corner withdrawal into grins. There is evidence in some cases that the causation of these movements has nevertheless changed little.

The most important general information communicated by protective responses is 'at what distance does a stimulus begin to appear dangerous'. This will depend in turn on computations of likelihood that the other animal involved in the interaction will approach it, or of approach by itself to the other. This has the interesting consequence that protective responses may be conspicuous, at one extreme, in very friendly displays and, at the other, in defensive threat. Their assumption by a high status animal close to another of lower status indicates that further approach is likely, in order to perform a response (e.g. play) which may be resisted without the superior at once attacking. (Protective responses will appear in confident attack, but only in the actual lunge.) At the other extreme, protective responses assumed at a great distance indicate the displaying animal to be of low status, and that it anticipates attack or other danger (and usually is likely to flee). Tinbergen (1959*b*) has already pointed out the apparent derivation of friendly displays from threat. If a fixed hierarchy of major variables is used, changes in causation at very high levels must be assumed; Tinbergen suggests an unexpected and undefined resemblance between 'feeling hostile' and 'feeling friendly'. At least for protective responses, a clear explanation for resemblances between the two situations can now be suggested, which does not depend on an evolution of the displays appropriate to one from those of the other.

Protective responses are conspicuous in greeting displays, which are most predictably evoked in primates by sudden mutual perception, particularly if the eyes meet. The animals behave, even although at a considerable distance, as if their faces had been thrust together. In higher New World monkeys, nearly all facial protective responses (with frown in place of scalp retraction) occur readily in greeting, whilst in Old World monkeys, grins and scalp retraction are best developed (Andrew, 1963*b*, who was wrong in describing the mandrill grin as a threat snarl instead of a friendly grin). Grins are further discussed later.

It is not clear how far such display components are, in causation, still protective responses, made conspicuous by the exaggeration of musculature, which reveals incipient movements that might be missed in more primitive forms. Such exaggeration of movement alone is obvious in such species as the black ape (*Cynopithecus niger*) and mandrill (*Mandrillus sphinx*) in the grin and eye closure which precede ordinary sneezing or follow ordinary yawns (and are very similar to the same movements given as a display component). Operant conditioning seems likely for some greeting grins, in particular the human smile.

4. *Vocalizations*

Andrew (1963 b), in arguing for a possible origin of mammalian vocalization from protective glottal closure, did not separate sufficiently clearly arguments about origin and about present causation; as a result too close a similarity in causation to protective responses was implied. Andrew (1962) pointed out that many primates (and other mammals) possess a main group of calls, which fall into a series of increasing intensity. Within the series one (low intensity) call may be given on recognizing sought-after food or in greeting a fellow, whilst another (more intense) call may be given in a frustrating situation, to a strange object or after loss of contact with fellows. Such situations will be recognized as amongst those already discussed as involving recognition comparison (with or without final mismatch). This close dependence of vocalization on periods of recognition comparison is important in communication: once alert responses are fully developed, they can change little with further extension of recognition comparison or increase of mismatch. Vocalizations communicate accurately a variety of such states. They can indicate that an animal is uncomfortable or lost, that it wants something but cannot obtain it, that it is likely to flee, or that it has perceived a novel stimulus. The parallel evolution of calls with comparable causation and function in the domestic fowl (Andrew, 1964) and probably other birds, suggests strong selective advantages for such communication. Further, since very prolonged mismatch may result in the development of immobility and unresponsiveness, calling may then be the only sign of the persisting central state (e.g. human crying, short wail of *Lemur*).

Other causal factors clearly must be considered. Not all calls are given in such a wide variety of situations, indeed some appear to be evoked by only a narrow range of stimuli. In some cases such calls can be shown to represent one out of a number of variants of a call with the sort of causation discussed above. Such variants can be produced by the superimposition of facial movements and changes in ventilation movements on a basic pattern of glottally produced sound. The slow extended expirations associated with developing immobility (below), or a sudden short expiration following startle are made audible in the tempo and amplitude of coincident vocalization (e.g. as moans or bark). Lip, tongue and jaw movements affect coincident vocalization by changes in the resonant properties of the oral cavity, if the sound provides adequate energy at the resonant frequency (Andrew,

1963*b*); grunts are suitable for this in animals between the size of a small monkey and a baboon. The most interesting case is that of the grunt modulations caused by lip smacking in baboons, which can convey in a call the likelihood of grooming, even if the tongue movements involved are too small to be visible. In this, and probably other instances (e.g. mouth corner retraction), the responses superimposed upon vocalizations appear to retain their original causation.

However, even when superimposition effects have been taken into account, causal factors other than facilitation during recognition comparison are clearly important for vocalization. The evocation of calls by sounds which resemble or are identical with them is common in mammals. It may affect complex calls used in long distance exchanges between groups used in maintaining territories (e.g. song of gibbons (*Hylobates*) and indri (*Indris*); chorus of howler monkeys (*Alouatta*)), alarm or mobbing calls (e.g. bark of baboons (*Papio* spp.; mobbing of *Galago*), calls given by isolated animals (e.g. wails of brown lemur (*Lemur fulvus*) or ring-tailed lemur (*Lemur catta*)) and calls associated with friendly interaction (e.g. grunts, brown lemur; grunts, which once started may suddenly build up to a general chorus, East African baboon). As a result an animal may be able to find out by itself calling where adjacent fellows are (since they are likely to answer) and often who they are. It may do more, since superimposed facial reflexes may reveal eating, lipsmacking and other behavioural states. The choruses which result from facilitative interaction are important in communication over long distances in mobbing (arboreal mammals), in maintenance of territories, and in some specialized instances like the calling of chimpanzees around a localized food source.

Finally, some reflexes associated with vocalization have themselves become visual signals. This is true of the contraction of the mouth sphincter muscle which accompanies violent expirations (perhaps because the *dilator naris* is a slip of the sphincter; perhaps as a quite separate reflex serving to regulate terminal resistance to airflow, and prevent too rapid a collapse of the thoracic cavity). The resulting lip rounding is conspicuous in many mammal vocalizations, and is particularly characteristic of primate threat, in which it accompanies violent expiration.

Mouth corner retraction similar to that already described as a protective response, and to movements associated with biting (below), occurs during high-pitched vocalization in so many mammals, that it must be regarded as a reflex of vocalization in its own right (perhaps with the function of protecting from vibration the great vessels of the neck and other structures). In evolution such grinning, like vocalization itself, may well be derived from protective responses (Andrew, 1963*b*) but this is not directly relevant to its current causation. The vocalization grin, at all events, is a conspicuous signal, and it is possible that some silent grins (e.g. human smile) indicate incipient vocalization.

5. *Locomotion*

In mammals, a crucial feature of locomotion is the gait which is selected. Situations which promote locomotion, but also make it likely that the animal will not move far, may result (e.g. in courtship in horses) in a speedy gait such as a trot (or even occasionally, a gallop) being performed slowly or on the spot. In primates, one equivalent is to bound on the spot by sudden limb extension, whilst retaining a firm grasp of the branch (advanced New and Old World monkeys, especially the stereotyped, vertical bobs of guenons (*Cercopithecus*)). Brachiators may swing but not let go.

Less obviously connected with locomotion are some movements and postures of the vertebral column. Lateral bending of the vertebral column serves to lengthen the stride in diagonal gaits, whilst arching in the vertical plane, followed by a sudden straightening during hind limb extension, adds power to the propulsive phase of a gallop. Both movements are most obvious in flexible bodied carnivores (e.g. Felidae, Canidae, Mustelidae), but they can be seen in primates which make considerable use of quadrupedal gaits.

Lateral bending of the lumbar vertebral column is often continued into the tail, which, in most mammals, makes conspicuous for communication the overall state of the vertebral column. In locomotion the tail swing probably serves to counterbalance a pelvic displacement in the opposite direction. Rather similar lateral bending also occurs in body shaking, and this too is probably reflected in tail wagging. The use of the tail in response to skin irritation probably originated in this way in ungulates. In both cases tail wagging can now be elicited in the absence of general movements of the vertebral column.

In many ungulates and carnivores (Kiley, 1969) tail wagging occurs when greeting a peer or inferior, as well as a superior (e.g. pig, cow, goat, dog), and in threat as well as in courtship. It is also obvious when the animal is trying to obtain milk during suckling in all these species. Wagging whilst hunting amongst undergrowth should also be noted in the dog. Kiley (1969) has shown for the same range of species that, in the absence of any social fellow (conspecific or human), non-reinforcement and searching for food both produce tail wagging. In all these situations, which involve periods of recognition comparison (with and without mismatch), tail wagging is probably caused by a facilitation of locomotion, itself caused by the fact that the animal is ready to perform particular responses appropriate to the activated central model.

Species with marked vertical flexibility of the back may show considerable vertical arching as a preparation for leaping (e.g. defensive threat of cat and great galago (*Galago crassicaudatus*)). Vertical tail tossing appears to bear the same relation to the arching and straightening of the vertebral column in leaping, as wagging does to the lateral bending of the diagonal gait. It is well-developed in the tail flicks of squirrels and tree shrews, which are given in mobbing displays, as well as in interactions with conspecifics.

The facilitation of locomotion, which causes the above displays, is sometimes accompanied by incompatible tendencies to move in two clearly specified directions. Generalized facilitation (here termed 'excitement') with no specific directional component also occurs (e.g. in the wild locomotion which follows release from cramped sleeping quarters). An interesting special case is the rising excitement shown by mangabeys (*Cercocebus*), gibbons and chimpanzees, accompanied by calls which rise to a crescendo just before a sudden end to the wild locomotion, since this occurs in response not only to a wide range of frustrating or attention-provoking stimulation, but to sounds like their own calls.

Locomotor display components thus communicate not only the likelihood of locomotion, but its probable intensity. Viewed more broadly they indicate that the animal is ready or waiting to respond, in a way which cannot be deduced unambiguously from response groupings which relate to attentional processes alone: alert responses, for example, may be given by an animal which is not prepared to move.

6. *Exertion, immobility and autonomic responses*

The autonomic responses involved in displays are usually treated as if they were a single stereotyped pattern occurring only in fear or conflict. It is argued here that this pattern (the familiar 'emergency reaction' of Cannon, 1929) is in fact associated with exertion in general, rather than restricted to situations involving attack or fleeing. Many of the reflexes involved also precede, and therefore predict, exertion, and it has been possible in consequence for some of them to become important in communication. The clearest evidence, both that appropriate reflex changes precede exertion, and that they may be associated with displays, has been given by Abrahams, Hilton and Zbrozyna (1960, 1964), who described in the cat a pattern of cardiovascular changes, which is obtained by stimulation in exactly the areas of hypothalamus, mesencephalic central grey and tegmentum from which defensive threat is obtained in the conscious cat; indeed defensive threat components often accompanied the cardiovascular changes despite anaesthesia. The pattern, which is not dependent on muscle contraction, involves active vasodilation in skeletal muscles and is functionally important in increasing their blood supply; mean arterial blood pressure rises nevertheless, thanks to cardiac acceleration, vasoconstriction in skin and intestine, and inhibition of the baroceptor reflex (Hilton, 1963). The crucial point here is that the pattern occurs, not only during defensive threat, but (together with alerting), as a first response to a new stimulus in a previously quiet animal, and in the first phases of defensive threat, when only alert responses have been elicited. The function of this pattern is generally agreed to be that of preparing for sustained or violent muscular exertion, and in man and dog it has been confirmed that the pattern appears in anticipation of a routine (and quite unalarming) bout of exercise (e.g. walking on a treadmill: dog) (Rushmer, Smith and Lasher, 1960). Adams, Bacelli, Mancia and Zanchetti (1968) have questioned these findings, after showing there indeed to be increasing muscle blood flow in the cat during defensive threat, but not in the anticipation of attack. It is doubtful if these results are relevant, since anticipation was behaviourally quite unlike the alerting discussed above (the animals flinched, flattened their ears and remained still); the blood flow measurements may well represent in any case interaction between vasodilation and increased resistance to flow due to sustained muscle tensing.

The idea of reflexes predictive of exertion thus seems reasonable. I would like to extend this further, and postulate reflexes, in general the functional converse of exertion reflexes, which precede as well as accompany immobility (i.e. generalized motor inhibition). An obvious example of immobility is freezing, which often begins with the animal crouched to leap, but can occur briefly in any posture, and sometimes passes into more relaxed immobility (Andrew, 1956a); the development of persistent immobility in forcibly-constrained animals (animal hypnosis) may be comparable. Such inhibition, usually in a resting posture, also develops in an animal which is persecuted by a superior (when movements may be punished) or tired or drowsy.

It may thus be useful to characterize the behaviour of an animal as lying at some point along a continuum from exertion to immobility. The reasons for shifts along the continuum will be extremely various, since they will reflect the likelihood of new responses as well as concurrent behaviour. Some effects are obvious (e.g. groupings such as locomotion, attack or copulation clearly require exertion). Others are less so: thus the possibility that general motor inhibition may be promoted when a specific response is withheld, is suggested by the fact that forebrain lesions (e.g. septal, hippocampal, frontal) which make freezing less likely also interfere with the ability to withhold specific operant responses when this is called for by a schedule of reinforcement (McCleary, 1966). Finally, it is important to note that the same stimulus may produce shifts in opposite directions at different times: a threatening conspecific, for example, may tend to induce exertion or immobility.

The main types of reflex which appear to identify where an animal lies on this continuum between preparations for exertion and for immobility are:

(*a*) *Respiratory*. Acceleration of ventilation in exertion and slowing in immobility are both made audible in vocalizations as well as being visible externally. Dilation of the external naris, normally confined to inspiration but sustained through expiration as well in vigorous ventilation, has been influential in the development of facial expression. Dilated nostrils are conspicuous in many mammals (e.g. due to pink vascularized interior in horses).

(*b*) *Thermoregulatory*. Although experimental evidence is largely lacking, observational evidence for the evocation of thermoregulatory responses by novel, significant or frightening stimuli is clear (Andrew, 1956*b*). It is here suggested that these are either cooling responses which prepare for exertion, or warming responses which prepare for immobility. These primary changes may be followed by normal corrections: shivering when chilled by nervous sweating is an obvious possibility. This should be contrasted with the suggestion (Andrew, 1956*b*) that changes in skin temperature following changes in blood flow (e.g. vasoconstriction) might evoke thermoregulatory responses. This may sometimes be a factor, but it is doubtful how effective it could be in the absence of changes in core temperature (Cooper, 1966).

Sweating, which is not part of normal thermoregulation, occurs in man and horse in response to frightening stimuli, and in man, to embarrassment and to a variety of conditioned stimuli (Andrew, 1956*b*; Sokolov, 1960). The original function of eccrine secretion was apparently not temperature regulation in most lines of mammals: thus the cow is apparently unusual amongst Artiodactyls, and man amongst Anthropoidea, in their marked use of sweating. Montagna (1965) suggested that a more general function of eccrine secretion is to moisten, and thereby improve the grip of, palms and soles (to which areas eccrine glands are confined in many carnivores, rodents and some primates). However, secretion in preparation for exertion would be just as appropriate for this function, which appears to be retained in man.

Panting and associated cooling responses are facilitated by frightening stimuli in birds (Andrew 1956*b*) and in dogs. Flushing in man, and perhaps also of the vascularized skin of gelada (*Theropithecus gelada*) and uacari

(*Cacajao rubicundus*), is another cooling response, which appears in a wide range of situations in which exertion is likely (e.g. in man: anger, embarrassment, excitement).

Warming responses are conspicuous in displays in which immobility begins to develop (see Fig. 1). Hinde (1955) pointed out that some passerines (e.g. *Fringilla*) assume resting attitudes when persecuted, and that

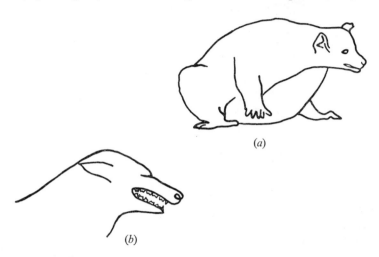

(*a*)

(*b*)

Fig. 1. (*a*) shows a ring-tailed lemur, which is grinning and giving a yip call as a result of the approach of a human being. (High pitched calls are usually accompanied by a grin in this species.) The animal has its right hand raised to make an aggressive thrust, but its general posture is that of a subordinate animal. Note in particular the way in which the large bushy tail is passed forward under the belly and back round the left shoulder. This is a warming reflex shown by cold or sleeping animals, which is also assumed very rapidly at the approach of a superior. The face shows protective ear flattening, but the eyes remain wide open.

(*b*) A generalized mammal showing more clearly the main facial reflexes of ear withdrawal and flattening, mouth corner withdrawal and lip retraction, and eye closure which are commonly associated together and given in response to noxious stimuli (potential as well as actual). Both drawings after Andrew (1963).

these may serve as submissive postures. Leg flexion and general feather fluffing, which are the most conspicuous features of such a posture, both serve to reduce heat loss. Piloerection in mammals has been little studied, and its thermoregulatory role is difficult to document. *Arrectores pili* may even be almost absent in some mammals e.g. lemurs (Montagna, 1962). Localized erection of both feather and hair appear to have a different causation and origin, and will be considered later. Postural warming responses, however, are obvious (e.g. huddles, with limbs clasped against trunk in many monkeys) in cold or tired mammals, and also appear in persecuted inferiors, just as in a dejected human being. The use of the tail as a wrapper around the body makes such behaviour especially obvious in long-tailed lemurs (e.g. Ring-tailed) where the posture may be assumed very rapidly (Andrew, 1963 a). In many mammals the tail may be tucked between the legs well before a full resting attitude is assumed, and this may be a warming reflex in causation as well as a protective one. Shivering is well

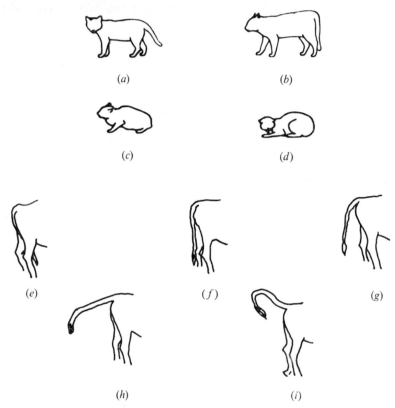

Fig. 2. (a–d) represent postures of the domestic cat which show systematic changes in postural tonus. The rise in limb extension, in posture (b), which is that of an animal likely to attack, by comparison with normal relaxed stance (a), is very marked. Postures (c) to (d) show progressive loss of tonus as (according to Leyhausen), fear becomes more intense; note that, in (d), the cat is lying rather than preparing to leap (after Leyhausen, 1956).

(e–i) show different tail elevations of domestic cattle, which provide another index of general postural tonus. The posture shown in cold, sick or frightened animals (e) is probably a special case, with active depression in a protective reflex. The relaxed position (f) is assumed when grazing or walking. In the remaining positions elevation becomes greater, the greater the involvement of active vertebral column extension in locomotion: (h) occurs in galloping and (i) in bucking and gambolling. The same tail positions also occur in conventional display situations. Position (g) is assumed when attending to a stimulus, and during threat and sexual behaviour, whilst further elevation may occur in the latter two instances (after Kiley, 1970).

known in man and horse in frightening and depressing situations; it occurs also in some other primates e.g. rhesus (*Macaca mulatta*) at similar times. It is my impression that it is associated with incipient immobility, but no good data are known to me. Pearson (1970) has shown for the guineapig in frustration situations an association of body tremors and tooth chattering with immobility periods; he feels that this characteristic rodent display may represent shivering, rather than biting movements.

(c) *Postural*. Postural reflexes (limb extension, anti-gravity tonus in vertebral column) are, of necessity, facilitated during locomotion to some extent. However, the relation of vertebral tonus to locomotion is complex: it usually reaches a maximum in diagonal gaits, since sustained tonus is replaced by alternate flexion and extension in gallops. Moreover, the same gait and speed of movement can be accompanied by high or low postural tonus. Such tonus differences are best regarded as revealing shifts along the exertion/immobility continuum (Fig. 2*a–d*).

Tail elevation reveals, in an exaggerated form for use in communication, the degree of vertebral anti-gravity tonus (Andrew, 1963*a*). This reaches a minimum when immobility is developing (tired, persecuted animals). In the horse, the tail is raised appropriately in ordinary locomotion as vertebral tonus rises, but maximum elevation is attained only in situations such as courtship, when other exertion responses are likely (Kiley, 1969). Schenkel (1948) showed that high elevation is a sign of confidence in the wolf (*Canis lupus*). This is true also, for example, of Artiodactyls (Kiley, 1969) (Fig. 2*e–i*) and macaques (Altmann, 1962). It should be noted that a confident animal is not necessarily one which is likely to attack: it is rather free, unconstrained by conspecific or danger, to initiate a wide range of responses. Equally, low confidence is not always accompanied by a tendency to flee. Limb extension gives very similar information. A confident gait and stance with increased limb extension is obvious in many primates, including man. Reduced limb extension, giving a crouched stance or slinking gait, is equally characteristic of tired or inferior mammals (see Fig. 3).

7. *Cutaneous secretion other than sweating, urination and defaecation*

Scent-marking displays, in which the secretion of specialized apocrine glands are applied to environmental objects or conspecifics, are widespread in mammals. A common condition, which may be primitive, involves similar scent marking, both of the female in courtship and of special points in the environment, e.g. opossum and tree shrew (*Tupaia* spp.). *Lemur* spp. have retained scent marking, despite the evolution of permanent societies. Male brown lemurs mark both scent posts and females, particularly vigorously when strange conspecifics are present. Glands occur on forearm and wrist, and around the anus. The conspicuous rubbing movements employed in marking appear to be initiated, at least in part, by irritation associated with secretion. At the same time, just as in many other mammals, olfactory stimuli are clearly important: great interest is shown during marking in the

Fig. 3. The two figures differ markedly in postural tonus. The one on the left (the 'Slopper') with rounded shoulders and legs which are not fully extended, can be seen to be unresponsive; he might in addition be depressed, drowsy or even ill. The Scout on the right is ready at any moment to respond. An appropriate facial expression would make him threatening, instead of merely confident. (Baden-Powell, 1929.)

scent both of the secretion and of urine. Many mustelids e.g. polecats (*Mustela putorius*) and skunks (*Mephitis*) produce nauseous scents in response to frightening stimuli, but it is not certain that such behaviour is best regarded as caused by fear: badgers (*Miles miles*), for example may secrete strongly during play (Neal, 1958).

Urination and defaecation are also facilitated in many mammals by frightening stimuli. This has commonly been attributed to 'general autonomic activation', but its relation to the autonomic reflexes discussed in the previous section is obscure. Some more elaborate displays based on such behaviour show similar facilitation (e.g. urine washing is common during mobbing in galagos), but others are certainly evoked by, and directed at, conspecifics or marking posts (e.g. micturition in dog, guineapig and rabbit). Olfactory stimuli, in particular the smell of urine and faeces, are certainly also important.

Scent signals thus act as indices of territorial possession, together with breeding status, and very probably, identity, which not only are effective in the absence of the animal but are also important in direct communicatory exchanges. The clearest experimental work is in mice where scent cues prevent attacks by males on females and juveniles, as well as initiating copulation with females (Mackintosh, pers. comm.).

8. *Grooming*; *shaking*; *stretching*

Grooming and shaking responses are to some extent separate causal groupings. Tongue, lips, teeth and paws are used in grooming to part, clean and smooth fur. Lateral shaking of head and body, accompanied in some mammals by piloerection, may terminate grooming and serve to settle the fur. It is also evoked by water or other substances on the body surface; rather similar but more violent movements occur when shaking free of an encumbrance or opponent.

Both grooming and shaking may occur at transitions between two different series of responses. Shaking responses occur particularly at the end of brief or incipient periods of immobility; in the chick this is common when food is unexpectedly absent from an accustomed food dish, and rapid locomotion usually follows. Shaking responses also follow immobility due to a strange situation. They are particularly conspicuous in ungulates as tail lashing and head shaking (Kiley, 1969); bill shaking is equally frequent and conspicuous in the domestic fowl at similar moments. Local piloerection may represent raising for shaking. Snorting (which clears the nostrils and may be regarded as a protective reflex), provides a sound signal at such times in many ungulates, and is an interesting parallel to the suggested origin of vocalization from protective glottal closure. Whatever their causation, such responses allow the identification of the moments at which a new behaviour sequence is about to begin.

Normal grooming is important in communication in many mammals. In primates, grooming of conspecifics dominates many interchanges, increasing in frequency with an increased tendency to seek bodily contact (e.g. for copulation, Michael and Herbert, 1963). Lip and tongue movements of grooming also appear before contact with the skin is achieved in a number of groups, in particular Old World monkeys. These have become very conspicuous lip smacking displays (Andrew, 1963 b). Causally they appear usually to continue to be grooming responses. By communicating intention to groom such displays indicate friendly intentions in a way complementary to the protective responses already discussed.

Stretching is usually a sign of waking from drowsiness or sleep, but it may follow sustained contraction of particular muscles. Yawning in baboons and macaques is a stretching movement, apparently associated with tensing the mouth sphincter muscle in threat.

9. *Bodily contact*

Two patterns can probably be distinguished in mammals as a first simplification, although there is great need for further data. Clinging and huddling in infant, and (in social species) adult, mammals has become important in communication in primates, in that it follows attack or novel stimuli, and allows exchanges involving comforting and being comforted.

The second pattern, rubbing flanks and sides of face along the substratum or a conspecific, is often associated with sexual behaviour. In the Virginian opossum, Roberts *et al.* (1967) found that such rubbing occurred on any furry surface before copulation (which was elicited by hypothalamic stimulation). The same movements may be used in scent marking objects in the environment as well as conspecifics. The causal relation between rubbing and copulation is still not clear; one interesting possibility is that stimuli from rubbing make copulation more likely.

10–14. *Attack, fleeing, hiding, defensive threat and copulation*

All of these groupings have been obtained as relatively coherent patterns, with a restricted number of components, by brain stimulation in mammals (e.g. attack in the Virginian opossum following hypothalamic stimulation, involves visual fixation of a rat, its seizure in the jaws, followed by stereotyped shaking: Roberts *et al.*, 1967). At the present stage of analysis, it is best to regard them as no more (and no less) important, both in the general causation of behaviour and in communication, than the groupings already discussed.

In primates, mouth opening for biting, cuffing forward or at the ground, and lunges forward of body and head all occur in displays as intention movements of attack. In many mammals, lip retraction occurs both in biting food and other objects and also before (and during) attack. The full movement often involves mouth-corner withdrawal (which allows the back teeth to be used), as well as lip retraction, and is almost or quite identical with the grin of the same species. This is obvious for the grin of friendly greeting and the snarl shown by some domestic dogs. Since this stereotyped pattern also occurs as a protective response (including the movements just before a sneeze), and as a vocalization reflex, it is clear that attempts to identify its evolutionary origin when it occurs as a display component have been premature.

In the cat, fleeing, in which the animal shows continual locomotion, often upwards, is obtainable by hypothalamic stimulation, and is clearly distinguishable from hiding behaviour, in which the animal shrinks, keeping close to the floor until it reaches a refuge (dorsomedial thalamic nucleus; Roberts, 1962). The same patterns are distinguishable in many mammals. Their interaction with attack and other response groupings deserves fresh moment-by-moment study. It is far from clear, for example, that the mechanism involved in fleeing need be involved in causing an attack lunge to be checked. At all events, incipient fleeing and hiding can often be recognized, and must often be of importance in communication.

In defensive threat induced by hypothalamic stimulation, the opossum opens its mouth widely, swings its head from side to side, growls and barks (Roberts *et al.*, 1967). The sites involved are discrete, and clearly separated from those giving attack or fleeing. Biting is impossible to provoke, and fleeing does not develop; there is thus, as has already been noted, little evidence that the pattern is caused by the simultaneous activation of attack and fleeing. In the cat, intense protective responses are the most characteristic part of the display, and it may best be regarded as the fullest and most intense development of this grouping, with super-imposed exertion responses. Erection of hair along neck, back and tail is so characteristic of defensive threat in cat and many other mammals that it must be mentioned. Birds, too, show feather erection along the neck in threat (e.g. ducks and geese), whose relation to thermoregulatory and other types of piloerection is obscure. The value in communication of defensive threat is obvious.

Intention movements of copulation, such as the raising and direction to one side of the tail in many female mammals and attempts at mounting in males, are clearly important in communication. In the Old World monkeys, in particular, the female pattern of presentation for mounting is now widely used by both sexes. In males, presentation is usually to a superior who may briefly mount (Altmann, 1962, rhesus); the causation is still obscure, but one function in communication is apparently to make friendly bodily contact more likely.

CONCLUSION

A recipient could be affected by the signals of a sender in a number of ways. At one extreme he might be affected only by one particular movement, sound or structure and ignore all others; at the other, he might attend, and respond to, the total combinations of signals, and the environmental situation in which they were emitted. Most observers are of the opinion that most mammalian communicatory exchanges approximate to the latter case, although single stimuli which appear to override nearly all other signals (e.g. induction of rolling in cats using extract of catmint) are also well known. Even so, it would be possible to explain most existing observations by postulating a specific different effect on the recipient of each of a large number of complex patterns. If this were the case, then the approach taken in this chapter would be useful only in deciding what is the information potentially available in a particular pattern, so that experiments can be designed to demonstrate what part of it is actually effective.

However, it seems more likely, in higher mammals at least, that analysis of signals by the recipient involves a number of steps, some of which may correspond to separate estimates of the states of groups of responses such as have been discussed here. Thus an animal might use the appearance of responses which, it has been argued, indicate processes of recognition, in order to decide the degree of significance (or novelty) of a stimulus perceived by the sender, before it could decide whether the sender had perceived prey or an enemy, or was responding to the recipient as a mate or as an infant. Equally, the likelihood and direction of future locomotion by the sender might be computed as a first step, whether or not a response relevant to the recipient was expected. If analysis of this kind is common, then it will be useful to take as units of study (in place of complex displays) such categories of signals as those which give information about recognition comparison, or estimated likelihood of danger to the face, or strategies of examining the environment, in addition to categories such as specific movements of attack.

We can only speculate as to the ways in which further analysis might proceed. In some exchanges, the recipient responds as if it had been specifically informed that it is about to be attacked, for example. However, in other cases it could well be that that recipient identifies a general state, such as that of confidence, in which the sender is unconstrained by the possible actions of the sender and other social fellows. Animal communication is so complex and involves so many levels, that it is usually possible for a worker to find examples of whatever he expects to find. One of the central arguments of this chapter is that if we adopt a single over-simplified theory of the causation of displays, the development of ideas about animal communication is likely to be constrained and distorted.

COMMENTS

(1) *Use of 'major' intervening variables in accounting for displays*

Most members of the group agreed that Andrew's insistence on conflict between individual responses rather than conflict between super-ordinated tendencies gave a more precise understanding of many signalling movements. Another example is provided by Smith's analysis of communication in tyrannid flycatchers, cited in the comments on Chapter 4. This is a refinement of the conflict theory valuable in many, but not necessarily all, contexts.

Whatever the level of the conflict, the tendencies postulated as being in conflict are only intervening variables – that is, hypothetical factors postulated to explain the relationships observed. The conflict theory of display can be used for a whole hierarchy of problems ranging from the analysis of the whole display behaviour of a species and the inter-relations of the patterns shown, to the bases of the individual patterns. By focussing on the latter simple answers can be obtained (cf. Andrew, p. 182), but sometimes at the cost of evading the former. On the other hand, and this is where Andrew's view comes into its own, the continued use of the more clumsy tools suitable for the initial problems can obscure understanding of the finer ones. For instance the earlier approach does bring understanding of the *average* direction of the flights of a mobbing bird, but not the direction of individual ones (see pp. 182–3): it is possible that a response tendency approach can provide the latter, but not the former.

However, even if the bases of individual display movements are being discussed, it is an empirical issue whether one approach or the other is more useful. Andrew cites Blurton Jones (1968) as showing that the great tit's forward threat occurs whenever full attack is prevented, whether by a physical barrier or through inhibition by fleeing. However Blurton Jones also found that another threat posture, the head-up, seemed to depend specifically on a tendency to flee, rather than merely on the blocking of attack. Blurton Jones concluded that some threat movements depend specifically on conflict between tendencies to flee, stay or attack, others merely on conflicting tendencies for locomotion in different directions, and yet others on conflict between one major tendency and any other which blocks it.

Andrew's view is that since some displays cannot readily be described in terms of conflict between two drive type variables, and since it is difficult to find *proof* that any displays are so caused, an analysis which does not employ such variables is preferable.

It is perhaps significant that discussion on this topic arises particularly when the expressive movements of mammals are under discussion. In lower vertebrates attention tends to be focussed on whole displays, while higher up the phyletic scale the parts of which the displays are constituted seem to be less highly correlated with each other. This may be related to a tendency (probable, but by no means established) from discrete towards intergrading and multidimensional signals higher up the phyletic scale, permitting a wider range of meanings to be conveyed. The topic is taken up again by Blurton Jones in Chapter 10.

(2) *The concept of stimulus contrast*

Some members of the group felt that the concept was used *post hoc* and was so flexible that practically any data could be explained by it. In addition, Horn (1967) argued that at least many of the findings on which Sokolov (1960) based his concept of neuronal models could be explained in terms of self-generated depression within the responding system, without the postulation of a matching system producing active gating. Andrew pays lip-service to Horn's viewpoint, but then continues his discussion using such terms as 'retrieval' of a 'central model' which are incompatible with that view. Andrew argues that this is justifiable because Horn did postulate what is essentially a central model in discussing temporal patterns, and one which had to be 'found' by the input – a process properly described as 'retrieval'. Furthermore, Andrew is concerned with the general problem of recognition, of which habituation is a special case.

Non-verbal Communication in Man

INTRODUCTION TO PART C

This section is concerned with non-verbal communication in man, but Van Hooff's chapter provides a link with Part B. He uses several sources of evidence, including comparative data from sub-human forms, to elucidate the nature of, and relations between, smiling and laughter. Not only does he make a convincing case for their evolution from particular expressive movements of lower primates, but he hints also at how these basic 'non-verbal' expressions may have been influenced by the development of human language.

Argyle and Blurton Jones provide some indication of the phenomena of non-verbal communication in adults and in children, discuss the methods used in their study, and indicate some of the current areas of controversy.

Eibl-Eibesfeldt, formerly a student of animal behaviour, stresses the cross-cultural uniformity of human expressive movements, and indicates some ways in which differences between cultures can be understood. Leach, an anthropologist, emphasises the superficial differences between cultures and looks for more basic relationships at a deeper level of analysis. Some resolution of the differing viewpoints in these two chapters is possible in terms of the phenomena selected for study. Grant, also formerly a student of animal behaviour, is concerned with the uses to which the non-verbal signals of patients can be put by the clinician.

The last two chapters, by Gombrich and by Miller, are concerned with non-verbal communication in the visual arts and the theatre. Although this might seem rather remote from the earlier chapters in the book, many of the themes which appeared in Part B occur again here.

8. A COMPARATIVE APPROACH TO THE PHYLOGENY OF LAUGHTER AND SMILING

J. A. R. A. M. VAN HOOFF

Laboratorium voor vergelijkende Fysiologie, Rijksuniversiteit, Utrecht

LAUGHTER AND SMILING: A REVIEW

Of all human expressive activities laughter has undoubtedly most fascinated philosophers and scientists from antiquity to the present. While it appears reasonable to explain most human expressive movements and postures as functional elements of the various forms of behaviour by which the individual interacts with its environment or as manifestations of a general or specific state of activation (Frijda, 1956), laughter, and crying too, seem to defy such an explanation (Lersch, 1957; Frijda, 1965). Both Plessner (1942, 1953) and Buytendijk (1948) set apart the two responses from the other emotional responses, because they are said to lack the relational reference, the symbolic character, that the other responses manifest. Especially in the case of laughter many authors have been baffled by its reflexoid stereotypy and automation on the one hand and the subtle spirituality of the stimuli that can release it on the other, and have considered it as a specifically human attribute (e.g. Koestler, 1949; Plessner, 1950).

Laughter: the releasing factors

The essence of the comical, the ludicrous, has been the main and sometimes the sole concern of many students of laughter. Common to their view is the emphasis on some contrast, unexpected change or contradiction in the situation. But even in 1898, Lipps had shown that not every contrast or change provokes laughter, and this character is therefore not sufficiently exclusive (see also Hayworth, 1928 and Dumas, 1937). Freud (1912) argued that laughter is associated with the 'gain of lust' (*Lustgewinn*) obtained when the tension due to inhibited tendencies (e.g. aggressive, erotic) is released in the morally innocent form of a joke. But it is beyond doubt that not all laughter can be explained in this way.

According to Bergson (1900) laughter occurs when one perceives that in a situation or an action an automatic, stiff and mechanical element is present instead of an adaptive and 'intelligent' one, the comical is 'du mécanique plaqué sur le vivant' or what by association is perceived as such. Again the question is whether all such cases necessarily provoke laughter, and also whether the causes of all laughter (for instance the pun) can be interpreted in such a manner.

14 [209]

It has also been emphasised that a certain emotional appreciation is involved, so that we consider an event or situation ludicrous when it has some awe-inspiring, big or solemn character that is contrasted with or degraded by, an element that makes us appreciate the thing as a trifle (Bain, 1859; Lipps, 1898). There is a 'contraste affectif descendant' (Dumas, 1937). Also sudden relief from strong tension may bring on laughter: whether it is laughter or a sigh of relief depends on whether the tension appeared justified after all (van den Dries, in prep.). In summary, we laugh if we do not take something seriously.

Some descriptions suggest that the ludicrous implies a certain temporal order (the 'big' preceding the 'small'). This is not necessarily so: the two incongruous impressions may also form simultaneously, or the attention may oscillate between them (as in cartoons: Koestler, 1949).

A few authors remark that for laughter to occur it is also of importance that the incongruous should be perceived in some social setting, for people are unlikely to laugh heartily when alone (e.g. McComas, 1926; Hayworth, 1928; Dumas, 1937; Frijda, 1965). There is diversity of opinion, however, as to the functional significance of the social factors.

Derivation and functional aspects of laughter

The earlier authors (e.g. Kant, 1781; Spencer, 1870; Freud, 1912) saw laughter merely as a means to discharge surplus tension or mental excitation, which accumulates if their adequate use is prevented. They imply that laughter restores the normal physiological equilibrium (see Koestler, 1949, for a modern representation of this point of view). Such views do not provide satisfactory answers to all questions. Why, for instance, has laughter taken its actual form; why don't we wriggle our toes? Is the discharge after a simple pun not disproportionate to the tension supposedly generated? Why do social factors facilitate the discharge? Is our species the only one with such a discharge system; if so, why can other species do without it?

Various authors have claimed that the elaborate form and varieties of laughter can be understood only if we assume that social selective pressures have contributed to its development (e.g. McComas, 1926; Dumas, 1937; Hayworth, 1928). McComas follows Darwin (1872) in regarding laughter in its most original form as the expression of joy, happiness and high spirits. Thus it appears early in childhood and later on it is especially characteristic of children's play. He assumes that it became so strongly developed in human beings because the long dependency in infancy required powerful signals that could release, control and reward care-taking by the parents and elder peers. Thus laughter, a signal of well-being, developed as the counterpart of crying, the signal of discomfort. Later on laughter comes to be associated with all kinds of pleasurable things and with maturity 'an intellectual element gives quite a different aspect to the problem'. Here is implicit the distinction between 'le rire de la joie' and 'le rire du comique', made later by Dumas (1937).

Strongly contrasting is Hayworth's (1938) theory that laughter developed originally in primitive humans as a vocal signal to other members of the group that they might relax with safety after some danger passed or was averted. It might stem – and here he follows Sidis amongst others – from the savage shout of triumph and the cruel mockery over a conquered enemy. The oldest form would be to jeer at other people we despise or hate, and might be comparable to the 'hassen' that is known from many birds (Goethe, 1939; see also Eibl-Eibesfeldt, 1967). Its function could be either to correct or to repel abnormal and non-conforming individuals. Gloating would be one of the early roots of the comic and the ludicrous. The genial aspect would derive from the binding of companions laughing together in a mutual realisation of safety.

Smiling

Many authors, most of them implicitly, regard smiling and laughter as patterns that differ only in degree, smiling being a less intense form, a diminutive of laughter. According to Spencer (1870) first the effectors with a low threshold (i.e. the facial musculature) are affected, giving the smile; when stronger excitations break through other muscles join in, giving laughter. According to others (e.g. Darwin, 1872; Hayworth, 1928; Koestler, 1949) laughter preceded smiling in the history of our species, the smile being a subdued version, a diminutive of the laugh.

Only a few authors regard the movements as qualitatively different. When the smile is combined with the typical staccato vocalisations of laughter, which resemble the alarm calls of other primates, Eibl-Eibesfeldt (1967) believes that an aggressive element comes into play. Buytendijk (1947, 1948) and Plessner (1950, 1953) regard both laughter and crying as a giving up of active control over one's own body and an abandoning of directed relationship to the environment. Smiling differs in that the active directedness to the environment is maintained; it is an active display of tranquillity and positive attitude. Plessner (1953) emphasises that although smiling lacks the energetic form of laughter, this does not mean that smiling therefore corresponds with a comparatively weak emotion; on the contrary, its specific emotion can fully absorb us. He notes, however, that laughter also often develops out of and trails away into a smile; often a smile may suffice where a laugh would not be out of place ('a little laugh').

According to Plessner, the reverse, however, is not always possible, and there are, in addition, intense smiles that do not tend towards laughter.

I have gone to some length in presenting these different viewpoints, without pretending to have given an exhaustive review. There is clearly quite a diversity of opinions, but nearly all are based only on general observations and philosophical reasoning. Apart from some developmental studies of the laughter and smiling responses in early infancy, to which we shall turn later, systematic ethological studies of these patterns in the older human primate have been long in appearing. Only recently have a number of ethological studies been started (e.g. Eibl-Eibesfeldt, 1968; Grant, 1968); some of these are concerned with nursery-school children (Blurton Jones, 1967; McGrew, 1969).

The purpose of this paper is to see to what extent the data acquired in recent ethological studies on non-human primates as well as some new data on man can clarify the issue further.

POSSIBLE PRIMATE HOMOLOGUES OF LAUGHTER AND SMILING

A comparative study of the facial displays of the higher primates (van Hooff, 1967) put forward the hypothesis that laughter and smiling could be conceived as displays with a different phylogenetic origin, that have converged to a considerable extent in *Homo*. The material that has become available since then, with a few exceptions still mostly of a qualitative nature, supports this hypothesis.

When searching for possible homologues or phylogenetic precursors of the expressive movements of smiling and laughter it is important first of all to see whether one can detect in the phyletic range of the primates morphologically similar movements that resemble our expressive movements more strongly, the closer in that range the species are to our own. Similarity with respect to the causal context and the functional aspects are also important criteria, but less easy to use since motivational shifts appear to take place much more readily during phylogenetic development than changes in the motor patterns (Baerends, 1959).

With respect to the present issue two primate displays come to our attention immediately. The first is the '*grin*'-*face* (van Hooff, 1962; Andrew, 1963) or *silent bared-teeth* display (van Hooff, 1967). The second is the *play-face* (van Hooff, 1962; Andrew, 1963; Bolwig, 1964) or the *relaxed open-mouth* display (van Hooff, 1967).

1. *The silent bared-teeth display*

This display is characterised by: fully retracted mouth-corners and lips, so that an appreciable part of the gums is bared; closed or only slightly opened mouth; absence of vocalisation; inhibited body movements and eyes that are widely or normally open and can be directed straight or obliquely towards an interacting partner (Fig. 1).

The *silent bared-teeth* display strongly resembles a class of *vocalised bared-teeth* displays, with which it shares the marked horizontal and vertical lip retraction. The vocalisations are mostly high-pitched and often loud. Depending on other (for instance temporal) characteristics these can best be denoted as screams, squeals, barks, geckers.

Fig. 1. *Silent bared-teeth* display by a 'submissive' crab-eating monkey (*Macaca irus*).

The *vocalised bared-teeth* display occurs not only in probably all primates (van Hooff, 1967), but also in most other mammals. It is phylogenetically one of the oldest facial expressions. In primitive mammals like marsupials (cf. Gewalt, 1966; Ewer, 1968), insectivores (cf. Herter, 1957), rodents (cf. Eibl-Eibesfeldt, 1957; Allin and Banks, 1968), primitive carnivores (cf. Hoesch, 1964; Pohl, 1967) and primitive primates (cf. Kaufmann, 1965; Epple, 1967) it is often the only facial expression. In primitive mammals the expiration is not always vocalised, but there may be a forceful 'spitting' or 'hissing' instead. Moreover, the lip retraction may not be very obvious, if the lips are not very prominent (cf. Ewer, 1968).

These animals characteristically show this intense vocalised display when they are subject to some threat or strong aversive stimulation. It is shown, for instance, in a situation of defence. Usually the actor manifests a strong or moderate tendency to flee (cf. Fig. 2). The display occurs especially when this tendency is thwarted, for instance when the actor simply cannot flee because it is cornered, or when other factors (e.g. the tendency to stay with youngsters or near some favoured possession) inhibit flight. In infants it may be a signal of discomfort.

With respect to the derivation of the display, i.e. to the original function of the different display elements that compose it, several possibilities exist that do not necessarily exclude each other. Strong expiration, lip retraction

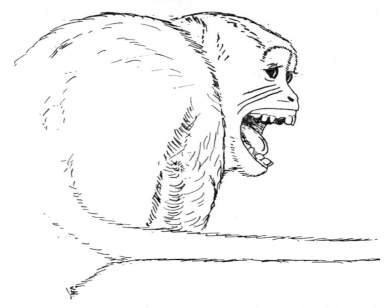

Fig. 2. *Bared-teeth scream* display in the crab-eating monkey (*Macaca irus*).

and other elements like tongue protrusion and horizontal headshaking are protective responses evoked by strong, aversive stimulation of the face and especially the oesophagus (Andrew, 1963). In the defensive posture the widely opened mouth and the baring of the teeth, which are usually accompanied by shrill barks or hisses, may be regarded as a preparedness to bite, should the attacker suddenly advance (cf. van Hooff, 1962). Given the strong vocalisation, the baring of the teeth may also be seen as a secondary effect of the vocalisation reflex, the muscles of mouth and throat region being tensed during strong vocalisation in order to protect the vibrating tissue (Andrew, 1963). The signal effect that the pattern elements undoubtedly also had may have led, during the course of evolution, to a generalisation of causes and a differential facilitation and ritualisation of the elements.

The intense *vocalised bared-teeth* display also occurs in the higher primates in identical situations. Here, however, it clearly may appear also in situations where the tendency to flee is not notably present and where other strong tendencies are thwarted. It may thus develop into a signal of general frustration and excitement (Andrew, 1963; van Hooff, 1962, 1967).

Of the elements of the compound display described, the horizontal and vertical lip-retraction and consequent teeth-baring has undoubtedly the lowest threshold. In many mammals that possess the *vocalised bared-teeth* display one may note that horizontal and, less frequently, vertical lip retraction occur in a fleeting manner, without the other elements which form the vocalised display. Moreover they may occur independently in the first and the last phases of the fully vocalised display.

In the more original arboreal types of higher primates (e.g. in *Cercopithecus*) a similar situation is found. A number of more advanced genera, notably *Macaca, Cynopithecus, Theropithecus, Mandrillus, Papio* and *Pan* (van Hooff, 1962, 1967), however, possess a distinct *silent bared-teeth* display ('grin', 'grimace'). Here, intense horizontal and vertical lip-retraction may be maintained for some time. Its occurrence in the course of the comparatively active social life of these genera indicates that it is a ritualised version of the low-intensity expression.

Whereas the *silent bared-teeth* display is a submissive gesture in most higher primates (for a review see also Spivak, 1968), a few species are now known in which the display can also be given by a dominant animal towards a subordinate (see Fig. 3) (*Macaca maurus, Cynopithecus, Pan* – van Hooff, 1967; *Mandrillus, Theropithecus* – van Hooff, 1967; Spivak, 1968). The context in which it then occurs suggests that it may function also as a reassuring signal or even as a sign of attachment in these species. Below I shall present some quantitative data that support this view for the genus that is nearest to us, the chimpanzee (*Pan troglodytes*).

In catarrhine monkeys reassurance and the expression of attachment are characteristically performed by the *lip-smacking* display. This expression has been evolved from the functional smacking which accompanies grooming (viz. when swallowing the particles which have been found) and perhaps also from the infantile suckling. The ritualised display differs from the functional smacking by its high, rather fixed rate and its typical intensity (see Chapter 4).

Particularly in the genus *Macaca* a combination of the jaw movements of smacking and the baring of the teeth has led to a new display, *teeth-chattering* (van Hooff, 1962, 1967). In many macaques, notably the more primitive long-tailed ones (e.g. *Macaca irus*) *teeth-chattering* is predominantly submissive and is performed by subordinate towards dominant group-members. However, in some of the more advanced short-tailed macaques, namely *M. sylvana* (Darwin, 1872; Lahiri and Southwick, 1966; van Hooff, 1967), *M. assamensis* (Osman Hill and Bernstein, 1969) and *M. speciosa* (Bertrand, 1968) this mixture of the *silent bared-teeth* display and the *lip-smacking* display can also be given by dominant animals towards

Fig. 3. *Silent bared-teeth* display during greeting by a 'confident' Celebes ape (*Cynopithecus niger*).

subordinates and between equals. The context suggests that it can serve as an expression of attachment and as a reassurement in these species.

In the apes *lip-smacking* plays a role of minor importance in comparison with macaques, baboons and related genera. I have never observed it in *Gorilla*. In *Pan* (the chimpanzee) it occurs only in the original context, namely in close connection with the act of grooming (van Hooff, in press). Only in *Hylobates* (the gibbon) can the *teeth-chattering* display be observed (van Hooff, 1967). In man, *lip-smacking* may have disappeared altogether, unless one regards the rhythmic tongue-protrusion that occurs in some cultures during courting as a last vestige of it (Eibl-Eibesfeldt, pers. comm.).

We may conclude that in the ascending scale of the primates leading to man, there is a progressive broadening of the meaning of the element of baring the teeth (cf. Andrew, 1963). Originally forming part of a mainly defensive or protective pattern of behaviour, this element becomes a signal of submission and non-hostility. In some species the latter aspect can become predominating, so that a reassuring and finally a friendly signal can develop. Correlated with this development is the tendency of *silent teeth-baring* to overlap functionally and, in some species also morphologically, with the *lip-smacking* display. At the end of the scale it has practically replaced *lip-smacking*. Our human *smile* appears to fit neatly at the end of this development.

2. *The relaxed open-mouth display*

Beside the *silent bared-teeth* display another display, the *play-face* (van Hooff, 1962; Andrew, 1963; Bolwig, 1964) or *relaxed open-mouth* display (van Hooff, 1967), is of interest. In the majority of primates it has much in common with the aggressive *staring open-mouth* display. It is likewise characterised by a rather widely opened mouth, and lips that remain covering all or the greater part of the teeth. It differs from the *staring open-mouth* display by the free and easy nature of the eye and body movements and by the fact that the mouth-corners are not pulled forward (Fig. 4). It is often accompanied by quick and shallow rather staccato breathing. In some species, the breathing may be vocalised (e.g. the chimpanzee). The vocalisations then sound like 'ahh ahh ahh'.

In all the primates in which it occurs the *relaxed open-mouth* display typically accompanies the boisterous mock-fighting and chasing involved in social play (Loizos, 1967). It can be regarded as a ritualised intention movement of the *gnawing* which is a characteristic part of the play of many mammals (see, for instance, Eibl-Eibesfeldt, 1957), and may function as a metacommunicative signal that the ongoing behaviour is not meant seriously, but is to be interpreted instead as 'mock-fighting' (Bateson, 1955) (cf. Fig. 5). In the chimpanzee the *relaxed open-mouth* display can easily be elicited by tickling, and many authors (e.g. Darwin, 1872; Foley, 1935; Kohts, 1937; Grzimek, 1941; Yerkes, 1943) were struck by its resemblance both in form and context with our laughter. The data suggest that both displays are phylogenetically closely related.

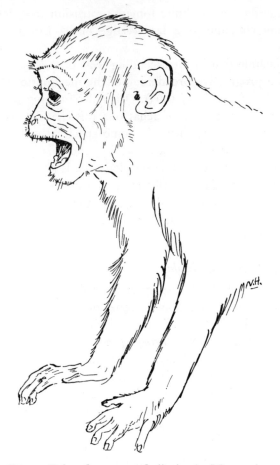

Fig. 4. *Relaxed open-mouth* display in *Macaca irus*.

POSSIBLE HOMOLOGUES IN THE CHIMPANZEE REPERTOIRE OF EXPRESSIONS

Before we discuss a phylogenetic hypothesis, a closer look at our nearest relative, the chimpanzee, may be illuminating.

Its *relaxed open-mouth* display (Fig. 6b) has been described extensively by several authors (e.g. Goodall, 1965; van Hooff, 1962, 1967; van Lawick-Goodall, 1968; Loizos, 1967, 1968). With respect to the *silent bared-teeth* display there is less information available on its exact role in social behaviour. I shall therefore present some data obtained by observing a semi-captive chimpanzee colony (eleven males and fourteen females; ages ranging from 3 to 16) living in a 10 ha enclosure at the Holloman Aeromedical Research Laboratory, New Mexico. In a recent study (van

Hooff, in press), I have described this group at some length, and I have discussed the methods of observing and recording. Behavioural interactions of pairs of animals were recorded in an appropriate short-hand writing. A great number of elements of social behaviour could be distinguished. I then determined to what extent these behaviour elements were causally related to each other.

Fig. 5. 'Gnaw-wrestling' (social play) by crab-eating monkeys (*Macaca irus*); the animal on the right shows the *relaxed open-mouth* display as an intention movement to gnaw.

Fig. 6 (*a*). *Horizontal bared-teeth* display in *Pan troglodytes*, the chimpanzee.
(*b*) *Relaxed open-mouth* display in *Pan troglodytes*.

The basic data are the behaviour protocols of the individual animals from which can be found how often each behaviour element is followed by each other behaviour element. Thus we can obtain the frequencies of transition between any pair of behaviour elements. Now if two elements had exactly the same causal basis, their transition frequencies with each of the other behaviour elements would be identical. The extent to which the causal bases of two elements in fact coincide can thus be assessed by comparing their frequencies of transition with the other elements. This is done by calculating correlation coefficients between the frequency patterns of these two elements: the coefficient will be $+1$ if the transition frequencies indicate identical causal bases, o if the causal bases are unrelated, and -1 if fully inhibitory relations are indicated. In this way a matrix of correlations between the behavioural elements can be obtained. The further analysis involves an assessment of the extent to which the matrix of correlation coefficients could be deduced from a smaller number of independent hypothetical basic variables. These basic variables may be referred to as 'systems', 'components' or 'causal categories', in part according to whether attention is focussed on the relations between the behavioural elements or the hypothetical causal factors which control them. The results of such a 'component analysis' depend on the situations studied, the types of behaviour sampled and, perhaps most important, on the extent to which one splits or lumps in selecting the elementary behaviour variables. These factors determine whether certain relations are expressed in terms of one general component or of more, independent, specific components. So these determine the 'hierarchical level' of the structure of causal relationships that one examines (van Hooff, in press). To the extent that the correlations can be deduced from the new categories, these can be said to explain the transition probabilities. The extent to which any one behavioural element is associated with each of the causal categories is described as its 'loading' on that category. This could vary from $+1$ (totally determined by that causal category) through o (totally uninfluenced) to -1 (inhibitory relations). The categories can be characterised by the behaviour elements with the greatest loadings on them. Thus if a category had high loadings from many elements, all associated with attacking another individual, it could be described as an aggressive category (or system or component).

Our research on chimpanzee social behaviour (van Hooff, 1970, and in press) yielded a categorisation of the behavioural elements in terms of a hierarchical structure of causal relationships. At a certain, rather high hierarchical level the fifty-three most frequently occurring behaviour elements appeared to be explainable in terms of five main independent causal categories. Of these the four most important categories could clearly be functionally interpreted and were designated as the 'affinitive', the 'play', the 'aggressive' and the 'submissive' system; the fifth could not and was given a neutral, more descriptive label: the 'excitement' system. At this high hierarchical level the *silent bared-teeth* display and the *relaxed open-mouth* display did differentiate clearly, as is manifest from their 'loadings' on the first four components; this indicates that they have different causal bases:

	affinitive	play	aggressive	submissive
silent bared-teeth	**0.65**	−0.19	−0.13	0.27
relaxed open-mouth	−0.12	**0.95**	0.04	−0.06

Moreover, of the *silent bared-teeth* display three types could be distinguished (see Fig. 7).

1. The *horizontal bared-teeth* display (**hbt**): strong horizontal and vertical retraction of the lips, so that the teeth and often the gums are maximally exposed; mouth closed (teeth meet more or less); practically no body movements; the partner is often looked at obliquely, indicating an evasive tendency.

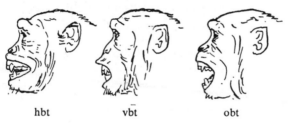

hbt   vbt   obt

Fig. 7. The three types of *silent bared-teeth* display of the chimpanzee, *Pan troglodytes*; from left to right: *horizontal, vertical* and *open-mouth bared-teeth* display.

2. The *vertical bared-teeth* display (**vbt**): no marked horizontal retraction, but there is vertical retraction, especially of the upper lip – it is lifted and often slightly protruding; again the teeth meet more or less; body posture and movements are relaxed; it is not a very conspicuous display.

3. The *open-mouth bared-teeth* display (**obt**): full horizontal and vertical retraction of the lips, so that teeth and gums show; mouth widely opened; movements calm and posture comparatively relaxed.

I have observed the *open-mouth* and the *horizontal* type in other chimpanzee groups as well; both have also been described by van Lawick-Goodall (1968) and Reynolds and Reynolds (1965) mention the horizontal type. So far I have seen the vertical type only in the Holloman colony, where it has also been recorded on film by Beckwith (pers. comm.).

I shall now consider a few additional data which illustrate the nature of these types. For each type of display the relative position of the actor and the reactor in the age hierarchy was determined. To this end the animals were divided into five age groups; animals of the same and adjoining age groups were regarded as equals. The results are given in Table 1.

Table 1. *Relative ages of participants in interactions involving silent bared-teeth display*

Actor	Reactor	**vbt**	**obt**	**hbt**
Elder	→ younger:	20	18	3
Equals:		5	15	14
Younger	→ elder:	6	12	23
Total:		31	45	40

The table shows that the *vertical* type (**vbt**) appears most frequently in the eldest members of an interacting pair, followed respectively by the *open-mouth* (**obt**) and the *horizontal* type (**hbt**). When the performer was the younger, the reverse was true ($p < 5\%$, two-tailed χ^2-test, in these and following cases). Only **vbt** showed a correlation with sex: out of a total of thirty-one recorded cases it appeared twenty-eight times in the five eldest males; sixteen of these times were in the eldest male.

Table 2. *Categories of behaviour preceding and succeeding the three types of* silent bared-teeth *display in the actor itself*

		Affin.	Play	Aggr.	Subm.	Excit.	Non-social
vbt	*prec*	11	0	3	0	1	16
	succ	11	1	0	0	0	19
hbt	*prec*	7	1	2	27	0	3
	succ	7	0	1	18	0	14
obt	*prec*	33	0	2	4	1	5
	succ	37	0	0	3	0	6

For each type of *silent bared-teeth* display I investigated which elements preceded and followed the display in the actor itself. Each of these preceding and following elements was grouped either in one of the five main causal categories in accordance with the results of the component analysis (see p. 220), or in a category including both non-social behaviour (as defined by van Hooff, in press) and the absence of any particular type of behaviour (as when an animal merely watched a partner). The results are given in Table 2.

The *open-mouth bared-teeth* display is typically associated with *affinitive* behaviour, and much more strongly so than the other two displays, with respect to both the preceding and following elements (for all differences:

$p < 1\%$). **Vbt** could be regarded as a low intensity form of **obt**, because it shows about the same relative association with the categories of social behaviour ($p < 1\%$), but it is more frequently preceded and followed by non-social behaviour. For instance **vbt** was shown a few times by an adult male upon meeting some youngsters, after which he simply walked on. **Hbt** differs from the other two, in that it is associated much more strongly with *submissive* behaviour elements ($p < 1\%$). The figures, though not significant, suggest that it is associated with a decrease in 'submissiveness'. Moreover, the display ends more frequently with non-social behaviour than it begins that way ($p < 1\%$).

For each type of *silent bared-teeth* display I also investigated which social behaviour elements (if any) occurred in the *reactor* both at the beginning of the display and at its end. Thus I tried to get an insight into the possible shifts in the partner's behaviour that are associated with the display. The results are given in Table 3.

Table 3. *Categories of behaviour occurring in the reactor at the beginning and at the end of each of the three types of* silent bared-teeth *display*

		Affin.	Play	Aggr.	Subm.	Excit.	Non-social
vbt	at begin	5	0	1	6	15	4
	at end	11	0	0	0	1	18
hbt	at begin	4	3	22	1	1	7
	at end	9	2	3	0	0	26
obt	at begin	22	0	2	4	2	12
	at end	39	0	0	0	0	5

Obt is quite definitely associated with *affinitive* behaviour in the partner; in this respect it differs from the other two displays ($p < 1\%$). Although there seems to be an increase of affinitive behaviour in correlation with all three *silent bared-teeth* displays, the only significant increase is that connected with **obt** ($p < 1\%$). In the case of **hbt** and **vbt** there is a significant increase of non-social behaviour (for all differences, $p < 1\%$). This is not so with respect to **obt**.

All three displays are associated with a decrease in *aggressive, submissive,* and *excitement* behaviour. In two cases the decreases are significant ($p < 0.1\%$): **vbt** is associated with a significant decrease of *excitement* behaviour, and **hbt** with a decrease of *aggressive* behaviour. **Vbt** and **hbt** each differ from the other two displays in the large frequency of respectively *excitement* and *aggressive* behaviour which precede them ($p < 0.1\%$).

With the data presented here the subject has certainly not been treated exhaustively, so that one cannot definitely conclude that the changes in behaviour of the partner are due to the displays. But it is very likely that the behavioural changes reflect the biological functions of the displays. With the *horizontal* and the *open-mouth bared-teeth* display the conclusions are in agreement with the qualitative evidence obtained from other groups. With the *vertical* display evidence from other groups is needed to establish whether or not it concerns an idiosyncratic trait of this group.

Fig. 8. An adult female chimpanzee shows the *open-mouth bared-teeth* display towards her child before embracing it. (After a photograph by Dr H. Albrecht.)

Nonetheless, the following conclusions seem justified. The *open-mouth bared-teeth* display is given by 'friendly' animals towards partners which usually are 'friendly' too. This tendency may even be strengthened: on a number of occasions two animals reciprocally showed the display and, for instance, then embraced (cf. Fig. 8).

The *horizontal* type is of a more submissive nature. Addressed to animals that tend to be aggressive, it may result in their leaving the displaying animal alone.

The *vertical* type seems to be a rather casual display by which mainly dominant males 'soothe' smaller animals, for instance, when these display the typical 'squat-bobbing' and utter the rapid 'oh-oh' vocalisations which are the typical representatives of the category of 'excitement' behaviour (van Hooff, 1970, in press); cf. Fig. 9.

None of the displays is clearly associated with play. This is in contrast to the *relaxed open-mouth* display: in 196 out of 224 independently recorded cases of the *relaxed open-mouth* display the partner responded with behaviour of the play category too.

Fig. 9. An adult male chimpanzee shows the *vertical bared-teeth* display towards two young females which are 'squat bobbing' excitedly in front of him.

A phylogenetic hypothesis of laughter and smiling

The data show that the chimpanzee *silent bared-teeth* display encompasses not only the phylogenetically old appeasing type, but also the more recent 'friendly' types. They confirm, moreover, that the chimpanzee in this respect is clearly at the extreme of the developmental range of the *silent bared-teeth* display in non-human primates (see p. 217).

Van Hooff (1967) has suggested that the similarity existing between this friendly version of the *silent bared-teeth* display and our human *smiling* on the one hand, and the similarity between the *relaxed open-mouth* display and our laughter on the other, each reflects a phylogenetic relationship. The human expressive movements certainly show an appreciable formal similarity with the displays mentioned. In most primate species that possess both displays they are, however, quite distinct, both in form and context. In the chimpanzee they have a low temporal association, and they belong even at a high hierarchical level to different behaviour systems (see p. 221). Such information as is available suggests that the situation is comparable in most primates. In *Erythrocebus patas* the *relaxed open-mouth* display is a common element of the behaviour repertory, whereas the *silent bared-teeth* display is absent and its place is taken by some gestural displays (Hall, Boelkins and Goswell, 1965).

The only exceptions to this rule amongst catarrhine primates, known to date, are *Mandrillus* (Schloeth, 1956; van Hooff, 1962, 1967), *Theropithecus* (van Hooff, 1967; Spivak, 1968) and perhaps *Cynopithecus* (unpubl.

Fig. 10. 'Gnaw-wrestling' (social play) by a pair of gelada baboons (*Theropithecus gelada*). Especially the male on the right shows strong teeth-baring, by turning its upper lip upwards and inside out. (After Spivak, 1968.)

observ.). These species possess the emancipated *silent bared-teeth* display (i.e. not only as a submissive but also as a reassuring signal). The *silent bared-teeth* face can also be noticed in playing *Mandrillus*; it then alternates with the true *relaxed open-mouth* face, which seems to occur more especially during actual play-wrestling. *Theropithecus* and *Cynopithecus* readily display the teeth and gums during play: we see here a mixture of the classical *silent bared-teeth* face and *relaxed open-mouth* face (see Fig. 10).

Human laughter and smiling also appear to shade into each other quite smoothly. They are undoubtedly highly associated temporally, and they are at least to a certain extent contextually interchangeable. From a purely morphological view-point our laughter can roughly be considered as an intermediate of the classical primate *relaxed open-mouth* display and the *silent bared-teeth* face (e.g. the chimpanzee open-mouth form), and the smile as a weaker form of it.

So, if the proposed homologies are correct, this means that laughter and smiling, though being of a different phylogenetic origin, must have converged and started to overlap considerably. Nevertheless van Hooff (1967) suggested that it might be possible to find situations where laughter and smiling are not interchangeable, and where specific causal and functional aspects which are in agreement with their different origins manifest themselves. Thus smiling might be typically used in the expression of sympathy, reassurance or appeasement (in accordance with the 'affinitive' *silent bared-teeth* display); while laughter occurs typically in the free and easy atmosphere of a comradely relationship where jokes come easily and everything is fun (in accordance with the 'playful' *relaxed open-mouth* display). Smiling in the latter context might be low-intensity laughter which, by causation and function, and perhaps also in form (see pp. 231–5), differs from the 'affinitive' smiling (cf. Plessner, 1942, 1953).

Practically no ethological research pertinent to this problem has been done, except for the work of Blurton Jones (Chapter 10, this volume). This author has grouped a number of social behaviour items of small children (2 to 4 years) by means of factor analysis. He found four categories of social behaviour patterns that correspond closely to the four most important ones I found during my chimpanzee study (van Hooff, 1970 and in press), namely 'rough and tumble play', 'aggression', 'crying' and 'social'. *Laughter* appeared to belong to the 'rough and tumble play' category, *smiling* to the 'social' category.

Another source of information is the judgement of these expressive movements by human subjects. Is there a difference in interpretation of the smile and laugh and if so is this in agreement with the phylogenetic hypothesis? This question I have investigated by means of the word questionnaire method.

THE JUDGEMENT OF LAUGHTER AND SMILING: AN INQUIRY

(a) Motivational categorisation of 'social words'

From a dictionary, ninety-nine adjectives and participles were collected, each referring to a certain social mood or attitude (for instance: 'afraid', 'affable', 'jovial', 'reassuring', 'nagging', 'roguish', 'meek', 'angry', 'ironical'). This collection was offered in a random order to a hundred subjects, who were asked to place each of the words in one of four categories, referring respectively to an 'aggressive', a 'submissive' or 'fearful', an 'affinitive' and a 'playful' mood or attitude. These four categories were selected because the component analysis of the structure of the social behaviour of the chimpanzee had revealed that practically all the social behaviour patterns of this species can be classified into these four categories (see above and van Hooff, 1970 and in press). A similar analysis of the social behaviour of *Macaca irus* (unpublished data) and qualitative evidence from the primate literature (see van Hooff, in press) suggest that these four main categories of social behaviour can be distinguished in many species. The factor analysis of the social behaviour of small children by Blurton Jones (Chapter 10) revealed corresponding categories (see above).

It is probable that in adult humans the categories 'playful' and 'affinitive' are less distinct, overt 'rough and tumble play' being a rare thing in them. However, irrespective of the hierarchical level at which these categories are independent, they can be used to find out whether laughter and smiling differ in the same respects as the *silent bared-teeth* and *relaxed open-mouth* displays differ in other primates.

For each word the inquiry yielded frequencies of scores in each of the four motivational categories, indicative of the social meaning of the word. Note that the motivation scores do not refer to an intensity of motivation, but to motivational 'purity' in terms of the chosen categories.

Table 4. *Frequency distribution of the scores for each of the four motivational categories*

Score magnitude	Aggressive	Submissive	Affinitive	Playful	Total
80–100	26	15	17	9	67
60– 79	6	7	3	1	17
40– 59	6	2	6	6	20
20– 39	5	7	10	4	26
0– 19	56	68	63	79	266
Total	99	99	99	99	396

The frequency distribution of the score magnitudes on each of the four categories is given in Table 4. The distributions are skew and tend to be U-shaped; the words tend to have high or low scores. There is a fair, though not equal, number of rather pure representatives of each category.

(*b*) *Association of the 'social words' and the expressions of laughter and smiling*

Besides the foregoing, four independent questionnaires were conducted during each of which about a hundred subjects were asked to establish for each of the words of our list respectively:

(1) whether the word indicates an attitude that can be accompanied by laughter ('possible context');

(2) the same with respect to smiling;

(3) whether the word indicates an attitude that is typically accompanied by laughter ('typical context');

(4) the same with respect to smiling.

So questions (1) and (2) reveal a possible association whereas (3) and (4) reveal a typical association. According to expectation, questions (1) and (2) yielded association-frequencies that were higher on the average than questions (3) and (4).

Smiling is undoubtedly more highly associated with the 'affinitive' mood than laughing. For, of a total of twenty-three words that had their highest score in the 'affinitive' motivational category, all had a higher association frequency with the smile than with the laugh in the typical context, and twenty-two did so in the possible context.

Laughter, by contrast, seems to be associated more highly with a 'playful' mood than smiling. For, of the fifteen words that had their highest score in the 'playful' category, twelve had a higher association frequency in the typical context with the laugh and three with the smile ($p < 1\%$). In the 'possible' context however, there were eight highest scores for the smile and seven for the laugh.

(c) Correlations between motivational scores and association frequencies

A similar picture emerges when the correlation coefficients are computed between, on the one hand, the association frequencies of both laugh and smile (in the 'typical' as well as in the 'possible' context) and, on the other hand, the motivation scores on the four categories. The results are given in Table 5; where relevant the significance of the difference between correlation coefficients has been given (as revealed by the $\bar{z}$-coefficient of Fisher, one-tailed; from Guilford, 1956).

The laugh and the smile are each relatively incompatible with both an aggressive and a fearful mood. The laugh is less negatively correlated with the aggressive than with the fearful mood and seems to differ in this respect from the smile. These trends, visible in both the 'typical' and 'possible' context, are however hardly significant.

The laugh and the smile are positively correlated with the 'affinitive' as well as the 'playful' category. They do differ however, in a number of respects. In accordance with the expectation, the differences manifest themselves particularly in the typical context. Thus the laugh is much more strongly correlated with the 'playful' than with the 'affinitive' mood, whereas the reverse, though insignificant, tendency exists with respect to the smile. The laugh is, moreover, significantly more strongly correlated with the 'playful' mood than the smile. By contrast the smile is significantly more strongly correlated with the 'affinitive' mood than the laugh.

(d) Evaluation of results

The results indicate that laughter and smiling are both primarily non-agonistic signals. In an affinitive mood a smile is much more likely and appropriate than a laugh. The latter characteristically marks a playful mood, but smiling in this situation is also likely.

Table 5. *Correlations (Pearson's r) between motivation scores and association frequencies. Correlations significant beyond the 1% level are in bold print. The significance of differences between pairs of correlation coefficients have been determined – where relevant – by means of the z̄-coefficient of Fisher, one-tailed (Guildford, 1956). The symbols used are:* ⟫⟩ *for* $p < 0.1\%$; ⟫ *for* $p < 1\%$; ⟩ *for* $p < 10\%$; / *for* $p > 10\%$

	Laugh typical		Smile typical	Laugh possible		Smile possible
Aggressive	−0.26	/	−0.44	−0.22	/	−0.33
	⌒		/	/		/
Submissive	−0.46	/	−0.43	−0.35	/	−0.26
Affinitive	0.15	⟪	0.53	0.23	⟨	0.42
	⩓		/	⌒		/
Playful	0.72	⟫	0.49	0.44	/	0.26

The specific aspects of the two displays may have been underestimated in this inquiry, since their overlap may in part be due to the fact that the subjects were left free to decide what was to be understood by a laugh or a smile. A check revealed that, although the subjects were rather unanimously of the opinion that vocalisation and the opening of the mouth are the main distinguishing features, they were not entirely agreed about the categorisation of the low intensity forms of the responses, especially of the 'diminutive of laughter', to be described on p. 232.

The results are thus in agreement with the phylogenetic hypothesis put forward above. However, they are also compatible with the view that laughter and smiling can be considered as expressions of different intensities of the same motivational state, if it is assumed that the categories of 'affinity' and 'play' reflect two ranges of the same primarily unidimensional scale, i.e. that it is not possible to distinguish independent specific causal factors. The dimensional analyses of Blurton Jones (Chapter 10) and van Hooff (1970, in press) show that this is unlikely in the child and the chimpanzee respectively. An indication that a unidimensional scale is also inappropriate for human adults can be found in the dimensional analyses of human recognition of emotions by Frijda (1968; pers. comm.). He found that the group of emotions recognised as pleasant comprised two clusters, which may correspond to our two positive categories. One was characterised by words like 'loving', 'endearing', 'gentle'; the other by words like 'gay', 'witty', 'amused', 'roguish' and 'coquettish'.

It is quite likely, though, that there is an implicative relation between the categories. Whereas the affinitive category (as indicated both by the chimpanzee data and by the human enquiry data) is connected with the establishment of a non-hostile or positive social bond, and the overcoming of social reserve, social play seems to require a social atmosphere that is relatively free of tension. Thus Kawai (1966) found that certain neurosurgical operations led to a reduction of dominance behaviour in adult macaques and to a reappearance of play. In introduction tests of chimpanzees (van Hooff, 1967) and irus macaques (unpubl. obs.) affinitive displays and interactions could be followed by social play behaviour; the latter was never first. So one could say that social play can occur at a certain level of familiarisation. Whether one regards the categories 'affinitive' and 'play' as differing by degree or in essence may therefore depend on the hierarchical level from which the motivational structure is assessed (cf. van Hooff, in press, for a discussion of hierarchy in this respect).

THE FORM OF SMILING AND LAUGHTER

In the foregoing I left without comment those views (see above) which, while regarding laughter and smiling merely as different degrees of intensity of the same phenomenon, imply that smiling is merely an attenuated form of laughter. Their similarity is indeed beyond doubt. During this investigation twenty students were asked to give a description of both laughter and smiling and to emphasise their differences; the vocalisations and the degree of opening of the mouth were unanimously declared to be the distinguishing features.

There are undoubtedly different forms of smiling (Grant, 1969; Brannigan and Humphries, 1969) and laughter. The human observer readily attributes a different 'meaning' to these different forms. The question is whether these different forms can be ranked in a unidimensional scale of increasing intensity. If this is not possible, it would be of interest to investigate whether the divergences are in accordance with the phylogenetic hypothesis set forth above.

Our research on this point is still in a preliminary stage. Although a detailed knowledge of the various forms of laughter and smiling (i.e. the way in which the different facial elements – see van Hooff, 1962 – can be combined) is still lacking, so that the number of dimensions along which the expressions may vary is still unknown, our qualitative observations on both nursery-school children and adults require us to consider at least two dimensions of variation.

First, there is a dimension which leads in its most intense form to the 'cheese' smile or *broad smile*. At low intensities only mouth-corner retraction may occur, the lips remaining closed. At higher intensities mouth-corner retraction becomes very marked; the mouth may open slightly and vertical retraction of the lips occurs leading to baring of the upper teeth and part of the upper gums and occasionally also of the lower teeth (see Plate 1 *a*. For a description of various types of smiles see also Grant, 1969). The *broad smile* may be seen, for instance, in greeting and has been recorded in this context by Eibl-Eibesfeldt (1968, and Chapter 11, this volume). It may then be accompanied by an eyebrow-lift and head-nod. It would seem that this type of smile, with active baring of the teeth, is associated particularly with the emphatic manifestation of a non-hostile, friendly attitude (e.g. during greeting, from adults towards children, when apologising, etc.).

At the extreme of the other dimension is the full-hearted *wide-mouth laugh* with relatively widely-opened mouth, mild to moderate baring of the upper teeth, mild retraction of the mouth-corners, characteristic 'ha ha' vocalisations, throwing backwards of head and a decreased tendency to maintain visual contact (Plate 1 *b*). At milder intensities it may appear in a diminutive form: the mouth is only slightly opened, and there is only slight retraction of the mouth-corners, and hardly any vertical retraction of the lips. The vocalisations also appear in a reduced form: they may consist of a series of chuckles or, at the lowest intensities, even of a little staccato unvocalised expiration, given through the nose if the lips are closed. Such casual chuckles and snorts occur frequently during normal relaxed social contact. In Plate 2 *a* a developmental range of such laughter is given. So, in the intensity range leading to the full-hearted *wide-mouth laugh* the silent *broad smile* does not necessarily occur.

The *wide-mouth laugh* is very characteristic of children's play. There it alternates with a typical expression that in its intense form is rare in adults. It is characterised by a wide open mouth, and only slight to moderate vertical and horizontal lip retraction, so that the teeth may or may not show.

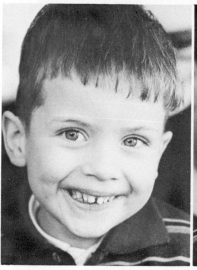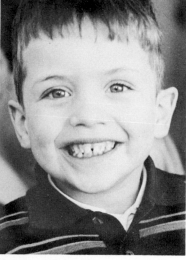

1 *a*   *Broad smiles* in a 5-year-old boy.

   b   Playful interaction in human children. Top: The active boy shows *play-face* with completely covered. teeth; the reactive girl shows low intensity *wide-mouth laugh*. Bottom: The girl reacts to the *pointe* with relaxed *wide-mouth laugh*.

(*facing p.* 232)

2a Left A *wide-mouth laugh* develops.

b Right The girl shows *broad-smile laughter* with intense baring of the teeth.

It is almost identical with the facial expression of the *wide-mouth laugh* but it is given silently without the vocalisations of laughter (see Plate 1*b*). Blurton Jones (1967), who first described it, noted its resemblance to the original primate 'play-face', as shown by, for instance, most of the macaques and the chimpanzee (cf. p. 217). It occurs especially in the anticipatory phase of social play both in the actor and reactor, to turn into vocalised laughter when the expected unexpected, the *pointe*, occurs. One

Fig. 11. Interspecific social play. Chimpanzee, in active role, shows *relaxed open-mouth* display, regularly accompanied by 'ah ah'-vocalisations. Boy, mainly in passive or reactive role, shows *wide-mouth laugh*.

can easily verify this by playing hide-and-seek or chase-play with a toddler. The resemblance with the chimpanzee is indeed striking (cf. Fig. 11). In the chimpanzee the staccato 'ahah' that may accompany the *relaxed open-mouth* display reaches its maximum also when some sudden move occurs or the grip is suddenly tightened during play. One can observe this best by playing with and tickling a tame chimpanzee.

It is clear that the variations within the smiling–laughter continuum can only be described in terms of a multi-dimensional model. A closer analysis is needed to reveal to what extent such expressive elements as the eyes (degree of opening; dynamics of looking), head posture (straight, slanting), vocalisations (relaxed, pressed, 'giggle') and various body movements can vary independently. It is conceivable that such variations could be related to changes in the general tendencies of withdrawal and aggression or to changes of more specific tendencies (e.g. nervous laughter, derisive laughter, etc.).

To account for the variations in the mouth and lip posture and the presence of vocalisation a model of at least two, and perhaps three dimen-

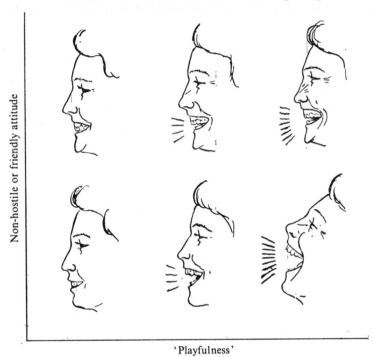

'Playfulness'

Fig. 12. Two dimensions of variation of the smile–laughter continuum. From bottom to top there is increased baring of the teeth. From left to right there is increased mouth opening and vocalisation.

sions is needed. In Fig. 12 a two-dimensional model is presented. The ordinate (vertical dimension) portrays the variation of the baring of the teeth; the abscissa portrays the opening of the mouth and the presence of the laughter vocalisation. The extreme morphological forms on the ordinate and the abscissa are respectively the *broad-smile* and the *wide-mouth laugh*. The *broad-smile laugh* of the upper right corner (cf. Plate 2b) can be well explained as a high intensity intermediate of the two forms. We could tentatively call the horizontal axis 'playfulness and mirth', and the vertical axis 'advertising of a non-hostile or friendly attitude'.

The existence of the forms that occupy the extreme places on both axes is a strong argument for the phylogenetic hypothesis; the *broad-smile* and the *wide-mouth laugh* appear to link up formally and probably also functionally with the *silent bared-teeth* and the *relaxed open-mouth* display respectively. In marked contrast with the situation in most other primates, however, the extremes *pur sang* are rarely seen in human adults. It is the intermediates and the rather difficult to distinguish low intensity forms that seem to dominate in everyday social interactions, or at least, there is frequent changing of these different forms into one another. Although it is impossible at this stage to make any quantitative assessment of this, it would seem that the extreme forms occur distinctly more frequently in children, and that in human adults there is a strong tendency to develop idiosyncratic stereotypes.

DISCUSSION AND CONCLUSION

The data from descriptive analysis support the view that laughter and smiling in the human species refer to a continuum of intergrading signals, the variation of which (at least with respect to such obvious characteristics as the mouth and lip posture, and the vocalisations) can roughly be described in at least two dimensions. The two extreme forms, the *broad-smile* on the one hand and the *wide-mouth laugh* on the other show formal, and most likely also functional, resemblance with respectively the *silent bared-teeth* display and the *relaxed open-mouth* display of the other primates.

The questionnaire data show also that the associations evoked in human subjects by laughter and smiling differ in the way expected on the basis of the proposed homology. Whereas an affinitive attitude, and the expression of positive intentions, is associated especially with smiling, a playful attitude is associated with both laughter and smiling (perhaps a bit more strongly with laughter). The questionnaire may not have adequately differentiated diminutive laughter and smiling. It would be of interest, therefore, to use this method to determine to what extent the morphologically different forms of smiling and laughter are judged differently by human subjects; this could be done by using photographs.

When the distinguishing features of laughter and smiling are considered in relation to the comparative evidence, especially of the closely-related chimpanzee for which quantitative data have been presented here, the agreement is obvious. Laughter then fits neatly in the phylogenetic developmental range of the *relaxed open-mouth* display, a metacommunicative signal, designating the behaviour with which it is associated as mock-aggression or play. Smiling fits well as the final stage of the development of the *silent bared-teeth* display. Originally reflecting an attitude of submission, this display has come to represent non-hostility and finally has become emancipated to an expression of social attachment or friendliness, which is non-hostility *par excellence*. The situations in which the *silent bared-teeth* display occurs have in common that a certain amount of uncertainty about the social relationship is overcome.

In normal human social interactions the forms of the smiling–laughter continuum which prevail can be considered as intermediate to those described as the typical extremes (this may well be more the case in adults than in children, but exact data are lacking). The amalgamation of the two original responses has even reached the point where their separate identities have almost been lost and are no longer easily recognised. I have pointed out that in this respect *Homo* differs from the majority of primates that possess both the displays under discussion. It is remarkable, however, that in a few catarrhine species, namely *Theropithecus*, *Mandrillus* and perhaps *Cynopithecus*, a parallel development has taken place (p. 226). The same seems to have occurred in some platyrrhine primates (cf. Andrew, 1963) and in the carnivores (Fox, 1970). These species also possess an emancipated *silent bared-teeth* display, i.e. an expression of non-hostility or even friendliness. Social play, one surmises, presupposes a social atmosphere that is free from too strong social tensions; the announcement by means of the *relaxed open-mouth* display, that the behaviour is to be understood as mock-aggression, parallels the *silent bared-teeth* display in that it implies a non-hostile attitude. This common element may have led to the further convergence of both displays. The proposed homologies and the implied levels of phylogenetic development are presented in the diagram of Fig. 13.

So far I have dealt with laughter and smiling as social displays, i.e. behaviour patterns that are released and/or directed with respect to one or more fellows. With a few exceptions (alarm barks, food grunts) the displays of primates convey information with respect to the actor's relationship to its social fellows. This certainly is true with respect to the *silent bared-teeth* and *relaxed open-mouth* display; indeed these displays are usually even released by and directed to the same partner. They have undoubtedly been shaped and ritualised within the frame of direct social interaction. Human laughter and smiling are undoubtedly also social in the sense that they occur especially in the company of others. Like many other human expressions, however, they often carry appreciable environmental information. We laugh when confronted with ludicrous things or situations, we may smile when confronted with something agreeable, tender or lovable. This emancipation of expressions, as compared with their homologues in the other primates, has undoubtedly been brought about by the selective pressures that have also influenced the development of human speech.

Many studies have been made of the early development of smiling and laughter in human infants (for a review see Laroche and Tcheng, 1963). Only a few authors have paid attention to the actual form of the expressions and specified what they understood by 'smile' and 'laugh'. Washburn (1929), who has given detailed descriptions, noted that the first expression to be generally interpreted as a social smile has an unusual shape. The mouth is widely opened, the lips are not much retracted, and the mouth

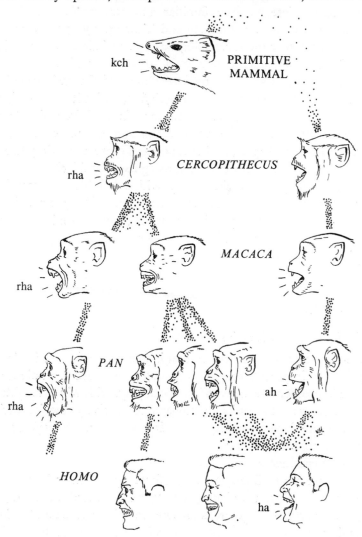

Fig. 13. The phylogenetic development of laughter and smiling as suggested by homologues in existing members of the phyletic scale leading to *Homo*. On the left is the speciation of the *silent bared-teeth* display and the *bared-teeth scream* display. The *sbt*-display, initially a submissive, later also a friendly response, seems to converge with the *relaxed open-mouth* display (on the right), a signal of play.

has a round aperture. Though mouth-corner retraction occurs soon after birth, the 'croissant' type of smile appears in social settings only at an age of 4 to 5 months. In the 'croissant' smile the mouth is not opened widely, but instead the mouth-corners are pulled back and slightly upwards. In true laughter and in the human 'play-face' the open-mouth posture remains. At about the time the *croissant-smile* appears, the human infant starts to differentiate between familiar and strange faces and smiling becomes selective (see e.g. Laroche and Tcheng, 1963; Gewirtz, 1965). One wonders whether the morphological differentiation corresponds with a motivational differentiation that is in agreement with the distinctions between laughter and smiling proposed above. Is the early *round-mouth smile* comparable to the *relaxed open-mouth face*, and is it the manifestation of a relaxed attitude when vital needs are fulfilled in which 'playful' interaction is appreciated? Does the *croissant smile* bear more an affinity to the *silent bared-teeth* display, and is it perhaps more a greeting display, in which the infant greets with relief the recognised familiar?

The present essay has certainly not provided definitive answers, but the comparative perspective in which the phenomena of laughter and smiling have been placed, and the questions that have come to mind as a result, may contribute to further clarification of the role of these displays in human non-verbal communication.

COMMENTS

1. *Social influences and laughter*

Radcliffe-Brown (1956) has stressed that in many societies, institutionalised joking plays an important role in abolishing friction from relationships in which it might otherwise arise. While social play, with which laughter may be associated, seems to *require* an atmosphere relatively free from friction, institutionalised joking serves to *produce* such an atmosphere (Ester Goody, pers. comm.).

Van Hooff comments:

'I agree with Radcliffe-Brown. In a former publication I have pointed out that man indeed uses laughter to "break the ice" in social gatherings, like a party, where there is *supposed* to be a relaxed and tension-less atmosphere.

Whereas in animals the expressions reflect the "true" development of the motivational attitude between social partners, man is often confronted with situations where convention imposes a certain social attitude upon him. This may have facilitated the use of laughter, more or less consciously, as a means of expressing or pretending that the actor wants to *engage* in a "playful" relationship, rather than expressing that the actor *is* in a playful mood; the mutual feed-back between the inter-actors may then help to really establish such a relationship. So this development may have contributed to the functional convergence of laughter and smiling.'

2. *Bases of laughter*

Van Hooff criticises the view that laughter is a means to discharge surplus 'tension' or 'mental excitation' on a number of grounds. It may be added that such a view is incompatible with current views on the nature of motiva-tion (see Preface to Chapter 7).

3. *Conflict in smiling and laughter*

Andrew comments:

'The division of human smiling into a silent "broad smile" corres-ponding to the silent bared-teeth display of chimpanzees, and slight smiles which precede laughter, is clearly a step forward, and preferable to the earlier lumping of all smiles by various authors, including myself. However, it remains to be demonstrated that it is correct to assume that complex displays, in which a number of parameters vary along continua, are best described in terms of two extreme fixed action patterns, whose varying superimposition produce all intermediate states. Such descrip-tions are possible (e.g. Lorenz on the dog and Leyhausen on the cat), but are attractive only if one accepts a theory of causation which postu-lates the interaction of two main causal variables, each of which when active almost to the exclusion of the other produces one of the fixed action patterns. Aggression and fear are traditional for the cases cited; I believe such a model to be unhelpful even then [Chapter 7] but it would be doubly so in the case of human smiles and laughter.

I agree strongly with van Hooff that there is no evidence of association between typical laughter and attack behaviour. Claims that laughter has aggressive motivation rest on assertions that it is usual during jeering or mocking by a group. No accurate descriptions of such behaviour have been published: I would predict that when they become available they will show that laughter is usual only when the human who is mocked is being treated as an object to be viewed or manipulated as the material of a joke, and that laughter disappears when attacks (e.g. pushes, kicks, stone-throwing) develop. Comparison with bird-mobbing should be abandoned until it is demonstrated that the calls of mobbing are caused (in any sense whatsoever) by aggression.

I am not convinced that human laughter represents a superimposition of the relaxed open-mouth display (play face) and the silent bared-teeth display. The former is quite recognisable in playful or joyful excitement in man as an independent expression.'

It may be added that the approaches of Andrew and van Hooff are not necessarily exclusive, but supplement each other. Andrew is interested in the causation of and information available in the elements of the displays, especially in their original and unritualised form, whereas van Hooff here concentrates on the social meaning of the whole pattern, and its evolution.

4. *Phylogeny of laughter*

Andrew comments:

'It would be best at present to consider the vocalisation of laughter and its associated facial reflexes as without close equivalent in the chimpanzee. In view of the evolution of language in the human line, it is likely that our ancestors would have differed more from the chimpanzee in their vocalisations than in other displays. Comparison with other 'infectious' calls such as grunt-choruses in baboons [Andrew, 1963] or chimpanzee choruses (despite their obvious differences from laughter) could be revealing. If laughter is considered separately from the broad smile, then associated mouth corner retraction may be related to the retraction which is a vocalisation reflex in a great many other mammals [Andrew, 1963], and the possibility remains that this is involved in the silent smiles, or smiles with chuckles, which precede laughter.

The assumption that the silent bared-teeth display (SBT) is more recent than the bared-teeth scream, and derives from the latter, is not yet clearly established. It seems more likely that mouth corner retraction as a protective reflex in the startle response (which is at least as ancient as mouth corner retraction as a vocalisation reflex) has given rise to the SBT. No doubt the two reflexes have overlapped and interacted through-out evolution, but the point is not an entirely trivial one. Once comparisons are made between the human smile ('broad smile'), and protective reflexes given in greeting, the need to postulate drastic changes in causation during evolution is greatly reduced.'

While Andrew's interpretation is certainly a plausible alternative, van Hooff comments that it is at variance with the fact that baring of the teeth is much more marked in a form of laughter, the broad-smile laugh, than in the 'deep' and louder, whole-hearted wide-mouth laughter. In the latter, retraction of the lower lip is often slight, suggesting that the muscles which are supposed to protect the throat are not much contracted. Van Hooff did not actually claim that the vocalisation reflex is older than the protective baring of the teeth, only that they are closely associated. He is confident, however, that *ritualised* teeth-baring as a display is more recent than the bared-teeth scream display.

Ambrose (1963) regarded laughter as a mosaic of smiling and crying. In a further discussion of this question in Chapter 10 of this volume, Blurton Jones finds that there is insufficient evidence to establish whether laughter is more closely linked to crying or to screaming.

9. NON-VERBAL COMMUNICATION IN HUMAN SOCIAL INTERACTION

MICHAEL ARGYLE

Department of Experimental Psychology, Oxford University

The discovery of the importance of non-verbal communication (NVC) has transformed the study of human social behaviour. Until quite recently social psychologists were baffled by the subtleties and intricacies of social interaction, and often analysed it in terms of how long encounters lasted, or who spoke most often. Now a new level of analysis has been opened up – the level of head-nods, shifts of gaze, fine hand-movements, bodily posture, etc. (as well as a similar detailed analysis of the verbal component). This kind of research started in the early 1960s; rather later we realised that ethologists, especially those doing field studies of primates, were using very similar variables. It looked as if human NVC was similar to animal social behaviour, and perhaps conveyed similar messages. Humans also have a verbal channel of communication, but other research showed that speech is accompanied by an intricate set of vocal and gestural non-verbal signals, which affect meaning, emphasis, and other aspects of utterances. This kind of research has led to fruitful collaboration with linguists.

RESEARCH METHODS

There are several different research strategies, corresponding to the conceptual assumptions of different groups of investigators.

(1) Those who think that non-verbal communication is a kind of language have tried to discover its elements and structure, rather than look for empirical laws and cause–effect relations. They have analysed short sequences of tape or film in great detail, but without statistical treatment (e.g. Scheflen, 1965).

(2) A second group of investigators has been concerned with the rules (i.e. implicit cultural conventions) governing verbal and non-verbal behaviour in different situations: this has been linked with an interest in the way people perceive or define situations and interpret NVC. This has led to rather informal studies of particular field situations and the analysis of etiquette books (Goffman, 1963), as methods of discovering the underlying, often unstated, rules. Another approach has been the deliberate breaking of conventions to show that they are there – for example students behaved at home as if they were lodgers, and moved their opponents' pieces at chess (Garfinkel, 1963). It is possible to do this kind of thing in a

more rigorous way: Felipe and Sommer (1966) carried out experiments in which personal space was invaded – an experimenter sat down close to 'subjects' on a park bench without explanation – the subjects all left very rapidly. Systematic research on the subjective interpretation of NVC has been done by social psychologists working on person perception.

(3) Social psychologists have developed a tradition of experimentation which consists of very well-controlled studies, conducted under very artificial conditions. Subjects may sit in cubicles by themselves, watch flashing lights and press buttons; often there is no verbal communication, no NVC, no real motivation, and there are no situational rules. Single cues are isolated, and their causes or effects studied, in extremely elegant experimental designs. However, the results obtained may be misleading in a number of ways. First, the results may be exaggerated. Argyle and McHenry (1970) replicated the previous finding that when people are seen briefly, the wearing of spectacles adds about 13 points to their judged IQ; however when they were seen talking for 5 minutes, spectacles made no difference to the judged IQ. Secondly, the results may be wrong: some results which have been obtained under laboratory conditions have not been confirmed under more realistic conditions. For example, the reinforcing of amount of speech or other aspects of verbal behaviour by means of head-nods and smiles has been found to work in the laboratory *only* when subjects become aware of what is wanted, but under field conditions learning takes place without subjects being aware of what is going on (Argyle, 1969). Another example is research on the perception of non-verbal cues, which has often been conducted under extremely artificial conditions. Subjects may be shown photographs of faces with different facial expressions and asked to judge emotions. Such experiments overlook the fact that facial expression is normally carefully controlled, and that changes in expression during interaction are often not due to emotions, but are part of the non-verbal signalling system that accompanies speech. There is a dilemma about experimental research which it is sometimes impossible to evade: in order to test certain hypotheses it is necessary to set up peculiar experimental conditions, but if this is done the results obtained may not be true to real-life behaviour. It may be impossible to achieve both internal validity and external validity at the same time.

The approach which is recommended here is to carry out rigorously-designed experiments, which test hypotheses, but which are carried out in realistic settings with clear meanings and conventions, and which contain all the main ingredients of ordinary social behaviour. There are a number of research procedures which meet these criteria.

Field experiments on unsuspecting subjects

Here one or more trained stooges approach a member of the public, who may be walking down the street, sitting in the park, etc. The stooges behave according to a standard procedure, and the subjects' behaviour is either recorded by them or an observer, or is filmed. Experimental variations are introduced by varying the behaviour or appearance of the stooges or features of the setting. Sissons (1970) for example made films of an actor who asked ninety people at Paddington station the way to Hyde Park; in half the interviews he was dressed and spoke in an upper-middle class manner, in half he appeared to be working class; the social class of the respondents was ascertained from a second interview.

Laboratory experiments which replicate real-life situations

Subjects are invited to the laboratory to take part in an experiment. They are then asked to take part in an interview or discussion or some other social situation, whose conventions are familiar to them. They meet real people, and real motivations can be aroused. One version is for them to meet a programmed confederate, as in the previous design. Another is for two or more genuine subjects to meet, and for some other feature of the situation to be varied, such as the distance between them, or the topic of conversation or the task. Rather more realism can be introduced by the 'waiting room' technique – subjects encounter programmed stooges in the waiting room.

Role-played laboratory experiments

There is no sharp line between the experiments just described, and those in which subjects are asked to pretend or imagine that they are in a certain social situation. In perceptual experiments they may be shown a video-tape and asked to imagine that they are meeting the person shown on the monitor. Other versions are less acceptable – when subjects are shown photographs, drawings or stick figures and are asked about the emotional states of those portrayed or the relations between them.

Statistical analysis of interaction sequences

These were discussed earlier. While such studies cannot always involve the manipulation of experimental variables, it is possible to test hypotheses from the data obtained – though the direction of causation may be ambiguous. Kendon (1967) for example found that speakers looked up at the ends of utterances. This led him to make a further analysis of what happened when speakers did *not* look up in this way: he found, as predicted, that there was a long pause before the other replied.

We will now describe the basic arrangements for those investigations which can be done in a laboratory. A typical observation room plan is shown in Fig. 1.

The experimental room is decorated to be appropriate for whatever situation is to be replicated; it should not be a white 'laboratory room'. Observations are made either by (1) TV camera and video-tape recorder,

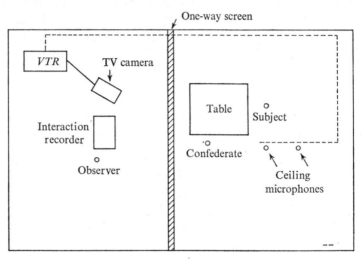

Fig. 1. Observation room arrangements.

(2) ciné-camera, (3) observers pressing buttons attached to some kind of recording equipment, such as a Rustrak recorder, or electronic counters to record number and duration of gazes, or other observable events, or by (4) audio-tape-recorder attached to ceiling microphones, to record length and duration of utterances. (See Fig. 1.)

THE MAIN NON-VERBAL SIGNALS USED BY MAN

Men use a number of different kinds of NVC, and it is convenient to classify them under ten headings. Each plays a distinctive role in social interaction; brief notes will be given on how each of them functions, with illustrative experiments. More details may be found in Argyle (1969).

Bodily contact

This may take a number of forms – hitting, pushing, stroking, etc. – most of which may involve a variety of areas of the body. There are great cross-cultural variations in the extent to which bodily contact occurs; in Britain and Japan there is very little, whilst amongst Africans and Arabs there is a lot. The most common bodily contact to occur in public settings in Britain is that involved in greetings and farewells. In most cultures bodily contact is far more common inside the family, between husband and wife and between parents and children. Even here there are tight restrictions on which part of the body may be touched by whom: Jourard (1966) found that male American white students were touched by their fathers only on the hands, though they were touched by opposite sex friends much more extensively.

Proximity

How close people sit or stand can easily be measured, but a considerable body of experimental work has yielded rather meagre results. It is found that people stand somewhat closer to people they like, and to those whose eyes are shut. However the differences of proximity involved are very small, a matter of 2–3 inches on average. There are much greater cross-cultural variations, in that Latin Americans and Arabs stand very close, while Swedes, Scots and the English stand much further apart. There are also consistent individual differences, but these appear to be unrelated to other aspects of personality, apart from a tendency for maladjusted people to be more distant (Lott, Clark and Altman, 1969). Porter, Argyle and Salter (1969) found that proximity communicates very little about an interactor: stooges who sat at 2 ft, 4 ft, and 8 ft were not perceived as different in personality. On the other hand when a number of people are present, proximity is found to reflect and probably communicate the relations between them (Kendon, pers. comm.). *Changes* in proximity communicate the desire to initiate or terminate an encounter: if *A* wants to start an encounter with *B* he will move closer, though this must be accompanied by appropriate gaze and conversation.

Orientation

This is the angle at which people sit or stand in relation to each other. The normal range is from head-on to side-by-side, and orientation can be assessed by asking a subject to meet a stooge or by asking two people to meet. Orientation has been found to vary with the nature of the situation – those who are in a cooperative situation or who are close friends adopt a side-by-side position; in a confrontation, bargaining or similar situation, people tend to choose head-on; while in other situations 90° is most common in England and the USA (Sommer, 1965; Cook, 1970). The main exception to this is that two close friends will sit head-on when eating. There are cross-cultural variations in that Arabs prefer the head-on position (Watson and Graves, 1966), and Swedes avoid the 90° position (Ingham, 1971).

Appearance

Many aspects of personal appearance are under voluntary control – clothes, hair and skin, while other aspects are partly so – physique and bodily condition. Furthermore much time, money and effort is put into the control of appearance, and this can be regarded as a special kind of NVC. The main purpose of manipulating appearance seems to be self-presentation, i.e. sending messages about the self. Thus people send messages about their social status, their occupation, or the social group they belong to, by wearing the appropriate costume – bank managers do not dress up like hippies. Appearance also conveys information about personality and mood – euphoric extroverts do not wear dark suits with black ties. Young women use all these signals, but their main concern is probably rather a different one – maximising their attractiveness as sexual objects. Appearance is meaningful only within a particular social setting where the significance of details of dress, hair or cosmetics is generally understood. Within modern cultures these fashions change extremely fast, so that simply being up to date becomes itself a main dimension of appearance.

Posture

In any given culture, many different ways of standing, sitting or lying are possible. To some extent posture has a universal meaning, like facial expression, but it also has a culturally defined meaning. There are conventions about the posture to be adopted in particular situations, such as church, dinner parties, etc. Posture is used to convey interpersonal attitudes: Mehrabian (1968) found that distinctive postures were adopted for friendly, hostile, superior, and inferior attitudes, and that these were perceived accordingly. Thus posture can be a signal for status; someone who is going to take charge sits in an upright posture (and in a central position, facing the others). Posture varies with emotional state, especially along the dimension tense–relaxed. This is of some importance since posture is less well controlled than face or voice, and there may be 'leakage', as, for example, when anxiety does not affect the face, but can be seen in posture (Ekman, 1969a).

Head-nods

We now come to the faster-moving non-verbal signals. A simple and seemingly minor signal is the head-nod, which plays a very important role in connection with speech. It usually acts as a reinforcer, in that when a piece of A's behaviour is followed by a head-nod from B, A tends to increase the frequency of that behaviour. Head-nods also play a crucial role in 'floor-apportionment', in that a head-nod gives the other permission to carry on speaking. On the other hand, rapid head-nods indicate that the nodder wishes to speak. Lastly, head-nods, like other bodily movements, are coordinated between two interactors, so that they appear to be taking part in a 'gestural dance'.

Facial expression

The face is a specialised communication area, which in non-human primates is used to communicate inter-individual attitudes and emotions. Much facial expression of emotion in humans appears to be culturally universal and largely independent of learning (Ekman, 1969b; see also Chapter 11, this volume). However there are considerable restraints on the expression of negative attitudes or emotions, so that spontaneous expressions are often concealed. However, some aspects of emotional expressions are very difficult to control – expansion of the pupils during arousal, perspiration during anxiety, and 'micromomentary' expression of concealed feelings (Haggard and Isaacs, 1966). Facial expression is also used in close combination with speech. A listener provides a continuous commentary of his reactions to what is being said by small movements of the eyebrows and mouth, indicating puzzlement, surprise, disagreement, pleasure, etc. A speaker accompanies his utterances with appropriate facial expressions, which are used to modify or 'frame' what is being said, showing whether it is supposed to be funny, serious, important, etc. (Vine, 1971).

Gestures

The hands are able to communicate a great deal: movements of head, feet and other parts of the body may also be used, but are much less expressive than those of the hands. Some gestures may indicate general emotional arousal, which produces diffuse bodily activity, while others appear to be expressions of particular emotional states, e.g. clenching the fists in anger. Gestures are also closely coordinated with speech and are made by a speaker to illustrate what he is saying, particularly when his verbal powers fail, or when objects of special shapes or sizes are being described. Hand (and head) movements may be closely coordinated with speech to indicate the internal structure of utterances, and to control the synchronising of utterances. Gestures can even replace speech, as in gesture languages.

Looking

During conversations each participant looks intermittently at the other, for periods of 1 to 10 seconds, for 25%–75% of the time: periods of mutual gaze, or eye-contact, are rather shorter. It is found that people look about twice as much while listening as while talking.

Looking plays an important role in communicating interpersonal attitudes and establishing relationships. The act of looking sends a signal to the other that a certain amount of interest is being taken in him, and interest of a kind which is signalled by the accompanying facial expression. Argyle and Dean (1965) postulated that amount of looking is a signal for intimacy, and found that people look more when another person is more distant, suggesting that looking and proximity can substitute for each other as signals for intimacy. This has been confirmed by later experiments; the effects of distance on individual gaze and eye contact are shown in Fig. 2.

Exline and Winters (1965) found that subjects looked more at people

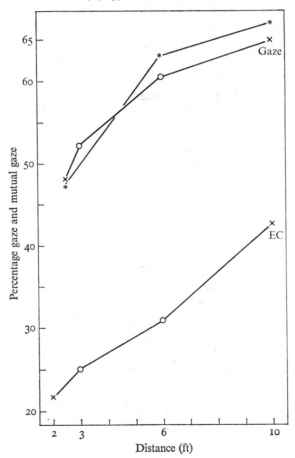

Fig. 2. Amounts of gaze at different distances from another person; gaze = individual gaze; EC = eye contact; * from Argyle and Dean (1965); ×, ⊙ – from Argyle and Ingham (in press).

they liked. However, looking can be accompanied by quite different facial expressions, and can signal aggression (as is common in animals), sexual attraction, or clinical interest for example.

Looking is closely coordinated with verbal communication. In the first place it is used to obtain information: feed-back on the other's responses while talking, extra information about what is being said while listening. In addition, shifts of gaze are used to regulate the synchronising of speech (p. 255). Gaze is used as a signal in starting encounters, in greetings, as a reinforcer, and to indicate that a point has been understood.

Non-verbal aspects of speech

The same words can be delivered in quite different ways by variations in pitch, stress and timing. Linguists distinguish between prosodic sounds which affect the meaning of utterances, and 'paralinguistic' sounds which convey other kinds of information (Crystal, 1969; see also Chapter 2 and comments on Part A, p. 91). Prosodic signals are pitch pattern, stress pattern, and juncture (pauses and timing) which affect the meaning of sentences, and are regarded as true parts of the verbal utterance (Lyons, pp. 52–3). Paralinguistic signals include emotions expressed by tone of voice, group membership expressed by accent, personality characteristics expressed by voice quality, speech errors, etc. These non-verbal signals are not closely linked with language, do not have a complex structure, and are similar to other expressions of attitudes and emotions. The emotional state of a speaker reading a neutral passage can be recognised from a tape-recording (Davitz, 1964). Thus anxious people speak fast and in a breathy way, i.e. with a high frequency distribution and with speech errors. A dominant or angry person speaks loudly, slowly and with a lower frequency distribution (Eldred and Price, 1958). There are speech styles which consist of other combinations of the same variables – the speech of surly adolescents, bright hostesses, etc.

THE DIFFERENT FUNCTIONS OF NVC

NVC in man is used to manage the immediate social situation, to support verbal communication and to replace verbal communication.

Managing the immediate social situation

Animals conduct their entire social life by means of NVC, and it appears that humans use rather similar signals to establish a similar set of relationships.

Interpersonal attitudes. These are attitudes towards others present – the main dimensions are found to be inferior–superior, and like–dislike. A superior attitude can be conveyed by (*a*) *posture* – body erect, head raised,

(b) *facial expression* – unsmiling, 'haughty', (c) *tone of voice* – loud, resonant, 'commanding', (d) *appearance* – clothes indicating high status and
(e) *looking* – staring the other down. The author and his colleagues
compared the effects of verbal and non-verbal signals for communicating
interpersonal attitudes. Typed messages were prepared indicating that the
speaker was superior, equal or inferior; video-tapes of a performer

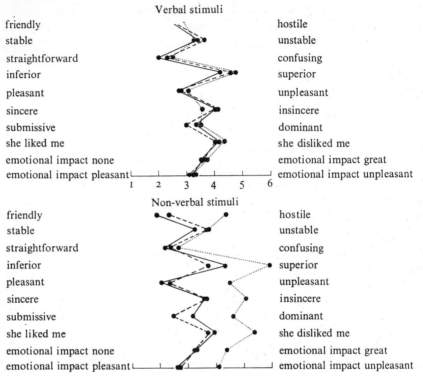

Fig. 3. Ratings of initially equated verbal and non-verbal cues for inferior and
superior attitudes when combined (Argyle *et al.* 1970). ——— = inferior, – – – = equal,
. . . . = superior.

counting (1, 2, 3 . . .) were made, conveying the same attitudes; the
verbal and non-verbal signals were rated by subjects as very similar in
superiority, etc. The combined signals were presented to further subjects
on video-tape – superior (verbal), inferior (non-verbal) etc., nine combinations in all, and rated for superiority. It was found that the variance due
to non-verbal cues was about $4\frac{1}{2}$ times the variance due to verbal cues, in
effecting judgements of inferior–superior (Argyle *et al.*, 1970). The
results of this experiment are shown in Fig. 3. Similar results were obtained in later experiments using friendly–hostile messages (Argyle, Alkema
and Gilmour, in press).

Emotional states. These can be distinguished from interpersonal attitudes in that emotions are not directed towards others present, but are simply states of the individual. The common emotions are anger, depression, anxiety, joy, etc. An anxious state, for example, can be shown by (*a*) *tone of voice* (see p. 251), (*b*) *facial expression* – tense, perspiring, expanded pupils, (*c*) *posture* – tense and rigid, (*d*) *gestures* – tense clasping of objects, or general bodily activity, (*e*) *smell* – of perspiration, and (*f*) *gaze* – short glances, aversion of gaze. Interactors may try to conceal their true emotional state, or to convey that they are in some different emotional condition, but it is difficult to control all of these cues, and impossible to control the more autonomic ones. Emotional states can be conveyed by speech – 'I am feeling very happy' – but probably statements will not be believed unless supported by appropriate NVC, and the NVC can convey the message without the speech.

Self-presentation. Information can be sent about an interactor's status, group membership, occupation, personality, or sexual availability. A person might want to be seen as an eccentric, upper-class, inventor, or as an important, left-wing, intellectual. This can be done by (*a*) *appearance* – especially clothes, (*b*) *NV aspects of speech* – especially accent, loudness, speed, etc. and (*c*) *general style* of verbal and non-verbal performance. A lot of self-presentation is concerned with role-distance i.e. showing there is more to a person than can be seen from his present role performance (Goffman, 1961). Again self-presentation *can* be done in words – 'look here young man, I've written more books about this than you have read' – but common experience suggests that it is not very effective. Self-presentation by NVC is basically a matter of using signals that are understood to stand for the real thing (Goffman, 1956), though it is clearly better to display the actual qualities where this is possible.

Thus in this context it may be suggested that NVC and verbal communication normally play two contrasted roles. NVC is used to manage the immediate social relationship – in much the same way as in animals; verbal communication is used to convey information connected with shared tasks and problems – though it is also used to give orders and instructions. However NVC *can* convey information, as in gesture languages, and the verbal channel *can* sustain interpersonal relations, as in informal chat, and it can communicate attitudes and emotions by the words chosen.

Sustaining verbal communication

Speech plays a central role in most human social behaviour, but many linguists do not always appreciate the importance of the role played by

NVC in conversation. However Abercrombie says 'We speak with our vocal organs, but we converse with our whole body' (1968:55), and Crystal (1969) and Lyons (Chapter 3 of this volume) recognise clearly the importance of non-verbal signals.

Vocal and kinesic NVC which affects the meaning of utterances

A person would not be accepted as speaking a language properly, indeed he would scarcely be understood, if he did not deliver his sentences in the pitch-pattern, stress-pattern and temporal pattern of grouping and pausing, proper for that language. We showed above that the meaning of sentences depends on their prosodic features. Similar considerations apply to kinesic signals, which can affect the meaning of a sentence by (1) providing the punctuation, displaying the grouping of phrases and the grammatical structure, (2) pointing to people or objects, (3) providing emphasis, (4) giving illustrations of shapes or movements, (5) commenting on the utterance, e.g. indicating whether it is supposed to be funny or serious (Ekman and Friesen, 1967).

Birdwhistell (1952) provided a system for recording 'kinesics' (bodily movements) on the analogy of the transcription methods used in phonetics; and Scheflen (1965) postulated that NVC has a hierarchical, three-level, structure corresponding to sentences, paragraphs, and longer sequences of speech. The method used by these investigators was to take a film or tape-recording of a fairly short sample of behaviour, and to make a very detailed but non-statistical study of what was happening.

Table 1. *Equivalent verbal and non-verbal units*

	Verbal	Non-verbal
1.	paragraph, or long unit of speech	postural position
2.	sentence	head or arm position
3.	words, phrases	hand movements, facial expressions, gaze shifts, etc.

(Cf. Condon and Ogston, 1966.)

It has been found that speech is accompanied by bodily movements; these are coordinated with speech in that a sentence may be accompanied by related hand or head positions. These movements also have a hierarchical structure, in which smaller verbal and bodily signals are organised into larger, and coordinated, groupings of both (see Table 1). Frame-by-frame analyses of small movements of hands, head, eyes, etc. in relation to speech, have provided evidence of 'interactional synchrony', i.e.

coordination of bodily movements between speaker and listener over periods of time corresponding to sentences, and even words (Kendon, 1970, 1971 *a*, *b*). However these movements are somewhat idiosyncratic, though there are cultural uniformities. Since most of them can be seen only in peripheral vision, their function is rather mysterious. Kendon (1971 *b*) suggests that in addition to displaying the structure of utterances they (1) make the speaker more interesting, and hold the listeners' attention, and (2) give advance warning of the kind of utterance that is to come.

Floor-apportionment. When two or more people are conversing, they take it in turns to speak, and usually manage to achieve a fairly smooth 'synchronising' sequence of utterances, without too many interruptions and silences. When people first meet, it is unlikely that their spontaneous styles of speaking will fit together, and there is a period during which adjustments are made – one person has to speak less, another has to speak faster, and so on. This is all managed by a simple system of non-verbal signalling, the main cues being nods, grunts, and shifts of gaze. For example, at a grammatical pause a speaker will look up, to see if the others are willing for him to carry on speaking – if they are they will nod and grunt. Just before the end of an utterance a speaker gives a rather more prolonged gaze at the others as is shown in Fig. 4. This shows the average amount of other-directed gaze by each person just before and after *A* stops speaking and *B* starts speaking, in a number of conversations analysed by Kendon. If this system fails, interruptions will take place, and there is a struggle for the floor (Kendon, 1967).

Feedback. When someone is speaking he needs intermittent, but regular, feed-back on how others are responding, so that he can modify his utterances accordingly. He needs to know whether the listeners understand, believe or disbelieve, are surprised or bored, agree or disagree, are pleased or annoyed. This information could be provided by *sotto voce* verbal muttering, but is in fact obtained by careful study of the other's face: the eyebrows signal surprise, puzzlement, etc., while the mouth indicates pleasure and displeasure. When the other is invisible, as in telephone conversation, these visual signals are unavailable, and more verbalised 'listening behaviour' is used – 'I see', 'really?', 'how interesting', etc. (Argyle, Lalljee and Cook, 1968).

Signalling attentiveness. For a conversation to be sustained, those involved must provide intermittent evidence that they are still attending to the others. If an interactor turns his back, or falls asleep, the others will assume that he has withdrawn from the encounter. To signal attentiveness interactors use (*a*) *proximity* – they are within the conventionally prescribed range, (*b*) *orientation* – they are appropriately oriented for the encounter in question, (c) *gaze* – they look at each other frequently, (d) *head-nods* – are used frequently by listeners, (e) *posture* – a listener adopts an alert, congruent posture, with slightly cocked head, or (f) *bodily movements* – these reflect the verbal and non-verbal signals of the speaker (see p. 254).

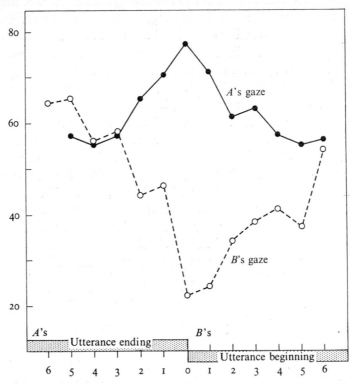

Fig. 4. Direction of gaze and the beginning and ending of long utterances. Frequency of other-directed gazes at half-second intervals before and after the beginning (broken line) and ending (continuous line) of long utterances. Pooled data from ten individuals, based on a total of sixty-eight long utterances (from Kendon, 1967).

Replacing verbal communication

Verbal communication may be impossible, or fail to work, for one reason or another, in which case NVC may take over.

Sign languages. The deaf-and-dumb language is a well known example. Simpler sign languages have been developed in a number of other settings when speech is impossible e.g. in broadcasting, on racecourses, in noisy factories, and between under-water swimmers. Some Australian aboriginal tribes have developed a gesture language, based on signs for objects and actions, that enables rapid communication and is used under certain ritual conditions (Brun, 1969). While English deaf-and-dumb language is based on letters of the alphabet, the American deaf-and-dumb language learned by the chimpanzee Washoe (pp. 37-45) is based on words.

Neurotic symptoms. Some clinicians believe that the symptoms of certain mental patients are a kind of NVC, used when speech has failed. Thus psychosomatic symptoms may be signals seeking attention, love and sympathy, or may be intended to control the behaviour of others (Szasz, 1961; see also Chapter 13).

THE PERCEPTION OF NVC

For NVC to have any effect it must be perceived, though it need not be consciously perceived. The term is used here to refer to more than the perception of size, brightness, etc. and to include cognitive activities of interpretation. Research here is concerned with the perceptual and cognitive activities of the *perceiver*, rather than the signalling of the sender. There has been a great deal of experimental work on social perception, but much of it has unfortunately been of a very artificial kind, so that rather little is known about perception during real social interaction. Another's NVC can be interpreted in at least four ways, which we will consider separately.

(1) *Interpretation as personality*

Since different people need to be treated differently, interactors try to categorise each other. They categorise one another in terms of age, sex, social class, occupation, and also in terms of personality traits. Individuals differ in the traits which they regard as important, and which will affect the way they treat others. Some people want to know primarily whether another is a Catholic or a Protestant, or whether he is Jewish, or how intelligent he is. All these categorisations are made on the basis of verbal and non-verbal cues, but NVC is of particular importance. The research findings in this area are of questionable validity, since they were obtained by such methods as looking at still photographs – e.g. in the studies of the effects of spectacles on judged IQ discussed above (p. 244). Physical cues which are used in this way include skin colour, hair colour and length, size of forehead, height, eye wrinkles, make-up, thickness of lips, size of nose, and shape of chin (Allport, 1961). These features may be taken as evidence of category-membership (e.g. being Jewish), or as a basis for inferring personality qualities (e.g. intelligence). It seems fairly clear, however, that individuals concentrate on certain categories or dimensions, and then assume that another possesses all the other stereotyped qualities associated with the race, class or other category indicated thereby.

17

(2) *Interpretation as emotions*

Information is needed on another's emotional state, and this can be inferred from his facial expression, tone of voice, and posture, as described above (p. 253). Whether or not a smile, for example, will be interpreted as an emotional state depends on the situation, the usual state of the person observed, and how well the observer knows him. Davitz (1964) found that there are considerable individual differences in the ability to judge emotions from tone of voice alone when neutral passages are read.

(3) *Interpretation as interpersonal attitudes*

(The same distinction from emotions is being made as before.) Whether another is friendly or hostile, or feels inferior or superior, can be judged from his posture, tone of voice, facial expression, etc. as described above (pp. 251–2). Tagiuri (1958) found that perceptions of liking–disliking were quite accurate – members of a group on the whole knew who chose them as friends, though 4% of others who chose them were seen as rejecting, and 9% who in fact rejected were seen as choosing: the difference is probably because negative attitudes are often concealed. A may be more concerned with perceiving how B perceives A, than with perceiving anything else about B – if A is a candidate being interviewed, or a performer in front of an audience for example. Argyle and Williams (1969) set up a variety of social situations and asked subjects 'To what extent did you feel that you were mainly the observer or the observed?', the answers being recorded on a seven-point scale. It was found that people felt observed when (*a*) being interviewed or assessed, (*b*) the other was older, (*c*) young females met young males; in addition some people felt observed most of the time e.g. insecure and submissive males. However, feeling observed was not increased when others looked more, though it was greater for those who themselves looked least.

(4) *Perception during on-going interaction*

As we have seen, during verbal communication each interactor needs intermittent information about the other's reactions, both for feed-back and to control floor-apportionment. The main cues are the other's shifts of gaze, head-nods, and small movements of the eyebrows and of other parts of the face. Most people are not consciously aware that they are receiving this information, but it is clear that they are making use of it from the lawful relationships found between these cues and other aspects of performance. This perceptual information is obtained mainly by means of intermittent visual scanning of the other's face.

There is a lot of research into the processes involved in the perception of non-verbal cues. Any of the kinds of interpretation listed above involves inference from a number of cues. When the cues are contradictory the observer does not normally average them, but rather opts for one and re-interprets the others. Individuals vary in the dimensions of personality they use, and in their sensitivity to various kinds of cue. While the recognition of emotional states is probably partly independent of specific experience, most of these perceptions depend on learning, and they can be improved by quite brief training (e.g. Jecker, Maccoby and Breitrose, 1965).

NVC IN SOCIAL INTERACTION

In social interaction there are streams of verbal and non-verbal signals which are closely inter-twined, and go in both directions. We shall discuss in this section some of the main kinds of sequential linkage involving non-verbal cues in the interaction process. Most human social interaction involves verbal as well as non-verbal communication, and most of the studies in this area have been concerned with both kinds of communication. We shall concentrate on sequences involving NVC here, though as shown above the two kinds of signalling are very closely related.

Response sequences

One approach is to study the statistical probability that some response of *A*'s will be followed by some response of *B*'s. One of the most common kinds of sequence is where *B* produces a similar response to *A*. It is found that if *A* smiles or nods his head *B* is likely to do the same (Rosenfeld, 1967); the same applies to posture, gesture, and to various aspects of verbal performance. Such rapid responses are unthinking and can be described as unconscious 'imitation': if an interactor becomes aware of what is happening he is liable to prevent himself from doing it. If this is a case of imitation, it should occur under the usual conditions for imitation, e.g. imitating people of higher status, who possess similar attributes, etc. (Bandura, 1962). This does not apply to 'reciprocity', which is a quite different kind of response-matching – the carefully thought-out and timed exchange of gifts or invitations, that is largely governed by the social norms about different kinds of relationship (Sahlins, 1965). Reciprocity is important in primitive societies, and differences have been found between social classes and cultures in the extent to which helping behaviour depends on reciprocity (Berkowitz, 1968).

Another important response sequence is that produced by reinforcement. Most research has concentrated on the non-verbal reinforcement of verbal behaviour, e.g. the effect of head-nods on frequency of expressing opinions.

The following non-verbal signals have been found to act as reinforcers – head-nods, smiling, leaning forward, looking interested, gazing at the other, and making encouraging noises (Williams, 1964). As argued above, it seems that, in the experiments conducted under real-life conditions, reinforcement was effective without awareness of what was going on (e.g. Verplanck, 1955). Other experiments suggest that interactors are not aware that they are giving reinforcements. Thus during interaction, all interactors are constantly giving and receiving reinforcement without much awareness.

We described earlier how a listener provides feed-back on what is being said. Careful study of the fine movements of listeners shows that they often make a continuous facial and gestural commentary on what is being said – what Kendon (1970 a) has called 'speech analogous movement'. Listeners may also produce 'movement mirroring', i.e. imitation of a speaker's bodily movements; this has been observed to occur intermittently, mainly at the beginning and end of utterances (Kendon, 1970).

It has been observed by Kendon (1971 a) that interactors coordinate their bodily movements in another way. For example if A turns towards B, B may turn slightly towards A; A has initiated a shift of attention towards B, or a desire for interaction with B, which B has recognised, and is indicating willingness to take part in. Another variety of this sequence is for A to look B in the eye, and for B to hold the gaze.

Skilled sequences of response

Social interaction cannot be analysed wholly in terms of S–R sequences, since each interactor has goals he is trying to attain: these consist of desired behaviour by others, or particular types of relationship or interaction. If the other does not behave in the desired way, continuous corrective action, directed towards eliciting the desired responses, is taken. For example if B is too dominant for A's liking, A may sit more upright, tilt his head back slightly, and speak more loudly. These sequences of responses have some of the characteristics of motor skills, such as driving a car. We have seen above how feed-back is sought at certain points e.g. at the ends of utterances. When another's face is invisible, interaction is found to be very difficult (Argyle, Lalljee and Cook, 1968). Another way in which social performance resembles a motor skill is in the integration of lower-order sequences, complete with feed-back loops, into larger units. Social performance thus has a hierarchical structure, where the smaller elements are habitual, while the larger units are subject to cognitive control and are influenced by rules, plans (in the sense used by Miller, Galanter and Pribram, 1960) and strategic considerations (Argyle and Kendon, 1967).

Equilibrium processes

However there is more to interaction than individuals responding to one another – those present must behave in a highly coordinated way for there to be any interaction at all. There must be coordination over (1) the content of interaction, i.e. the nature of the activity, (2) the role-relations e.g. whether a candidate is being interviewed for a job or whether he is assessing the firm, (3) how intimate the encounter is (warm–cold), (4) the dominance relations (inferior–superior), (5) the emotional tone (anxious, serious, happy etc.), (6) the proper sequence of acts (questions should lead to answers, gestures should be responded to), and (7) the timing and amounts of speech (who shall talk most, and when). To work out a pattern of interaction between two or more people where such synchronising occurs requires some rapid group problem-solving; this problem-solving is carried out mainly by the use of minor NV cues. Small attempts may be made at intimacy or dominance, with careful study of how the other reacts, so that these can be withdrawn if his reaction is too negative. To sustain the interaction a sufficient supply of rewards must be delivered to the other in order to keep him in the situation. Encounters often begin with a period of informal chat, the purpose of which is probably to enable some degree of synchronising to be established. Little information is exchanged during this period, but the interactors are able to emit signals conveying inter-personal attitudes and other NVC and to begin building up an equilibrium.

Once equilibrium has been established it proves very resistant to change, as is found in interaction between psychotherapists and patients over long series of sessions (Lennard and Bernstein, 1960). For some combinations of people, equilibrium is very difficult or impossible, e.g. if all want to dominate, or one wants to be intimate and others want to be formal and distant (Schutz, 1958).

SOURCES OF VARIATION IN NVC

Interaction in different group settings

We mentioned earlier that there are different rules of behaviour in different situations – for eating, at seminars, in church, etc. Here the tasks to be performed are quite different, and lead to totally different patterns of NVC. The social conventions about each of these situations specify what NVC may be used. A further source of variation is between such basically different forms of grouping as families, groups of friends and work groups. Sociological data suggest a number of differences in the communication systems of these groups (Argyle, 1969). Surveys by Argyle and Little

(1971) have found great differences between the kinds of NVC and social interaction in these settings.

The family has a biological basis and is found in some form in birds and mammals. The parents are drawn together originally by sexual motivation, children depend on them for food and protection, and are socialised by them. The family is a group with a definite social structure in that the behaviour and relations between the main members are similar in all cultures with some inter-cultural variation; there is a standard pattern of behaviour between father, mother, older and younger sons, older and younger daughters. The members of the family live together and have joint tasks to perform in connection with eating, sleeping, care of children and maintenance of the home. Social interaction has a special quality – there is more intimacy, aggression and affection; there is more bodily contact, less formality and politeness than in most other groups; the members know each other extremely well, and every act is loaded with meanings and associations.

Work-groups are found in all men and some animals. Work is done in groups because cooperation and complementary skills are needed for many tasks. There are also social satisfactions in work, and a great deal of purely sociable verbal and NVC occurs. The pattern of social interaction however is very much determined by the technological arrangements, and the formal ways in which men are related in the work-flow system. The main forms of NVC for work purposes are (*a*) the rate or method of working acting as a signal to others; (*b*) in coordinated task activity such as two-man sawing, physical movements are also social acts; (*c*) helping; (*d*) guiding, by means of bodily contact; (*e*) gesture language, where noise or distance prevents speech; (*f*) non-verbal commentary on work performance, by raised eyebrows, or physical blows and (*g*) non-verbal accompaniments of verbal messages, indicating for example whether they are to be received as 'advice' or as 'orders'. The non-verbal signals for sociable purposes are as described earlier – communicating interpersonal attitudes, etc. (Argyle, 1971).

Friendship groups are one of the main forms of human groups and also occur in some of the primates; the members are brought together solely by interpersonal motivations, rather than to perform any particular task. Adolescent groups play an important role in socialisation; adults spend some of their leisure with friends. Non-human primates play with and groom their friends; humans may have common interests to pursue or they have to invent activities which will generate the desired form of interaction – dancing, eating, drinking, walking, etc. There is more self-presentation than in the home, there is more attention to appearance, behaviour is more formal and polite, greater efforts are made to preserve a synchronising flow of interaction, and clearer signals of positive interpersonal attitudes are communicated.

INDIVIDUAL DIFFERENCES IN NVC

There are of course individual differences in the use of non-verbal signals. For example people vary in their preferred *proximity*: this can be measured by asking subjects to talk to the same person, or even to a hat-rack, and the variations are found to be consistent across different situations (Mehrabian, 1968). There are also characteristic expressive styles in the use of *bodily movements* (Allport and Vernon, 1932), and in *looking* behaviour, lengths of utterances, etc. (Kendon and Cook, 1969). The same is probably true of all the other non-verbal signals, though the variation in their use between different types of situation must be borne in mind.

Certain groups of non-verbal signals are commonly found to occur together. For example, there is the pattern of warm, friendly, affiliative behaviour which includes looking a lot at the other, close proximity, smiling, leaning towards, etc. An individual with strong affiliative motivation will seek out situations where he can behave in this way, and will try to establish friendly relations in many situations.

However any individual will use quite different non-verbal behaviour patterns in different situations, and may be friendly, hostile, inferior or superior, on different occasions, just as he is able to play a number of different games. An individual's performance in different situations is not very consistent (Mischel, 1969). The behaviour he emits does not depend on his personality alone, it also depends on (1) the rules governing particular situations, (2) the role-relations of those present, i.e. he will behave differently to males and females, and to those who are older and younger, and (3) the personalities of the others. We have recently found that an individual has a limited number of social performances, each of which is used for a range of social situations and relationships: 'personality' could be regarded as the sum of these performances (Argyle and Little, 1971).

THEORY AND EXPLANATION

Three main theoretical approaches to the explanation of human NVC can be distinguished, though each is concerned with rather different aspects. We shall outline a fourth approach, which incorporates features of the other three, and adds considerations from social psychology.

First we must refer again to the three main functional types of human NVC (see pp. 251–7). (1) *Signals used to manage the immediate social situation.*

These are similar to the signals used by animals: they have no reference to specific objects but only to the state and intentions of the organism. (2) *Signals used to sustain verbal communication.* These do not occur in animals, are used in close coordination with speech, are fast-moving, and probably have a complex temporal structure. (3) *Replacing verbal communication.* Gesture languages are developed when language is inconvenient, are often based on language, but may be independent of it. These three kinds of NVC are different in a number of ways, for example on Hockett's 'design features' (cf. Thorpe, pp. 27–35), and a different kind of theory will probably be required for each.

Biological evolution

Social behaviour in the lower animals consists of the stereotyped production of non-verbal responses; in mammals and especially the higher primates the system is more 'open' and has to be completed by socialisation experiences (Part B of this volume). In either case social behaviour consists of the emission of non-verbal signals, using parts of the anatomy adapted for the purpose, in order to achieve ends of biological importance – collective defence of territory, mating, rearing offspring, gathering food and so on. It may be safely assumed that this entire pattern of non-verbal signalling and social behaviour has emerged in the course of evolution because it contributed to the survival of individuals or groups. There is evidence that the signals for different emotional states are much the same in all human societies and occur also in blind children, so they are presumably little affected by experience (Eibl-Eibesfeldt, Chapter 11, this volume). It is likely that the same is true of the signals for interpersonal attitudes. We have argued above that NVC in humans is used to control the immediate social situation, and shown that it is more effective than speech for doing this.

Several authors have stressed the similarities between human and animal social behaviour (Morris, 1967). How far can this type of explanation be pressed? There are two main limitations to it, corresponding to the next two theories. (*a*) Despite the similarities between cultures in emotional expression, there are extensive differences between cultures in other aspects of NVC. Particular interest has been taken in the Japanese and Arabs, whose signalling systems are very different from those in Europe or the USA. For example Arabs stand closer, touch each other, face head-on, use a lot of eye-contact, and speak louder (Watson and Graves, 1966). A given signal can mean quite different things in different cultures e.g. hissing in Japan is a sign of deference. As Darwin recognised (1872), while the expression of the emotions in man appears to be unaffected by experience, many other non-verbal signals that we use, such as nodding and shaking the head, clearly are not. (*b*) One of the main differences between men and apes is our use of language: most of our social behaviour involves language. Some animal gestural signals are used to send messages of a similar kind, and much of human NVC is used in close connection with the emission, reception, or control of speech. This is particularly true of head-nods, shifts of gaze, and hand movements, which are used quite differently from the way animals use them.

The structural-linguistic approach

A number of investigators have studied NVC in a similar way to that in which linguists study a language. They have tried to find out whether there is a hierarchical structure, in which smaller units are grouped to make larger ones, and if there are rules of sequence and composition.

We have seen that prosodic vocal signals and certain kinesic signals are closely coordinated with language, affect the meaning of utterances, and form part of a total communication system with a definite structure. Other non-verbal signals, such as direction of gaze, also operate in a structural way, for example in governing the synchronisation of speech and in negotiating greetings. Language, in order to communicate at all, must follow certain grammatical rules; the meaning of a word depends partly on the other words with which it is grouped: similar considerations may apply to NVC, and NVC may have some kind of syntax, governing both its own sequences, and the links with language. The detailed syntax for this kind of NVC has yet to be worked out in detail, and it may not be so strict or detailed as the syntax of language itself. While no explicit theory of the origin of this aspect of NVC has been offered, it seems to be assumed that it is part of a wider system of communication and social behaviour which also includes language proper; thus it could be based on the same neural structures and have evolved at the same time as language. An alternative explanation might be that the verbal communication system and its accompanying physical structures are primary, and that non-verbal signals are inevitably acquired in order to make communication feasible – to control synchronising, provide feedback, etc.

Sociological approaches (*Symbolic interactionism and ethnomethodology*)

In many situations there are definite rules governing what shall happen, for example at committee meetings, in church. In other situations there are implicit rules which are less obvious, but which become obvious when they are broken. In its emphasis on rule-following this approach is similar to the previous one: however where the structural-linguistic approach is concerned with the rules which NVC must follow to coordinate with speech and to communicate, the sociological approaches are concerned with the rules, at a more macroscopic level, governing styles of behaviour and sequences of events in particular situations and settings. This explanation in terms of rules is quite different from explanations by empirical laws: behaviour is rule-governed when people think that this is the proper thing to do, or if they are shocked when the rule is broken. There are different rules, in a given culture, for behaviour at a lecture, a seminar and a party. Barker and Wright (1954) carried out a sociological study of a small Midwest town, and concluded that there were 884 'behavior settings'. Goffman (1963) has analysed in detail some of the rules governing NVC in American middle-class society, and suggests that all or most of social behaviour can be explained in terms of these rules. These rules have developed slowly, as part of cultural history, because they are useful. Their explanation would lie in showing how they had been useful to particular groups. There would thus be no universal laws of behaviour, merely a large number of arbitrary rules for particular situations. An alternative approach is to look for universal laws that are affected by dimensional aspects of situations – the number of people present, their formality, etc.

Sociologists have also emphasised the subjective meanings given to non-verbal signals by the culture and by particular groups. For example, in his account of 'self-presentation' Goffman (1956) maintains that people manipulate the impressions others form of them by clothes and gestures which have certain meanings. In his account of 'body-gloss' he observes how people make their behaviour comprehensible to bystanders by sending signals which indicate the motivation of their behaviour (Goffman, 1971). The importance of non-verbal acts which have culturally defined and publicly shared meanings is shown in rituals and ceremonies, such as marriage. While this approach emphasises the meanings given in a particular culture, there is no reason why it should not be able to accommodate meanings of a universal biological origin. On the other hand Goffman has tended to emphasise the *different* meanings which the same act may have in different social settings.

A social psychological approach

The contribution of the biological approach must be recognised – there is an unlearnt basis to emotional expression, and there is a basic physiological equipment on which all NVC depends. It must be accepted that most human social behaviour involves speech, that much NVC is used to accompany speech, and that this NVC may depend on similar structures to speech or have been learnt as part of the skill of verbal communication.

The contribution of the sociological approach must also be recognised – there are considerable cultural variations in many aspects of NVC, and the same signal can have different meanings depending on the culture and the situation. To all this however a social psychological dimension may be added.

(1) Systems of interaction build up in particular social settings, groups, and cultures, to deal with particular communication needs.

(a) *Managing the immediate social situation.* Interpersonal attitudes and emotions are expressed freely in the family, but are carefully restrained outside, where more fragile relationships have to be sustained. On the other hand there is virtually no self-presentation in the family, where it is pointless. In some situations it is prevented or controlled by the wearing of uniforms.

(b) *Sustaining verbal communication.* Non-verbal signals play a number of important roles in sustaining verbal communication under normal conditions. If these conditions are changed in any way there are adjustments in the NVC: for example when visual cues are removed, as on the telephone, auditory cues replace them for purposes of feed-back and floor-apportionment. When the videophone becomes available, a different set of adjustments will have to be devised and learnt.

(c) *Replacing speech.* Systems of gestures are developed in situations where speech is impossible.

(2) When a system of communication has been established in a culture or a social group, it has to be learned by new members. This may take place through imitation, combined with the reinforcement of successful performance (Mischel, 1969). There may also be trial-and-error learning: McPhail (1967) found that adolescents engage in a lot of 'experimental' behaviour, much of which is awkward, aggressive and quite unsuccessful, but that this is replaced by more effective 'mature' behaviour after the age of 17 to 18. Learning by explanation and understanding is not common in the case of NVC, which is usually emitted and received in a spontaneous manner, below the conscious threshold. However, in groups of adolescents there is often discussion of such minutiae of social performance, and in social skills training and in psychotherapy insight is given into NVC, so that it is brought more under conscious control.

CONCLUSION

NVC in humans consists of signals similar to those used by animals. However human social behaviour uses a second channel – language. We have suggested that there are three distinct kinds of human NVC, which have different origins and modes of functioning. (1) Some NVC is used to communicate attitudes and emotions and to manage the immediate social situation. This appears to be very similar to animal communication, though there are cultural variations in the signals used, and situational rules governing their use. (2) NVC is also used to support and complement verbal communication. This is found only in man, is coordinated with speech in a complex way, and appears to be part of an overall system of communication, with complex rules of sequence and structure. (3) NVC is also developed to replace language, as in gesture languages. Some animal communication is of a similar character, and human gesture language can be taught to chimpanzees (pp. 35–47). The outstanding problems in the study of human NVC are: (1) further understanding of the functioning of particular non-verbal signals, and how they are perceived; (2) the elucidation of the second kind of NVC, how exactly it is fitted to speech, and whether there is some kind of overall grammar; (3) further analysis of cultural differences and similarities, and the conditions under which NVC is learnt; (4) further knowledge of the processes under which new systems of NVC develop in a group or culture.

Acknowledgement. I am grateful to Dr Adam Kendon for his detailed comments on this chapter.

COMMENTS

(1) *Definition and classification*

Much of the current work on non-verbal communication in man is concerned with establishing that the behaviour of one individual is affected by the non-verbal behaviour of others, and with describing and classifying the signals and effects involved. Questions of 'intent to communicate' are not yet important (cf. MacKay, Chapter 1). Although Argyle is concerned with situations in which both verbal and non-verbal components may be important, the focus in this chapter is on the latter, and the relationship between verbal and non-verbal components is important only in some cases (Lyons, Chapter 3). Thus the classificatory systems which Argyle uses tend to cut across those discussed in Part A. For example, what he calls 'managing the immediate social situation' involves in part non-vocal elements unrelated to speech, but paralinguistic and prosodic features also contribute, and so indeed does the choice of words in the verbal component. 'Sustaining verbal communication' involves primarily prosodic and paralinguistic features, but expressive movements and postures quite unrelated to the verbal content also play a role. And when sign languages are used to replace verbal communication, although wordless they must be based on words ('anthroposemiotic' – see Sebeok, cited in the Comments on Part A).

(2) *Description*

Further discussion of techniques for observing and recording inter-personal interactions is given by Hutt and Hutt (1970).

(3) *Paralinguistic and prosodic phenomena*

An earlier bibliography is given by Mahl and Schulze (1964).

(4) *Neurotic symptoms*

See Grant, Chapter 13, this volume.

10. NON-VERBAL COMMUNICATION IN CHILDREN

N. G. BLURTON JONES

Department of Growth and Development, Institute of Child Health, University of London

INTRODUCTION

In this subject the smooth-flowing descriptions that one finds in the literature bear little relationship to what is actually documented. This is particularly true of the development of non-verbal communication and its effects on development in the first year of life. For instance, several authors have concluded, or suggested, that smiling plays a vital role in the development of a child's attachment to its mother, and hence in the development of social behaviour. These, often incidental, remarks have come by repetition to be regarded as proven fact. Consequently this review attempts to integrate the observations of earlier workers with more recent work, rather than to re-report uncritically discussions of theories which have hitherto appeared in the literature as if they were established facts. Much of the early work is not easily accessible and it may be useful to have a contemporary evaluation and summary of the more notable studies.

The major topics covered in this review are: the little-known 1–2-year-old and his surprisingly extensive social behaviour; the early studies of laughing, smiling and crying on the well-studied 3–5-year-olds, and the recent more wide ranging studies of their other non-verbal communication patterns. Discussion centres on 'intention' and some misconceptions in the use of feedback models of communication, and more briefly on the contrast between the approach advocated by Andrew (Chapter 7) and that of many other students of displays. Where not otherwise specified, my remarks concern studies on white English or American children.

Methods employed in studies of non-verbal communication in children

Recent studies on non-verbal communication in children, such as those of Grant (1965, 1968, 1969), Brannigan and Humphries (1972) (and in an applied context by Currie and Brannigan, 1970), and Blurton Jones (1967 and in press), have used children as a handy brand of human subject that behaves rather than talking about behaving. This enables facial expressions and sounds to be investigated in terms of the situations that give rise to them, the behaviour they coincide with or lead on to, and the changes in behaviour of the reactors which follow them (see pp. 103 and 219). To these workers the 'meaning' of a facial expression means two quite separate things: first, the causes of the facial expression, that is, the factors which make the child emit the message; and secondly, the effects of the expression on other children. By contrast some early studies of children, like traditional studies of adult non-verbal communication, were concerned with a search for undefined 'meanings' of facial expressions, as when subjects had to name the meaning of a photograph or posed expression. These have slowly changed into a study of the degree of agreement over the meaning of words describing emotional states. I have argued elsewhere (Blurton Jones, in press), as also has Vine (1970), that these studies confound several different features of the response to facial expressions. Fortunately very little of this work was done on children, so that it need not be discussed at length in this paper.

Some observational studies attempted to categorise facial expressions and noises on one side or other of a nature–nuture division. The best of these studies was that on the development of smiling in blind children by Thompson (1941), which enforced the more integrated approach to development which was later seen in the studies of smiling by Freedman (1964). In recent years there has been increased interest in non-verbal communication between the infant and its mother in the first year of life (e.g. Ambrose, 1961; Wolff, 1963; Robson, 1967; reviewed by Vine, in preparation). Much of this has relied on measuring the efficiency of standard test situations in eliciting responses from the child. Very little has involved data on the actual performance of the mother and the child. Some such studies are now in progress (see e.g. Richards and Bernal, 1972).

NON-VERBAL COMMUNICATION IN THE SECOND YEAR

This is the least known, but most exciting, year of childhood, yet there has been almost no systematic work on it since that of Bridges (1933). Bridges' study, though clearly based on ample and able observation, is reported in a completely non-quantitative way and concerns children in an institution whose interactions may be expected to differ in some way from those living in families. Ainsworth (1969) and Schaffer and Emerson (1964) covered the 12–18-month period in their studies of attachment, but report no systematic observations which allow us to extract data on non-verbal communication. The systematic field studies by Anderson (1972) of urban children out with their mothers in a park provide almost the only source. Fortunately Anderson had paid special attention to a number of the gestures shown by children in this situation.

Bridges (1932 and 1933) describes the development of social interactions of children from the first month to 2 years old. By 15 months their interactions are basically the same as those of 2-year-olds and one may expect the sounds and facial expressions that characterise these interactions to play some role even at this early age. Bridges describes how 15–18-month-old children (and younger ones) held 'babbling conversations' with each other, complete with gesturing, but with no meaningful words. It would be fascinating to see whether the devices for deciding who talks when, that have been described in adults (Chapter 9), are present in these 'conversations'. The children in Bridges' study were in constant contact with other children though presumably in limited contact with adults, so that the institutional setting may have exaggerated the development of social behaviour as well as increasing the opportunity to observe it. As early as 4 or 5 months of age babies attended to crying children and by 7 or 8 months they would smile and reach out to a baby in a neighbouring cot, apparently in response to its cooing sounds. At 11–12 months they imitated each other's sounds and by 13–14 months mutually laughed and smiled a great deal. Bridges says that at this age they began to play more in twos than alone, 9–10-month-old babies put in the same play-pen explored each other in the same way as they explored themselves, and by 11 months they would hold themselves up by the play-pen fence and chase each other around, laughing as they did so. Suddenly at 11–12 months they began to respond to another child taking an object they were holding by crying, and soon after this they would hold on and 'squeal'. Biting, hitting and hair-pulling occurred in this situation by 14–15 months, but biting quickly disappeared under social pressure.

Gestures in the second and later years

I shall describe some of the findings of Anderson (1972) on this age group, along with data on the same gestures in older children. As many as half of the children that Anderson observed were seen to take an object to the mother, but they did not necessarily wait for her to inspect it. Blurton Jones (1967) and Connolly and Smith (1972) found this to be a common response of older children to a stranger, while Blurton Jones and Leach (1972) recorded it as common in greetings to both mother and other adults when they came to fetch a child from a play group. Leach (1972) also found it common in interactions of $2\frac{1}{2}$–4-year-olds with their mother. It thus appears that this behaviour has some social significance over and above the object given, and possibly even independent of the response to the object. Some child psychiatrists say they often find it useful to begin interaction with a child via some object.

   Pointing was a common response to novel events and always involved looking at the mother, stopping and standing still, and then orienting the body to face half way between mother and object. Anderson remarks that the person a child looks at when pointing is one of the best indicators of who its mother is – as good as or better than approach or proximity. Blurton Jones and Leach (1972) found much the same in 2–4-year-old children in greeting situations: pointing was one of the behaviour patterns most specific to the mother as opposed to other adults.

Waving in the 12-month-old child is an indication of imminent inter-action and not a symbol of departure. It indicates that the child will be near its mother very shortly. If it should leave after a wave it is for a much shorter time than if it does not wave. Anderson stresses that waving in 1–2-year-olds often occurs in situations where its significance is difficult to understand, for instance when the child is facing away from its mother. It sometimes leads to striking the mother. In 2–5-year-olds the same sort of loose rather rapid waving occurs in rough and tumble play, in 'temper-tantrums' and when queuing, for example, to go down a slide. It may be that waving reflects a very generalised kind of activation. Blurton Jones and Leach (1972) found that waving to the mother at separation was charac-teristic of the older children in their sample, that is those over 3 years old.

The arms-raised posture, where both arms reach up in front usually above shoulder height, is common in 1-year-olds and leads to the mother's lifting and carrying the child. Anderson observed it most when the mother was moving. He made the very important observation (discussed by Bowlby, 1969) that children of this age are unable to follow a moving mother, and he regards this as a perceptual inability rather than a loco-motor inability. The same posture occurs in older children: Blurton Jones and Leach (1972) were able to show that its effectiveness is not solely due to the approach that goes with it, and that it is equally effective in children who have cried at separation and children who separate readily. In greeting situations it often occurs with smiling, but unlike smiling its frequency is greater in children who cry at separation.

Anderson also describes a number of inconspicuous gestures which have been investigated in older children by Grant (1968, 1969) and McGrew (1969, 1972). We have no data to suggest that they actually affect the behaviour of other people but they do emit information about the behaviour that the individual is likely to show subsequently. Putting part of the hand in the mouth is found by Anderson to occur when a 1-year-old is not making a long sortie from its mother and not engaged in any clearcut activity. The child is quite likely to be stationary and when he does move it is likely to be towards his mother. If other adults are around, a child with his hand to his mouth will respond by watching rather than by interacting.

(Out of doors Anderson finds fear of strangers to be almost non-existent, and it is very common for the toddlers to approach strangers.)

Touching the back of the head with the flat of the hand occurs during pauses before changes in direction of movement, and when the child stops to look at objects or people, if this is followed by movement towards the mother. The same situations also often give rise to thumb-sucking and 'ear-flipping' (rhythmical movements of the fingers on the ear), with the arms in a position commonly adopted during sleep. Grant (1968) concluded that head-grooming in nursery-school children and in adults, occurred in situations of alternation between approach and retreat. Blurton Jones and Leach (1972) found thumb-sucking at separation from the mother to be more common in children who sometimes cried at this time. McGrew (1969) found 'auto-manipulation' to be commoner in agonistic–quasi-agonistic interactions – combinations of rough and tumble play and aggression (Blurton Jones, 1967) – than in non-agonistic ones. It was commonly seen to be performed by a child after losing a fight or in approach–avoidance conflict. Grant (1969) distinguishes several forms of automanipulation and finds differences in when they occur: this would clearly repay further study.

Some other workers, less in touch with current animal behaviour research, have reported on so-called 'displacement activities' in human behaviour. Where these workers treat 'displacement activity' as an explanation, for example Kehrer and Tente (1969), this influence of ethology has been for the worse. But it has been beneficial in those cases where it has led workers to describe hitherto undescribed behaviour, much of which may well act as signals during interaction (e.g. Berg, 1966; Mitchell, 1968). Mitchell (1968), working mainly with adults, discussed displacement activities from the point of view of a dermatologist. There are instances where tension or anxiety appears to exacerbate or bring about skin disease through increasing the amount of scratching and rubbing. More straight-forward symptoms, such as fissuring of the thumb resulting from sucking, are also described. Mitchell lists thirty-seven different activities concerning the body surface which give him the impression of resulting to some extent from elicitation of tendencies to flee or attack. Of particular interest is 'scratching and rubbing the head, particularly the vertical and occipital scalp', also described by Grant and by Anderson (see above). Mitchell also lists 'grinding the teeth', also described (incisors, during sleep) by Every (1965). Clenching and grinding the molars appears to be common in adults but has not been described in children, although banging them together with the mouth open has been observed. Tooth-grinding during sleep has been observed in young children. This reviewer finds it hard to regard this as having a signal function and doubts the causal (but not the functional) connection with aggression which Every claims for it.

Kehrer and Tente (1969) describe a large number of movements occurring in their 'frustration' tests on children sitting on hospital beds. (The reader should not discard sitting on a bed as being an artificial situation, simply because in our culture it is usual to sit on a chair. In many cultures most sitting is done on the ground.) Some of the commonest movements in this situation are not described by other authors: for example, 'kneading the feet', which they illustrate with photographs, will instantly remind any parent of innumerable forgotten observations of this behaviour. Sucking was also common in their tests. Unfortunately their descriptions are confused with classifications taken from the animal work of the early 1950s and it is impossible to tell whether they observed any of the patterns described by Grant (1965, 1968, 1969), Anderson (1972), Berg (1966) or McGrew (1969). Their list summarising 'human displacement activities' is slightly more usable.

Early studies of so-called 'nervous habits' in older children have also described some of these movements and many others in addition. Koch (1935) described 'nervous habits' only in terms of the parts of the body involved: her data are therefore not directly comparable with those of Grant, McGrew or Anderson, but they suggest a close similarity to the behavioural situations in which these authors found these inconspicuous gestures to occur.

By the time a child can walk it has a rich repertoire of gestures, and engages in several kinds of interaction. Unsystematic observation suggests that the repertoire of facial expressions and noises is comparably complete: it would be particularly interesting to see how their occurrence in child–child interaction compares with that in older children. Child–child interaction amongst children living in families is found by Anderson (pers. comm.) to be common, though not as continuous as at later ages. Reports by the mothers of children at our play group also suggests that child–child interaction is far more common in 1-year-olds than is usually assumed. The slow and jerky appearance of these interactions between 1-year-olds implies that there is a deficit in communication, but it is not known how far this is due to incomplete development of non-verbal communication or how far it is due to lack of language (or at least a mutually understood language). Comparison with deaf children would illuminate this question, but complete studies seem unlikely to be practicable until computer systems for analysing sequences of behaviour become more widely available.

NON-VERBAL COMMUNICATION IN 2–5-YEAR-OLDS

This is the age group in' which most of the recent work has been done. The following summary draws on work, much of which is about to be published, by zoologists who have turned to studying displays and inter-actions of humans. Accounts published so far are those of Grant (1965, 1968, 1969), Blurton Jones (1967) and McGrew (1969). Most of the work involves direct observation and photography, with investigation of the situations and sequences of interaction in which gestures and facial expressions occur. The differences between this approach and the traditional approach to studying adult expressions (e.g. Osgood, 1969) have been described briefly by Grant (1969) and Vine (1970).

Like the neonatal period, this age group probably attracted recent studies because of the accessibility of the children at nursery schools. There were also many earlier studies in nursery schools which covered the general occurrence of crying (Landreth, 1941), smiling and laughing. Recent studies (Grant, 1969; Brannigan and Humphries, 1972; Blurton Jones, 1972) have gone into greater detail on the entire range of facial expression, and have concentrated on problems of causation. Grant (1969) and McGrew (1969, 1970) have given detailed descriptions of postures and gestures which may have signal effects. Children of this age have also been used in studies of the ability to name the meaning of expressions (references in Vine, 1970). The ability develops rather late but provides no evidence against the supposition that non-verbal communication is already working in 3-year-olds. This supposition should however be tested further, for instance by looking at the responsiveness of children of various ages to different signals. Although occurring at lower frequency, most of the signals described for 3- and 4-year-olds have also been seen in 2-year-olds, and the gross aspects of the interactions with other children are organised in the same way as those of 4-year-olds (Blurton Jones, 1972). The very rich repertoire of facial expressions and gestures of nursery-school children suggests that highly complex non-verbal communication may be going on at this age. But so far studies have concentrated on the causation of these signals, at the expense of studies of their effects.

The effects of such signals can be studied by careful analyses of sequences of interaction, if other concomitants are held constant by statistical partial association tests. In a relatively simple situation Blurton Jones and Leach (1972) were able to show that the arm-raised posture that children direct to their mothers does affect the likelihood of the mother touching the child. In child–child interactions a great many variables, such as relative size and weight and position in the peck-order, would also have to be taken into account. The use of trained adult actors may be a successful technique for testing responses to facial expressions, provided that their behaviour is also recorded to show that they do in fact give the signals that were planned and not others. Buhler and Hetzer (1928) used adults posing angry expressions and sounds in their tests and claimed that children in the first year responded differentially to the adult signals. It was not established how far this was due to simple differences such as the volume of sound, or the absence of normal smiles, head on side, and other friendly gestures. In any case the responsiveness to facial expressions may develop through such influences.

Children show some of the 'speech markers' that have been described for adults. But there seem to be many differences and it may be partly these differences that makes it so easy to tell when a child in a game of 'tea-parties' or 'picnics' is speaking and behaving as its parents would in this situation.

Michael and Willis (1968) investigated the use of twelve common gestures in 4- to 7-year-old white American children. Their main aim was to investigate age, sex and social class differences, and the results are parallel to those on verbal ability. The middle-class children were better at transmitting gestures on request and at interpreting them. The first year at school had a very big effect. The gestures are not described in their paper other than by the words they were 'used to indicate' – e.g. go away, come here, yes, no, be quiet, how many, how big, etc. Some of the gestures with no generally expressed verbal equivalent which occur in this age group were described in the earlier section on gestures in the second and later years.

Crying

Crying has not attracted much attention from researchers after the newborn period. Apart from Landreth's (1941) study, the available information is mostly incidental to studies of other aspects of behaviour. For instance Schaffer and Emerson (1964) used crying on separation as their measure of the child's 'attachment' to its mother. They found that crying commonly occurs in this situation, and that its frequency and intensity increased up to 18 months when their study finished. Blurton Jones and Leach (1972) observed crying on separation from the mother at a play group to be common (especially in boys) at ages $2-2\frac{1}{2}$ years, but rare in children over $2\frac{1}{2}$. McGrew (1972) reports that crying in this situation was very unusual in children joining a play group at age 3 years. Children at play groups may not be representative because mothers who know that their child will not separate may not attempt to start them at a play group until later.

In 3–5-year-olds at nursery most authors seem to agree with Landreth

(1941) that falls and quarrels are the commonest situations for evoking crying. Landreth found boys to cry more than girls both at home and at school, which might reflect a persistence in the sex differences in 'irritability' and ease of comforting which have been described in newborns. Alternatively it may reflect the greater frequency of fights and of climbing and rough play that most authors describe for boys. This would also explain the apparent disagreement between Landreth and Dawe (1934) who found that girls were more likely to cry during quarrels than were boys (though the latter report should be treated with caution because (a) Dawe did not define quarrels or explicitly discriminate them from rough and tumble play and (b) because event recording from a whole group of children may mean one only notices the noisiest or most prolonged disputes). Ding and Jersild (1932) and Blurton Jones (1967) both remark that crying seems commoner in social situations than non-social, stressing their impression that cries per fall are greater when the fall is brought about by another child than when the child falls on its own. Many parents also comment on the apparent inhibition of the child's crying after a fall by an adult laughing, and Blurton Jones (1967) felt that this applied to falls during rough and tumble play.

Possible effects of crying on other children have been discussed by several authors. Bridges (1932) warned against assuming that small babies cry in response to others crying when they may merely all be simultaneously approaching their next meal-time, although she does suggest that in institutions they may learn that crying by other babies is a cue to the arrival of feeding bottles. Among nursery-school children crying can elicit the so-called 'nurturant' behaviour of older children which includes a noticeable 'hand-on-the-back' gesture, bending down, and gentle talking with the head on one side (Blurton Jones, 1967; McGrew, 1970). The loss of muscle tone and inactivity, often including sitting on the floor, which Blurton Jones (1967) described in the nursery school can also be seen in the presence of the mother and perhaps has some relationship to the inability of under-3-year-olds to follow a moving mother, as described by Anderson (1972, and in Bowlby, 1969).

Laughing

Most early studies of nursery-school children (Enders, 1927; Ding and Jersild, 1932; Kenderdine, 1931) agree that laughing is most commonly seen in association with vigorous movement and mobile play. Smiling was found to be common in this situation but also in less mobile and more obviously social interactions. Ding and Jersild (1932) disagree with Ames (1949) over the relative importance of 'social approach' and 'gross motor behaviour', but this seems (there is too little description to be sure) to result from the use of different definitions of laugh (smiling being the residue) or from Ames's restriction of the category 'gross motor behaviour' to non-social behaviour. Anyway the uncertainty does not allow one to conclude that Ding and Jersild's New York Chinese subjects differed from Ames's white American children.

All authors agree that laughing occurs only in the presence of other children and that smiling also is predominantly found in the company of others. Kenderdine (1931) and Justine (1932) examined the effects of various contrived situations on laughter and confirmed that moving about was a more frequent situation for laughing than were, for instance, situations which they 'realised were socially unacceptable' and 'humorous situations' (which presumably means situations which the experimenter found funny). The misfortunes of others did not elicit laughter: if anything, they elicited watching.

There is disagreement over differences in the frequency of smiling and laughing between different ages, but none of the studies report on how long the children had been in the group in which they were observed, which McGrew (1972) has shown to be related to the amount of interaction. Ding and Jersild (1932) also mention features of group composition as possible influences on the amount of smiling and laughing. These authors also remark on laughter occurring when 'romping about' and at the bottom of a slide. The latter is a situation where the children in our play group often laugh, and quite often pile on top of each other or chase each other back to climb to the top of the slide.

Recent studies confirm the association between laughing and motor activity (Smith and Connolly, 1972; Blurton Jones, 1967). Blurton Jones (1967) stresses the association of laughing and open-mouthed smiles with running and jumping, hitting at others, wrestling and chasing in a temporal grouping of behaviour which he calls 'rough and tumble play'. In a recent study Blurton Jones (1972) found that factor analysis showed these items as loading strongly on the same factor. Smiling (without wide open mouth or laughing) loaded most conspicuously with talking, giving objects to children, receiving objects and pointing at things while looking at a child. One should remember that this social, active behaviour gives rise to many more specific situations, resembling those used by Kenderdine and Justine, which can fit with several of the classical theories of laughter as listed by Ding and Jersild. These findings provide independent support for some of the conclusions of van Hooff (Chapter 8).

Ambrose (1963) describes the beginnings of laughing during the first year. He regards laughter as a mosaic of components of smiling and crying or, in more general terms, of stimulus-maintaining and stimulus-terminating behaviour. He argues that the rapid alternation and super-imposition of diaphragm and abdominal muscle activity shows simul-taneous arousal of breathing responses to pleasurable stimuli and those typical of crying. Blurton Jones (1967) commented that there was a continuum in the form of the sounds between laughter and screaming in nursery-school children. Although screaming is quite different from crying, Ambrose's kind of argument might still apply, and indeed a contrast of screaming and laughing would fit better with Ambrose's argument that the situations which elicit laughing in a baby combine the fear-evoking stimulus of rough handling with the pleasure-evoking stimulus of the familiar parent. However, one would expect more obvious variation in the sounds and facial expressions if Ambrose's argument was correct. One should see an even distribution of all grades between screaming and avoid-ance and laughing and cooing, especially when laughter first appears in development. It seems likely that changes of the sort Morris (1957) termed 'typical intensity' (Chapter 4) may have occurred during evolution, limiting the variation of the movement pattern. There is, in fact, insufficient evidence to decide whether laughing develops from a mosaic of smiling, cooing and crying, or from a mosaic of smiling, cooing and screaming, or by some other means.

The facial expression which accompanies laughter (but which also occurs with no vocalisation) is identical in form to the 'play-face' of van Hooff (1962), although it can also include forward movement of the lower jaw. This expression, and sometimes laughter, is evoked by tickling and by peek-a-boo games, and also by chasing as soon as the child is mobile (by whatever means).

Ambrose's interpretation of laughter fits well with many later features of the occurrence of laughter and play-faces, but it leaves one to wonder why these 'fearful' situations should be sought out and be as reinforcing as they appear to be. Since rough and tumble play has been persistently ignored by psychologists there is little information about its origins or later development. The existence of laughter and the play-face as charac-teristic signals associated with this behaviour suggest that it is an important piece of social behaviour that may involve important non-verbal communi-cation. Its widespread taxonomic distribution makes it difficult to conclude anything about its function from comparative studies. Its effect on peers is usually greater than its effect on mothers, which suggests that it has no function specifically concerned with mother–infant interaction distinct from the one concerned with interaction with peers and other adults (there are many species in which the mother will not play with her infant).

Smiling

Ding and Jersild (1932) describe smiling as occurring less in violent activity but more in 'non-verbal social contacts, embraces etc.' Ames describes smiling as occurring during 'verbal social' and 'social approach' interactions with other children and with the teacher. In so far as one can tell what these terms refer to, they would appear to fit with the findings of recent studies on nursery-school children. Smiling also occurs in children's interactions with their mother, especially during the greeting associated with approach and touching the mother. It is also included in the studies of children at a play group by Leach (1972) and Blurton Jones and Leach (1972). The likelihood of an individual showing a smile in greetings is not related to whether the child at some time cried when its mother left it at the play group. Children who smile a lot have mothers who smile a lot, and are likely to smile at the teacher when they arrive. Our present view is that in this age group smiles indicate the likelihood of 'social' interactions (giving, showing, talking) with the mothers which vary independently on the short-term time scale, and quite possibly during development, from the proximity-seeking, clinging behaviour to the mother.

At some ages Ames found more smiling to the teacher than to children, but does not give the number of non-smiling interactions. This is unfortunate because some observers get an impression that many children smile more often in interactions with the teacher than with children and that this gives rise to errors in teachers' ratings of 'sociability'. Information on this would also be useful for investigating the idea that smiling may have some features of 'appeasement' displays. So far the best evidence on this comes from the current studies of Krebs (pers. comm.) who found that children low in the peck-order smiled more when initiating an interaction with high-ranking children than high-rankers when initiating interactions with low. However, smiling in response to smiles is commoner in those high in the peck-order. These points suggest that the appeasing situations found to evoke the fear grin and similar displays in other primates may not be so far removed from the situations which evoke human smiling. Robson (1967) reports that children from 4 months on may react fearfully to a sober-faced mother, suggesting that maternal smiling may have a reassuring effect on the infant. The situation is thus perhaps comparable to that of many primates (Altmann, 1962) where the same movement may have both appeasing and reassuring functions (see Andrew, Chapter 7). It may be that it simply reduces the tendency to flee or attack without necessarily increasing the tendency to interact in any other way, and it may also be that the causation of smiling is relatively independent of that of other kinds of interaction. In other words, the tendency to smile may not be related to tendencies to approach, or flee, or hit or embrace, etc., but, just as in a baby, be simply a response to the front view of a face. Investigations of smiling and laughing are unlikely to progress much further than this until quantitative studies which distinguish the fine categories of smile described by Grant (1969) and Brannigan and Humphries (1972) are available (see Figs. 1 and 2). The still finer divisions, and the emphasis

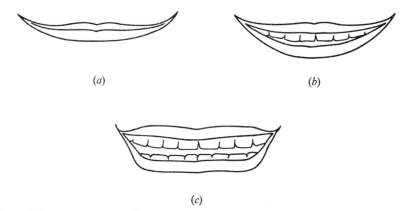

(a)   (b)

(c)

Fig. 1. Three common smiles: (a) simple smile, (b) upper smile, (c) broad smile. (From Brannigan and Humphries, 1972.)

on the separate occurrence of the various components, in Blurton Jones's (in press) catalogue of criteria for describing facial expressions opens the way to the application of the approach of Andrew (Chapter 7) to smiling and other complex expressions of children.

Among adults, and sometimes among children, one sees smiles (in addition to the exaggerated 'coy' ones of Washburn, 1929) which seem false. Some features which give this impression are: a wide smile with no wrinkling round the eyes (normally wrinkling has become conspicuous by the time the smile is wide), no narrowing of the eyelids, and a disproportion of raising the upper lip to retracting the mouth corner (sometimes accompanied by squaring the lower lip, without moving the jaw forward, unlike the 'oblong smile' of Grant).

Aggressive interactions

There seems to be agreement among modern workers both as to what is most properly labelled aggressive behaviour, and as to which gestures and facial expressions accompany it. However, none of these workers have yet published the quantitative evidence for their conclusions. Some of the expressions may also be found in non-aggressive and non-social situations, and this gives useful information about how they arise in the social ones. Such cases also raise some fundamental points about the classification and interpretation of expressions which are taken up in detail later in this

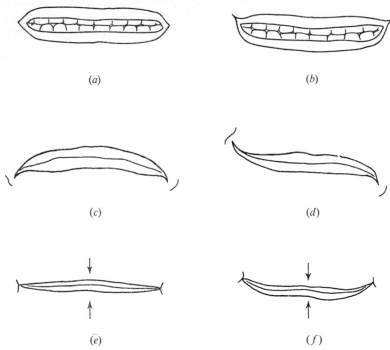

Fig. 2. Interactions in the form of units: (*a*) oblong mouth, (*b*) oblong smile, (*c*) mouth corners down, (*d*) wry smile, (*e*) tight lips, (*f*) compressed smile. (From Brannigan and Humphries, 1972.)

chapter. The examples given below indicate the degree of detailed description which is necessary for studying facial expressions, with all their diversity and subtlety.

Grant (1969) lists nine items under the heading 'aggressive elements'.

'Lips forward. The lips are apart and pushed forward. The mouth is open to some extent. This unit is frequently associated with "Look at" and "Aggressive frown" and forms part of the full aggressive display.

Small mouth. The corners of the mouth are drawn in towards the centre so that the mouth appears small. There is little or no pushing forward of the lips as there is in "Purse".

Tight lips. The lips are pressed tightly together.'

Grant says that the last two elements are very closely associated with actual hitting in children, as opposed to the 'aggressive frown' and 'lips forward' which more commonly occur simply as displays. However, tight lips is also common in many other situations of physical exertion (see Blurton Jones, in press).

'Intention bite. The lower jaw is pushed forward and the lower lip dropped to expose the lower teeth. This element is associated with aggressive activity and has been observed to lead directly to biting the other child.'

Grant says that there seems little doubt that it is an intention movement of this bite.

'Sneer. The centre of the upper lip is drawn up to expose the teeth. Quite often this unit is one-sided, only one half of the lip being drawn up. This drawing up of the lip results in the wrinkling on the bridge of the nose.'

Grant also distinguished a very abbreviated form of this element as 'lip up', where the lip is raised and dropped immediately. 'Sneer' lasts an appreciable time, say a second or longer. He says that the situations in which they occur are the same.

'Head forward. The head is brought forward towards the other person. This is a distinctly slower movement than is seen in the following element, and the position is likely to be held for some considerable time.'

This is also described by McGrew (1970).

'Threat. A sharp movement of the head towards the other person.

Chin out. The chin is pushed forward, tilting the head back and stretching the neck. [See also Blurton Jones, in press, and McGrew, 1970.]

Wrinkle. The skin on the bridge of the nose is wrinkled by much the same movement as "sneer". However, the teeth are not exposed. The lips usually remain together and there is no rolling out of the top lip.'

Grant also mentions the oblong smile as occurring when 'some agonism is present' and Blurton Jones (in press) has argued that the squared lower lip (which Grant expresses as 'corners of the mouth round slightly') and opposition of the incisors which distinguish this smile are seen sometimes in fights (Blurton Jones, 1967) but also in non-social situations which involve physical danger, or risk of impact.

Grant describes an 'aggressive frown' in which the brows are drawn together and down in the centre. This is distinguished from frowns which involve eyebrows sloping up in the centre and some simultaneous raising. Blurton Jones (1967, 1972) found that this 'aggressive frown' is indeed associated with hitting other children. However, a similar eyebrow position, with the addition of m. orbicularis oculi contraction, occurs during crying. The 'lips forward' element described by Grant is also described by Blurton Jones (in press), who observed it during fights, usually accompanied by shouting: it may give aggressive shouts their particular quality. 'Fixate', described, but ill-defined, by Blurton Jones (1967, 1972), seems to be the equivalent of 'gaze fixate' in McGrew's catalogue and compounded of 'look at' and 'threat' in Grant (1969). Perhaps the precise duration of gaze is important in identifying this impressive but subtle pattern of 'cold angry stare'. A number of arm and body gestures have also been identified in recent studies of children's quarrels. Certain combinations of face direction and eye direction may also be important.

NEGLECTED AREAS OF NON-VERBAL COMMUNICATION IN CHILDREN

Non-verbal communication by sound

I have already described some of the work on crying and laughing but there are other ways in which the sounds children make may transmit usable information. Besides language there are (a) the non-verbal features of language 'the way they say things' rather than 'what they say', which include for example the differences between talking, shouting, singing, and (b) an undescribed range of non-verbal noises: screams, sighs, grunts, giggles, babbles, imitations of machines, etc.

Wasz-Hockert et al. (1968) found that 'pleasure cries' do not begin until about 3 months. They are associated with smiling and looking at an adult who may be stimulating the child 'with gentle movements' or making 'soft noises', and were recorded after feeding and changing, often when the child was held by an adult. The categories 'pain cry' and 'hunger cry' are still distinct at this age and remain so through the first half year, and probably much longer.

Workers on the development of language have stressed the importance of children's babbling during the 6–18-month period as an antecedent of language. They emphasise the importance of adult responses to this behaviour in aiding the development of language. Babbling is thus assumed to be an adaptation with the function, not of signalling information about the infant's immediate or imminent state, but of ensuring that the infant receives the right sort of outside stimulation to aid language development. Some workers have regarded this and other little-understood behaviour, especially of newborns, as evidence that the nervous system is unorganised and puts out a mass of 'random' behaviour which differentiates during development. But besides the insuperable problem of defining 'random' in vacuo, there is increasing evidence of organisation and differentiation early in the baby's life. In any case, the precise ways in which babbling aids development may need to be investigated individually. In the past a variety of functions have been ascribed to it: it has been assumed to provide practice involving self-teaching, a stimulus to parental speech which can be learned and imitated, and an emitted response awaiting parental shaping by operant conditioning: it may also be a stimulus for parental imitation which gives additional information (air-transmitted not bone-transmitted sound) about the effects of the movements the baby has just made. Some of these mechanisms have been shown to operate in the learning of song by birds (see Chapter 6) and these parallels have interested students of the development of language.

Little seems to be known about the precise social, emotional and behavioural contexts in which babbling occurs and misconceptions abound. Sometimes (Bridges, 1933, and pers. obs.) babbling occurs between 1-year-old and 1-year-old, sometimes it is done while alone, sometimes in the company of parents, sometimes during meal-times and sometimes while playing. Single words, or idiosyncratic words not in the dictionary, are used with pointing, with approach, clinging, and a whining voice which stops when the parent responds by changing the situation. 1-year-olds also shout restricted sounds alone or at the sight of a friend far away, and make high-pitched double screams while running about vigorously.

Recent studies of nursery-school children have not concentrated on sounds, but they do describe some common noises. Besides laughing and crying, Grant (1969) describes squeals, short and high-pitched sounds during 'exciting games' and, usually with crying, sighs, sobs and sniffs. Blurton Jones (1967) describes a kind of scream which seems characteristic of a situation where something has been taken from a child by another, and which seemed very effective in getting responses from the teacher: there may be a relationship between this and particular facial expressions and 'whining' speech (Blurton Jones, in press). There is undoubtedly a huge field awaiting investigation here.

Non-verbal aspects of speech also have been little investigated in children though they are conspicuous, as they are also in mothers speaking to their children. Some early studies related indirectly to this topic. Hattwick (1932) showed that there were differences in the mean pitch of spontaneous speech by children, between occasions when they were talking to themselves, to another child or adult, in a group, or shouting to someone from more than ten feet: the range of pitch did not differ between these situations. Perhaps pitch range reflects the amount communicated while mean pitch reflects the kind of audience. In another study Hattwick (1933) showed that published songs for children were set in a much higher pitch than the children used when they sang these songs outside the teaching situation. One wonders whether the song writer confused the high-pitched voice with which he would talk as an adult talking to children with the pitch of the children's voices themselves. Fairbanks and his collaborators (1949*a*, *b*) showed that both for boys and girls aged 7 and 8 there were as many and as long 'voice breaks', sudden drops in pitch held to be typical of adolescent boys, as there were in adolescents. They concluded that these are not 'exclusively sex-linked phenomena'. This leaves the possibility that they have a function not just to communicate the sex of the speaker, but some variable feature of his behaviour.

Communication by structures

In the biologist's sense of communication (signal function), structures are as eligible for discussion as are behaviour patterns. Besides the bare face and elaborate facial musculature, several anatomical features of children have been suggested to have a signalling effect, with the implication that this may have contributed to the selection pressure for their evolution. Few of these suggestions have been followed up systematically either by careful comparative study or by tests of their effects. One of the rare examples of an attempt to do this is the study by Gardner and Wallach (1965) of Lorenz's (1943) suggestion that the head shape of human babies is important in eliciting parental behaviour. Gardner and Wallach limited themselves to asking adult subjects to say which of two profiles was more babyish. Variation along a scale towards greater cranium to face proportions went with classification as more babyish, even beyond the proportions of a real baby.

With careful recording of mother–child interaction it should be possible to see if there are effects on the mother of the size of the buccal pad (which gives the rounded appearance to the cheeks of children up to about $3\frac{1}{2}$) which can be separated from differences in the child's behaviour or age. Some of the other suggested signalling structures, like eye-whites, tear secretion on the eyeball, changes in pupil size, everted lips and blushing would be harder to test. Testing for the effects of the absence or weakness of adult features like secondary sex characters would also be scarcely practicable. The significance of the pale inconspicuous eyebrows of children might be easily tested with the aid of the mother's make-up kit. The 'pouch' beneath the eye which Grant (1969) describes as swelling in certain situations may well have an effect in signalling, but more needs to be known about the structure and its taxonomic distribution to establish whether it is a special adaptation.

Communication by odours

As far as this reviewer can discover, little or no work has been done on communication by odour in man, perhaps because it is a subject which arouses such strong feelings in our culture. This of course indicates that it may have a great communicative importance. No studies of odour seem to have been done on children, but remarks that mothers sometimes make, for instance that babies smell very nice, or that babies' faeces are not offensive until the baby changes to solid food, suggest that there is a field for research here.

Non-verbal communication from adults to children

Non-verbal communication in children must include the responses to signals as well as the production of signals. In interactions between parent and child there may be signals emitted by adults to children which they do not direct to other adults. These may or may not be limited to mothers and fathers. It is widely held in the 'advice to mothers' kind of writing, and in text-books for people working in child care, that babies and children are very sensitive to non-verbal signals and even to such information as that conveyed by the muscle tone of the arms that hold them. This may well be true, but the rather small body of data (e.g. Buhler and Hetzer, 1928) behind these statements is certainly inadequate.

The high-pitched voice of adults talking to children has already been mentioned. It can also be heard from 4-year-olds talking to 1-year-olds. In the same situation both adults and older children talking to smaller children are liable to bend at the waist, sometimes even to kneel or crouch down, and to hold the head on one side (as they do to peers if asking for something) and/or to wave the head slowly from side to side.

The ways in which mothers touch children may also communicate something about the probable responses of the mother. To move a naughty child they grasp it round the wrist and pull it more or less forcibly, whereas to keep one that is in a well-behaved mood close they take it hand-in-hand. Mothers and close relatives often stroke their child's head, usually from the crown back over the occipital area, while it stands beside them. This seems to be a brief signal of 'recognition and affection' in a situation of continuing contact where smiles other than open-mouthed 'play' smiles are unlikely. Whether it in any way relates to the behaviour of a child stroking its own head in conflict situations is open to speculation. The popular phrase 'a pat on the head' seems not to refer to something mothers do, though it may be done by other adults.

Some mothers carry a child horizontally under their arm if it has provoked their disapproval, yet still has to be carried. Rheingold and Keene (1965) have described the commonest modes of carrying seen in Washington, DC. Theirs seems to be the only quantitative study, and it is difficult to know whether to take statements in the cross-cultural literature like 'the infants are carried on their mothers' backs' as meaning that this is the major or the only form seen, or just whether it was seen occasionally and struck the author as so different from his own culture that he felt bound to report it.

Some features of the mother's touching behaviour reflects her ability to respond to the infant's signals rather than being themselves signals to the child. One who hugs, then boisteriously throws in the air, a child who approached with oblique eyebrows and 'droopy' mouth, instead of lifting it then holding it steadily until it lifts its head to look around, is responding quite inappropriately. The same would be true of one who responds to an approach that involved jumping up and down, open-mouthed smiles and flapping arms (held out towards her and above shoulder height) with lifting and sustained holding instead of by more boisterous close contact.

DISCUSSION

Recent direct studies do not seem to have progressed far enough to allow any general conclusions to be drawn about non-verbal communication in children. But two methodological points which may affect the course of these studies do need discussing. The first is the role to be played by control theory in studying the mechanisms of non-verbal communication. The second, on which I can reach no clear conclusion, is the difference between the interpretations given by Andrew (as in Chapter 7) and those given by other students of displays.

The role of control theory

In recent studies of causation of facial expressions it has been assumed that concepts like 'intention' were either unimportant for the explanation of expressions of young children, although possibly important in adults, or were altogether irrelevant to scientific explanations of behaviour. This latter assumption ignores the interactions between behavioural studies and control theory which are represented in works such as Miller, Galanter and Pribram (1960) and MacKay (1965), and which are used in a context nearer to the subject matter of this paper by Bowlby (1969).

For workers using purely observational techniques there has as yet been no attempt to show how a goal-directed piece of behaviour or an intentional act may be distinguished from any other during observation of the uncontrolled natural situation. But there are natural occurrences which give distinct impressions one way or the other and it is the duty of workers using direct observation to make clear what sort of occasions these are.

Many mechanisms in animal behaviour have been described which have the effect of finding a goal. Ways in which the behavioural mechanism selects routes to the goal can be of several kinds: simple repetition of the first activity, sequential arrangements of alternatives, alternatives occurring at different strengths of stimulus (a threshold system), alternatives evoked by particular specific aspects of the goal discrepancy (which do need to be specified when behaviour mechanisms are described this way) or thwarting environment, or combinations of these.

In non-verbal communication of children, no systematic studies of goal directedness seem to have been done, but at least one kind of simple and informative experiment is possible, although it partly concerns verbal communication. I have tried, on a few occasions, asking the teacher who runs our research nursery not to answer or respond when a child calls to her. I then recorded the subsequent behaviour of the child. The techniques for getting attention seem to relate closely to the time since the first call. The child calls again, then calls louder, then with pitch higher and higher as time goes on (provided that other factors inhibit approach, e.g. inability to get off the top of the climbing frame), the facial expression changes at the same time, showing squared upper lips and oblique eyebrows. If approach is possible the child may approach and move into line with the teacher's face, then tilt his head on one side and continue speaking, and next he may touch the teacher. In comparable interactions with children they are more likely to touch each other, and I have even seen a child grasp the head of another and turn its face towards its own face. My impression is that these alternative strategies are basically arranged on a scale related to increasing time and probably to increasing strength of whatever stimulus initiated the call, but that they are also related to the status of the reactor and to factors affecting the likelihood of approach to the reactor (e.g. Can the bricks be left without them falling? Was the child about to approach to show the reactor an object anyway? etc.).

Interest in control theory could lead to an increased emphasis on the effects of non-verbal signals as well as on their precipitating causes. It should also lead to analysis of the relationship between the effects and the causes. But when one tries to do this in detail it becomes clear that while control theory may set a useful framework it must not be allowed to gain a spurious air of explanation. Its use by some behavioural scientists makes it clear that they think they have an explanation rather than a framework. Advocates of the application of control theory to behaviour tend to stress the identification of the goals of behaviour as the main aim of causal analysis. Having identified the goal it is too easy to act as if the general theory of control systems explains, or removes the need to discover, the condition which sets the goals, and the rules by which discrepancy between feed-back and goal is measured (evaluator) and is related to a course of action (selector). The mechanisms involved in the evaluator and the selector, to use MacKay's (1965) terms, are usually left undescribed. In any but the most boringly simple behaviour these are the major questions about causation which active research workers investigate. Because of this there is actually no more conflict between the approach of workers on causation like Tinbergen (1959), Andrew (1957, 1963), Hinde (1958), Tugendhat (1960) and Blurton Jones (1968), and the concepts and questions raised by control theory than there is between students of causation of behaviour and students of the evolution of behaviour. Control theory may play a role very closely analogous to the role of the theory of evolution by natural selection in clarifying the ways of thinking about behaviour. Both theories provide helpful frameworks of the type: 'the task of this machine is xyz, when you have shown how the machine performs this task you have succeeded'. Research thus becomes more tidily goal-directed! But control theory models perform a confidence trick when they make us think that we have explained the behaviour. Where we think we have an explanation we are more often really saying 'the system does its job very well'.

Control theory tells us that behaviour does its job, which biologists knew anyway, but it does also tell us usefully that the job is more likely to get done if the relevant behaviour will only be switched off when the job has been done. But it seems to me impossible to regard the gaps in control theory approaches to behaviour as being trivial details.

Implications of size of behaviour items for study of causation and function

Andrew (Chapter 7) challenges the now traditional approach to the study of causation of displays, which is the approach used by recent workers on non-verbal communication in children. His challenge is based on the widespread occurrence of small parts of display movements outside their generally accepted social context. He challenges the assumption that the combination of parts is a better starting point for a study of the causation of displays than are the individual components. Certain points about the different 'kinds' of smile illustrate this very general problem about describing behaviour, the problem of selecting the size of units to describe and record. This is important because, as Blurton Jones (1968) and Andrew (Chapter 7) have suggested, this selection may limit the number of possible interpretations of the behaviour which results from investigation of its occurrence.

It is worth pointing out that this is no reason for giving up direct observation of displays. It does not affect the advantages of the detailed descriptions and definitions used in the modern studies as contrasted with earlier studies of children's behaviour and facial expressions. The early studies (and their contemporary descendants) used rating scales or categories of behaviour which were not fully described and which do not enable one to identify the behaviour being recorded. Modern studies of displays, whatever the size of unit employed, describe and define these units in a way which allows replication of the studies. Whatever one's view about the causes of the different forms of smile, there can be little doubt that it is valuable to distinguish them and to investigate separately their roles in interactions. If children respond differentially to different forms of smile by their mothers there is little future in looking for effects of a general category of maternal smiles on the behaviour of the child. Conversely, the forms of smile shown by children may greatly influence the responses of other people to them.

Grant's 'lip in smile' and 'oblong smile' are useful examples for discussing the problem raised by Andrew. Grant lists 'lip in smile' as a separate item but says it 'may well be a combination of upper smile and bite lip'. Andrew might well prefer the latter phrasing rather than the implication that 'lip in smile' was somehow different from the combination of the two components. Whether this distinction is important depends on whether we imply that the factors which elicit 'lip in smile' differ in any way from a mixture of the factors which elicit 'upper smile' with the factors which elicit 'bite lip'. In many cases it will prove extremely hard to tell whether there is a difference. But the classification may force one into the assumption that there is. If one sets about examining the situations and the behavioural sequences in which 'lip in smile' occurs one may find some very clear correlations with behaviour and base an interpretation on these. But these correlations may not show up, or different correlations may result, if one examines 'upper smile' and 'bite lip' separately. Andrew argues that the first interpretation will be entirely erroneous, while Blurton Jones (1968) implies that it leads to an incomplete interpretation rather than an erroneous one.

Much depends on the phrasing and interpretation of the conclusions. If one concluded for example that 'lip in smile' was a display with appeasing function which occurs when approach and avoidance are simultaneously elicited, this would not be jeopardised by showing, for example, that lip-in is caused by situations of danger while mouth-corner raising is a fragment of the behaviour involved in making quiet ('friendly') noises (Andrew, 1963). That oblique eyebrows can be produced by making someone look upwards against a bright light does not negate the finding that the commonest situation in which oblique eyebrows occurs in children is before or after crying, and that factors which bring about crying are in some sense a cause of both crying and oblique eyebrows. It may be that the occurrence of oblique eyebrows in bright light gives us a clue as to how oblique eyebrows comes to occur in the crying situation.

The evolutionary concept of ritualisation (see Cullen, Chapter 4), and the changes in causation called 'emancipation' by Tinbergen (see 1959, and earlier papers), give the possibility of maintaining that the social display is nonetheless quite independent of the non-social equivalent, except for their 'final common pathway'. So too does Andrew's suggestion that displays may acquire novel causation by learning. However, ritualisation requires to be demonstrated, and as Blurton Jones (1968) showed, this is very difficult.

The advantage of Andrew's approach is that it forces one to look at many more details of organisation of behaviour than does too ready an acceptance that a display is ritualised. While I accept most of the criticisms that Andrew makes, I do not feel that his viewpoint is as different from mine as might appear, nor necessarily as different from that of others investigating displays as he believes. Nor do I feel that Andrew's approach completely invalidates that of Tinbergen (1959). Tinbergen has an explicit, operational methodology. Andrew's methodology is not well described and at times leads one to very non-operational concepts which are as much evasions of the question as ritualisation can be. For instance, Andrew often leaves unanswered the question of how particular anticipatory responses come to be elicited in certain situations. To say that a display is a piece of thermoregulatory behaviour and the animal knows it will get hot if it fights, is no more useful and has more bogus explanatory aura, than saying that the display is a response to the causal factors which also produce attack.

The different sources of displays which Andrew lists are not necessarily independent of each other, or of the animal's imminent behaviour. Andrew himself provides an example in the elicitation of protective responses: 'This will depend in turn on computations of likelihood of approach by the other animal in the interaction, or of approach by itself to the other.' Here Andrew is beginning to put up explanations of the observed associations between displays and other social behaviour. These explanations imply that the association is partly due to internal factors and is not simply a fortuitous association of external stimuli. The significance of external stimuli must depend partly on the state of the animal, which itself may be influenced by the other external stimuli. Thus stimuli which affect the propensities to approach, groom, or attack, may then affect decisions as to which stimuli are classified as 'potentially dangerous to sense organs' or as a 'mismatch', or as 'startling'. It becomes increasingly difficult for this writer to distinguish Andrew's position from that reached by his own experimental study of great tit displays (Blurton Jones, 1968), or by his analysis of teeth-chattering in stump-tailed macaques (Blurton Jones and Trollope, 1968), which Andrew evidently ignores in favour of the more orthodox data on associations between facial expressions and attack. The difference in view-points may be reduced to the question of whether research will progress best by starting at the level of what Andrew calls 'major variables' or the level recommended by Andrew. Much of the difficulty of these discussions of causation may arise from the unsuitability of describing complex causal networks by a word like 'cause', which leads one to seek 'the cause' of a piece of behaviour (cf. the use of 'direct cause' in Andrew, Chapter 7), and from the unhelpfully concrete feel of names of 'motivations'.

None of these arguments should obscure the importance of distinguishing those displays that occur in a wide range of situations from those which occur in a narrow range. Nor do they excuse us from trying to explain why some occur in a more restricted range of situations than others do.

The difference between interpretations of displays by Andrew and by others (e.g. Tinbergen, 1959) also affects the question of responsiveness to displays. There is an apparent contrast between traditional views and the view that the reactor must relate many separate items together into a relatively unique, very informative pattern, from which not only imminent action but also information as to its strength, duration and probability of occurrence can be deduced. Traditionally one has assumed that displays might be responded to in a relatively 'automatic' way, influenced by relatively few variables in the situation or the behaviour of the actor. This is in part because assumptions about response to displays have been modelled on studies of sign stimuli. The seeds of an approach to the possibility of more complex kinds of reaction were present in Tinbergen's (1951: 81, and also earlier papers) discussion of heterogeneous summation.

A parallel difference in approach can be seen between workers on non-verbal communication in children, and some of those (see Duncan, 1969) working on adults. The latter argue that some non-verbal signals have no unique meaning and get their meaning largely from the context in which they occur. But this view is in danger of depending on an anthropomorphic (introspective?) view of meaning. It would be rash after seeing Andrew's paper to assume without serious investigation that a gesture accompanying conversation had no cause common to all occasions on which it occurs. Nor can one assume that their effects are not constant without examining data on their effects. But if one only tried to express their meaning in a verbal equivalent (usually arrived at by no better evidence than the observer's own conscious reaction), as do some of the investigators of non-verbal communication in adults, it is highly likely that these will be found to vary greatly. A crude example may be useful. Tail-wagging is done by dogs in many situations and Andrew had suggested a possible 'low-level', 'physiological' cause (arousal of running). But if we wish to translate its meaning into a verbal statement by the dog (the equivalent of what some researchers do) we have none that makes sense to us in all these situations. Also, if we wish to take other features of the situation into account we can give a meaning which says more about this specific occasion but less about tail-wagging. Tail-wagging may 'mean' 'take me for a walk' (when the dog wags while holding his lead in his mouth), it 'means' 'thank you for my supper' while he is eating, or 'pleased to meet you' while sniffing another dog. Whether tail-wagging has a constant meaning or not depends on how we define meaning.

COMMENTS

1. 'Displays'

Blurton Jones is concerned both with movements adapted in evolution for a signal function, and with movements which are potentially able to convey information. He uses the word 'display' primarily to refer to the former, implying a parallel with the display movements of lower species (cf. Andrew, Chapter 7).

2. Goals

In his discussion Blurton Jones discusses the different ways in which goals may be achieved. But in doing so he includes such strategies as 'simple repetition of the first activity', and implies that any stimulus which brings a sequence to an end can usefully be described as a goal. This would seem to stretch the goal concept to a point where it is no longer useful for the type of analysis advocated by MacKay in Chapter 1 (see also Comments to Part A).

3. Application of control theory

In the discussion at the end of his paper, Blurton Jones expresses some reservations about the usefulness of the theory of control systems for the study of behaviour, largely because, in his view, its use often involves 'a spurious air of explanation'. MacKay stressed that an 'information analysis' of the type *he* advocated was intended as a framework to suggest new questions rather than to offer explanations in the absence of data. It is concerned with elaborating questions about the 'evaluator' and 'selector' just as much as with identifying criteria of goal-directedness. Thus to MacKay, Blurton Jones' reference to the gaps in control theory approaches suggests a misconception of their function; while Blurton Jones asserts that no one ever does actually bother to elaborate questions about the 'evaluator' and the 'selector'.

11. SIMILARITIES AND DIFFERENCES BETWEEN CULTURES IN EXPRESSIVE MOVEMENTS

I. EIBL-EIBESFELDT

*Arbeitsgruppe für Humanethologie am Max-Planck-Institut für
Verhaltensphysiologie, Percha, Germany*

Is there a signalling code – a language without words – common to all men? The question has been much discussed and contradictory statements have been published. As long ago as 1872 Charles Darwin pointed out certain similarities in the expressive behaviour of men with different cultural backgrounds. He interpreted these as being due to characteristics inborn in all men, but this opinion has been challenged repeatedly. For instance Birdwhistell (1963, 1967) has advanced the hypothesis that no expressive movement has a universal meaning and that all movements are a product of culture and not biologically inherited or inborn.

Many anthropologists indeed were so struck by the cultural diversity, that they considered the culture-independent invariables to be negligible. Thus Andree (1899) wrote:

'One could fill easily a book with the enumeration of the various greeting customs in different nations. The scientific value of such an undertaking however would be trifling. One would encounter an incredible diversity of more or less unexplainable peculiarities and one would wonder about the delicate etiquette of the greeting patterns and the waste of time involved' (translation from the German original).

Biologists, biologically-oriented anthropologists and psychologists, on the other hand, have again and again emphasized the basic similarities of human expressive behaviour (e.g. recently, Ekman, Sorenson and Friesen, 1969), but a mutual understanding of the two points of view is still to come. This is partly because human behaviour has, so far, been inadequately documented, as Frijda (1965) has emphasized. Indeed we can search our large film libraries in vain for systematically collected documents on human expressive movements. These libraries contain a rich documentation of cultural activities (baking, weaving, pottery, etc.), but unstaged documents about human social behaviour (greeting, flirting, hugging of children by their mothers and the like) have rarely been collected. There have, however, been some exceptions, such as the films of Sorenson and Gajdussek (1966), Gardner and Heider (1968) and Bateson and Mead (1942). Gardner's film contains a rich documentation of various types of social behaviour, but this is included in a monograph and interspaced with many other scenes. And, to judge from the index, the films published by Sorenson and Gajdussek are similar. However, a person interested in greeting would like to find in one place films dealing solely with the activity he is interested in. This type of film library, containing documents on the social behaviour of man, still does not exist and, worse still, most of the social behaviour patterns have not even been documented cross-culturally. This is to be deplored, since social behaviour leaves no fossil tracks and once we have missed the chance to document the greeting behaviour of a native tribe at its first encounter with people of another civilization, the opportunity may be gone for ever.

This situation encouraged us to start a programme on the cross-cultural documentation of human expressive behaviour (Eibl-Eibesfeldt and Hass, 1967). We work with angle lenses that allow us to film people without their being aware of it. This is a prerequisite for any documentation of natural undisturbed behaviour. Even natives that are not familiar with the technique of filming get restless when a camera is pointed directly at them. But if the camera does not point towards them, they are not bothered by it. Of course, those close to the camera show interest in the proceedings, but this fades after a while and if one stays for a time in a village, one can document the intimate events of family life in people who would run away if the camera were pointed towards them. The technique works even in Europe and at close distance. People may notice the opening of the angle lens, but the assumption that a camera pointing in one direction is filming in that direction is so strong that very few people realize what is going on. We film most of the events in slow motion (48 frames per second). Sometimes, however, we use the technique of speeding up by filming at 2–7 frames per second.[1] This allows us to record longer-lasting events as a whole, e.g. a ritual, a family resting on the beach, a flirting couple and the like. If the speed of filming is chosen appropriately the movements of the body, head and limbs can be followed, although the people seem to move very fast to the later observer. Such documents allow us, for example, to count the number and duration of contacts between persons, to measure the distance they move apart and to observe the sequences of patterns involved. It is a valuable technique for collecting data for statistical analysis as well as for documenting the total pattern of an event.

A prerequisite for the later evaluation of the collected film documents is a detailed commentary accompanying every shot and stating in what context each pattern occurred and what the person did before and after the film was taken. Only in this way is an objective motivational analysis possible. It is actually the method ethologists employ when studying animals: that a specific posture signals threat or courtship is after all not assessed only from the posture, but also from its statistically significant recurrence in certain situational and sequential contexts (see Chapters 4 and 8, this volume). Data collected in this way reveal upon comparison many detailed similarities between the different cultures, as will be demonstrated in a few examples.[2]

The similarities in expressive movements between cultures lie not only in such basic expressions as smiling, laughing, crying and the facial expressions of anger, but in whole syndromes of behaviour. For example, one of the expressions people of different cultures may produce when angry is characterized by opening the corners of the mouth in a particular way and by frowning, and also by clenching the fists, stamping on the ground and even by hitting at objects. Furthermore, this whole syndrome can even be observed in those born deaf and blind (Eibl-Eibesfeldt, 1970a).

The similarities involve in addition minute details of the behaviour patterns involved. I shall discuss one example at length. When greeting over a distance people smile and nod; and if very friendly they raise their eyebrows with a rapid movement, keeping the eyebrows maximally raised for approximately $\frac{1}{6}$th of a second. I have filmed this so far in Europeans, Balinese, Papuans, Samoans, South American Indians (Waika, Quechua) and in Bushmen, and have observed it in a number of other groups (Eibl-Eibesfeldt, 1968, Plate 1 a–j). In all these cases the pattern signals readiness for contact, as can be deduced from the contextual and sequential analyses. In Central Europe the eyebrow flash is used mainly as a greeting to good friends and relatives, but if people are reserved they do not use it. This holds true for representatives of primitive cultures as well. In New Guinea I found that Papuans of different tribes initiated a greeting towards me, or responded to my greetings, with an eyebrow flash. In the village Ikumdi of the Kukukuku people, however, I did not get an eyebrow flash at all. This village was first contacted seven months before my visit and the government patrol at that time ran into trouble. The natives had shot one of their native porters, and in revenge the patrol had burnt some huts and broken the shields and war clubs of the natives. This was apparently the reason for their being reserved towards me, the second visitor. Greeting was restricted to a nod or a nod and smile at the most. However, eyebrow flashes were used when they were greeting each other (Eibl-Eibesfeldt, 1968).

In some cultures the eyebrow flash is suppressed. In Japan, for example, it is considered as indecent. In Samoa, by contrast, it is regularly used in greeting and also as a general sign of approval or agreement, when seeking confirmation, and when beginning a statement in dialogue. We use the signal in approximately the same situations, though we perform it less readily in a greeting encounter. We use it in addition frequently during flirting, when strongly approving, when thanking, and during discussions – for example, when emphasizing a statement and thus calling for attention.

We are normally not aware that we use this signal, but we respond strongly to it in greeting situations. We smile back and often answer with an eyebrow flash. However, if we are not familiar with the person we experience embarrassment. In an experimental set-up subjects addressed with an eyebrow flash then looked away, either by shifting the eyes or even by lowering or turning the head. Only those very familiar with the experimenter returned the eyebrow flash. (Unpublished experiments by Mario von Cranach.)

It is interesting to note that this signal has previously escaped the attentions of scientists. At any rate I have been unable to find any description of it in the literature. And yet, because the eyebrows emphasize this eyebrow flash, women give much attention to this region, often colouring the upper eyelid (in contrast) and thus making the signal more conspicuous.

We mentioned several situations in which eyebrow flashes of approximately the same stereotyped form occur: greeting, flirting, approving ('yes'), seeking (asking) confirmation, thanking, and emphasizing a statement (calling for attention). Can we find a common denominator for all these? In flirting and greeting the initiator certainly asks for contact and signals at the same time readiness to accept contact. When approving, for instance during a conversation, the eyebrow flash again signals acceptance of ideas and suggestions. It is a 'yes' to a social interaction, just as in the situations of greeting and flirting. The factual 'yes' by eyebrow flashing, as observed in many Polynesians, is probably derived from this. When thanking we again accept not only the present, but also the social bond symbolically initiated by this event. Finally in emphasizing a statement and seeking confirmation we seek contact and approval. So the basic common denominator is a 'yes' to social contact, and it is used either for requesting such a contact or for approving a request for contact.

By looking for other contexts in which eyebrow raising occurs we get hints as to its possible phylogenetic origin. People regularly raise their eyebrows during surprise and hold them in this position for a while. The same movement pattern occurs during a conversation when people ask questions, for example: 'What did you say?' In both cases people attend, opening their eyes to perceive better: the eyelids are opened and the eyebrows raised in connection with the opening of the eyes. Finally we raise the eyebrows during disapproval, indignation, and when we look at a person in an admonishing way, a pattern reminiscent of the threat stare of a number of infra-human primates. Again this admonishing look at the same time signals attention.

As a hypothesis I would propose that the eyebrow lift of surprise – originally part of the opening of the eye – was the starting point for the ritualization of several 'attention' signals. Some of these can be grouped together as the friendly attention signals, as represented by the eyebrow flash, and are mostly given in combination with nodding and smiling. Further evidence that the starting point for the evolution of these friendly attention signals was the surprise reaction is the fact that surprise is often involved in meeting somebody: the utterance 'Ah, it's you' when meeting in Central Europe is regularly accompanied by an eyebrow flash.

In its more generalized meaning the movement signals, as stated previously, a 'yes' to social contact, either requesting contact or approving a request. Smiling and nodding are added in these situations, whereas the derived factual 'yes' does not demand these additions, although they sometimes accompany it, especially the nod. The contact and approval-seeking eyebrow flash, as when emphasizing a statement, can, but need not, be accompanied by nodding and smiling.

The brow-lifting which accompanies indignation can also be derived from the surprise reaction. It can be interpreted as surprise concerning a misbehaving group member. It obviously signals to the group member that he has attracted attention, that he is being looked at but not in a friendly way. There is also a continuous stare emphasized by the eyebrows being held up, which is quite distinct from the eyebrow flash, for in the latter the impression of staring is avoided by the rapid lowering of the eyebrows following their raising. The expressions of disapproval and arrogance are related to indignation, but here the contact is rejected. The brows are kept raised, but in addition intention movements of withdrawal are added. The head is lifted in a backward movement and the eyelid is lowered, thus cutting off contact. In some cultures this pattern is used to express a factual 'no' – for example in the Greek (p. 303).

With the exception of the derived expression of arrogance (and 'no'), the eyebrow movements signal to another person that he or she is being looked at. This is necessary, since man cannot tell over a distance with certainty from the eye movements alone, whether or not he is being looked at (von Cranach, in press).

It is of interest to note that very marked lifting of the eyebrows occurs in some kinds of Old World monkeys, notably the macaques, baboons and mangabeys (van Hooff, 1967). In some species the effect is enhanced by a strong colour contrast between the upper eyelids and the surrounding skin. Accentuation may also be achieved by crests of hair (black ape) or eyebrows (patas monkey). This display element may have been selected as an indicator of visual contact. Depending on possible accompanying expressive elements and situational factors it may be interpreted as a threat or a positive signal.

Fig. 1 represents the hypothetical evolution of eyebrow movements into signals in man.

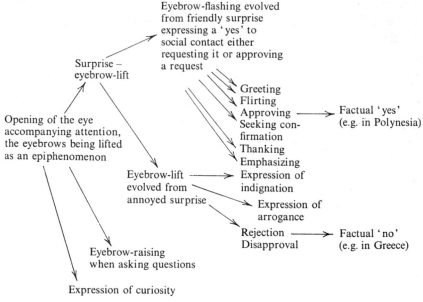

Fig. 1

Besides the eyebrow flash, which we have discussed more extensively, a number of other behaviour patterns occurring during greeting are similar in detail in different cultures. In addition to the striking formal similarities, similarities in principle are numerous (Eibl-Eibesfeldt, 1968, 1970b, c) (see p. 306). Some patterns of greeting, such as embracing and kissing, are apparently very old, since they occur also in chimpanzees (van Lawick-Goodall, 1968). The homologies of our smiling response are discussed by van Hooff in Chapter 8.

Another complex of behaviour patterns which is similar in a diversity of cultures is that of coyness, embarrassment and flirting. One pattern of embarrassment is the hiding of the face or just the mouth behind one hand. I have filmed this in Europeans as well as in Samoans, Balinese, Africans, Papuans and Waika Indians (Plate 2a–d). The pattern seems to be derived from hiding, since in its less ritualised form it can be observed in children (Plate 3a–c). This pattern also occurs in flirting girls. In addition the latter demonstrate ambivalence between flight and approach motivation – for instance by turning away and at the same time looking at the partner. Very often these patterns occur in successive ambivalence: the person looks at the partner, lowers the head, looking away and sometimes turning away, and finally looks back again (Plate 4 *a–c*). Giggling and hiding of the face behind the hand indicate a real conflict. The pattern can express itself in eye language alone: a girl looks at her partner, then lowers the lids thus cutting contact, and looks away.

These patterns of coyness which occur in flirting can be interpreted as ritualised ambivalence between flight and approach tendencies. In addition we can observe patterns inviting contact. Besides turning toward, looking toward, and the already mentioned eyebrow flash, tongue movements can occur. By playing with the tongue, Waika girls flirt with young males, and vice versa. In Central Europe also the pattern can be observed but it is loaded with sexual meaning and therefore considered as indecent.

So far I have emphasized similarities in human expressive behaviour between cultures. The list could be continued with many other facial expressions and gestures, for instance the patterns expressing grief (sagging of the shoulders, facial expression). But there are also differences to be considered. These may involve not only minor details, but sometimes also the basic patterns. An interesting example is the expression of yes and no. In many cultures people say 'yes' by nodding their head and 'no' by shaking their head, as Central Europeans do. This is for example the case with the Waika Indians, Samoans, Balinese, and Papuans. The Ceylonese, however, have two ways of saying yes. If one asks a factual question the answer 'yes' consists of nodding (example, 'Do you drink coffee?'). If, however, agreement to do something is expressed, the Ceylonese sway the head in slow sideways movements. The head is tilted slightly during this movement and it is very different from our no-shaking. This occurs, for example, when I ask the person 'Will you join me for a cup of coffee?'. 'No' in Ceylon is always expressed by headshaking, in the way we do it.

Another example of cultural differences in these basic expressive movements can be observed in Greece. 'Yes' is expressed with a nod as we do, but when saying 'no' the person jerks his head back, thus lifting the face. Often the eyes are closed and the eyebrows lifted for a while. When the 'no' is strongly emphasized one or both hands are lifted up to the shoulders, the palms facing the opponent. They may also say 'Okhi' or just click their tongues.

Here we are certainly confronted with cultural ritualizations. Interestingly enough, however, it is not the movement pattern itself which is traditional, but just its use in a particular situation. The movement occurs cross-culturally as a gesture of refusal and disagreement in a social context (see indignation, p. 301), though not everywhere as the plain statement for

'no'. It may, however, occur more widely when the 'no' is more emotionally loaded. For example, Central Europeans use it when expressing 'for
heavens sake, no!' Then the head is jerked back, the eyelids are lowered
and the hands raised with the palm facing the opponent. The raising of the
eyebrows which often occurs with the lowering of the eyelid signals
indignation. Furthermore, most of these patterns are part of the gesture
of pride, and both have the common origin of being intention movements
of withdrawal and refusing social contact. In an interesting parallel,
people of some cultures use the eyebrow flash (see above) for expressing
'yes' (e.g. the Samoans): as we have seen, this signal occurs cross-culturally
and signals readiness for social contact. In each case a signal that had had a
more specific function acquired the more general meaning of 'yes' or 'no'.
The use of these signals for the plain statement is, however, relatively
rare, and nodding and headshaking are far more widespread in this context.
Nodding and eyebrow flashing are often combined, the most common
sequence pattern being: lifting the head, eyebrow flashing and nodding.

The origin of headshaking and nodding was discussed by Darwin, who
suggested that shaking originated from food refusal. When the baby is
satiated it refuses the breast by turning its head away. It could indeed be
that this refusal gesture becomes ritualized into a 'no' by emphasis and
rhythmic repetition of the movement. It could, however, also be derived
from a shaking-off movement, which is part of the behavioural repertoire
of birds and fur-bearing mammals. In man it often accompanies a shiver.
Deaf- and blind-born children also refuse by headshaking. Nodding can be
interpreted as ritualized submission (Hass, 1968): we nod regularly during
conversation as a gesture of reassurance, so to speak, submitting to the
ideas of the speaker.

These examples show that cultural variation can result from the use of
the available, probably inborn, patterns in slightly different ways. In
addition there are numerous gestures which are culturally ritualized both in
pattern and meaning – for example the method of saluting by tipping the
rim of the hat or by lifting the hat, the latter being said to have originated
from the lifting of the helmet as an expression of trust. It is interesting to
note that a number of these culturally developed patterns show similarities
in principle in different cultures. This suggests that the acquisition of these
expressive movements may have sometimes been guided by phylogenetic
adaptations involving specific learning dispositions. Whether these are for
example 'innate releasing mechanisms' biasing the perception of the
individual, or drive mechanisms channelling behaviour in particular ways,
has yet to be explored (Eibl-Eibesfeldt, 1970 b, c).

1 The eyebrow flash, a universal expressive pattern used during greeting at distance, each example being illustrated by two photographs showing the lowered eyebrows at the beginning and the maximally raised eyebrow during greeting. (a) and (b): French; (c) and (d): Balinese; (e) and (f): Samoan; (g) and (h): Waika Indian; (i) and (j): Papuan (Huri tribe). (1 (a) by Hans Hass, others by author. All the photographs in this and following plates are copied from 16 mm movie films.)

(a)

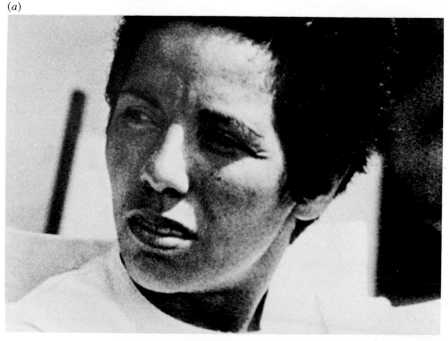

(b)

(c)

(d)

(e)

(f)

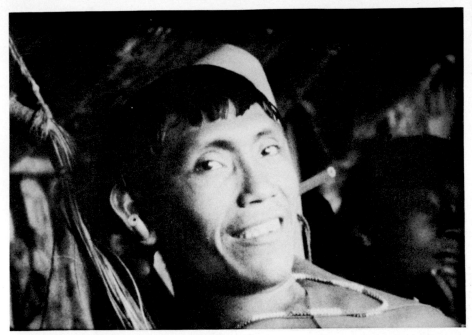

(g)

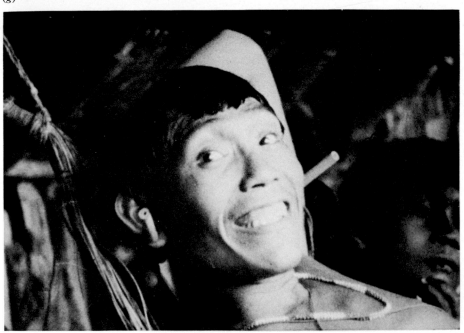

(h)

(i)

(j)

2 The hiding of the face during slight embarrassment: (*a*) Samoan girl after being teased about her strong interest in a photograph showing a male (she still holds the photograph in her hand). (*b*) Balinese girl after being given a compliment about her beautiful hands. (*c*) Turkana male when addressed and complimented for his nice head-dress. (*d*) Waika warrior when teased by his companions for openly flirting with a girl. (Photographs by author.)

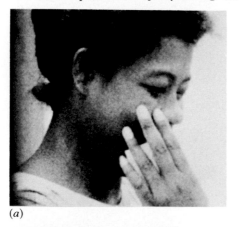

(*a*)

(*b*)

(*c*)

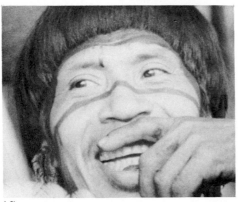

(*d*)

3   Hiding as a part of coyness behaviour. The 3-year-old girl was washing her hands. When asked by the author to show them she giggled, said no, then showed them and immediately hid herself as shown in the sequence. (Photographs by author.)

4   Turkana girl flirting after having been given a candy for which she had thanked. The sequence illustrates successive ambivalence (see text). (Photographs by Hans Hass.)

DISCUSSION

There are detailed cross-cultural similarities in both the meaning and the patterning of expressive behaviour. It is highly improbable that these similarities are due to chance. It remains to discuss the various possible ways in which they could have arisen. Many of the expressive behaviour patterns in man are certainly passed on by tradition. This is clear with patterns that are unique to one culture, such as the lifting of the hat. Often the historical origin of such patterns, their spread and the way in which they are learnt during ontogeny, can be followed. However, whether there are universals that are culturally traditional still needs to be explored. Since those behaviour patterns that are culturally learnt vary between cultures (for example, the development of dialects and languages – Erikson, 1966), it seems less probable that expressive patterns which occur as universals are culturally learnt. It is more likely that their universality is due either to common conditions in early upbringing channelling learning in a common manner, or that they are inborn. If Darwin was correct in suggesting that our headshaking as a gesture expressing 'no' develops from a sideways turning of the head as a gesture of refusing the breast after satiation, then this could well be an example of how a similar gesture can be learned in so many different cultures independently. In a similar way the refusal gesture of warding off with the open palm might be acquired from the pushing away of unwanted objects. There are many cases, however, where it is difficult to see how a complex pattern, as for example a facial expression, or a whole syndrome of behaviour, could have been acquired in this way.

Recently I have systematically filmed and studied the behaviour of deaf- and blind-born children of different ages. The results show that the basic patterns of facial expression are present in these highly deprived individuals. They laugh, smile, sulk, cry, show surprise, anger and the like (Eibl-Eibesfeldt, 1970a). The probability that they acquired all these facial and gestural expressions by learning is practically nil. Of course, one could imagine that shaping processes could take place accidentally: a smile, for instance, could be rewarded when it occurred the first time. But it would have to be an orderly pattern and not a mere grimace to release affection. And although shaping could be a faint possibility in the case of smiling, it is difficult to see how an anger syndrome (which would not be rewarded) could possibly be shaped in this way.

Another argument, which I once encountered in discussion, was that the deaf- and blind-born individuals may have acquired information about facial expressions by the sense of touch. However, I know of three deaf-

and blind-born who are thalidomide children, born without arms: they cannot acquire information in this way. They nevertheless exhibit the normal repertoire of facial expressions. Finally, I may mention that the typical facial expressions are also shown in those deaf- and blind-born who are so much brain-damaged that even intensive trials to train them to hold and guide a spoon fail. It is difficult to imagine how they could have learned social expressions without any deliberate training. If anyone insists in such cases on the learning theory, the burden of proof for such an improbable hypothesis lies on his side. It seems more reasonable to assume that the neuronal and motor structures underlying these motor patterns developed in a process of self-differentiation by decoding genetically stored information. That would mean that the motor patterns in question are phylogenetic adaptations, which I assume they are. For a detailed theoretical discussion of the concept of phylogenetic adaptations see Lorenz (1965) and Eibl-Eibesfeldt (1970a, b).

The cross-cultural comparison revealed also similarities in principle. These could be explained in several ways. One possibility which has to be taken into consideration is the possible influence of phylogenetically acquired receptor mechanisms (innate releasing mechanisms) on perception. Experiments of Ahrens (1953) and Fantz (1967) with babies, and of Hess (1965) and Coss (1969) with adults, demonstrate that we react strongly to two horizontally presented dark spots and even more strongly when a central dark spot is surrounded by a brighter iris-like circle. Depending on the size of the dark spot ('pupil') sympathy or antipathy (small circle) is released in adults. People react differentially to these characteristics, without being able to tell what they are responding to (Hess, pers. comm.). If it is true that we respond innately with fear to a stare or to eye-patterns resembling a stare, then this alone could bring about the independent learning of patterns of shifting the glance, or of cutting off the stare by lowering the lids and briefly looking away, during friendly social contact, without further imitation or advice.[3] The widespread use of eye-patterns as protective devices (Koenig, 1970) could also be explained by an inborn perceptual structure.

Another perceptual structure may be responsible for the widespread use of phallic figurines as guards (Wickler, 1966; Eibl-Eibesfeldt and Wickler, 1968; Eibl-Eibesfeldt, 1970d). These figurines show a male genital display, which is reminiscent of the phallic display of many infra-human primates (Ploog, Blitz and Ploog, 1963; Wickler, 1966). In vervet monkeys, baboons and others, some males sit at the periphery of the group 'on guard': with their backs to the group they display their genitals, which are often conspicuously coloured. This display is addressed to members of other groups and serves to aid in spacing. If a member of another group approaches, erections occur in the guards. The display probably derived from a mounting threat (Eibl-Eibesfeldt, 1970*b*) which became ritualized into a postural display. By behaving thus the males become territorial markers.

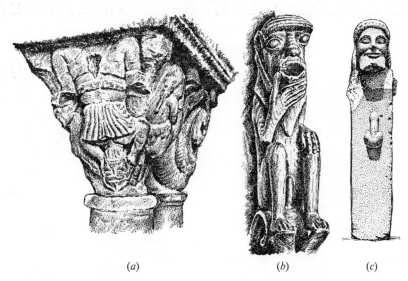

(*a*) (*b*) (*c*)

Fig. 2. Phallic European sculptures: (*a*) capital of a column in the cloister of St Rémy, France. The penis of the male figure (standing head-down) has been chiselled away. (*b*) 'Gähnteufel' sitting below the pulpit in the Cathedral of Lorch (Germany). (*c*) Herme of Siphnos (Greece). (After Eibl-Eibesfeldt, 1970*d*, and Wickler, 1966.)

It is hardly mere coincidence that male phallic displays as well as mounting occur in very different cultures as an aggressive display, though they are sometimes expressed only verbally. Nor does it seem a mere coincidence that guards, scaredevils and gargoyles in very different cultures are shown in phallic display. We find such figures in Europe, Japan, Africa, New Guinea, Polynesia, Indonesia and ancient South America, to mention just a few examples (Figs. 2, 3, 4 and 5). In modern Japan phallic amulets are used for protection and as good luck charms (Eibl-Eibesfeldt, 1970*d*). These similarities indicate that a perceptual structure, probably of sub-human primate origin, guides man, when he produces such guards.

In addition we can be sure that a number of the basic expressive patterns of man are phylogenetically acquired.[4] Many motor patterns develop independently of example. The degree to which such patterns are expressed, either as slight intention movements or in full intensity (or sometimes not at all) may, however, vary between cultures. In the Japanese, for example, the eyebrow flash is considered as being indecent and is therefore rarely seen. Here the cultural filter is evidently repressive.

Occasionally the argument has been advanced that the rich variety of our expressive patterns makes it unlikely that they are phylogenetically fixed. A similar argument was once put forward by Schenkel (1947), who studied the expressive behaviour in wolves and was struck by their high

(b)

(a)

(c)

Fig. 3. (a) Scarecrow from field in Bali; (b) phallic scaredevil from Bali (front and side view); (c) house guardian from Nias. (After Eibl-Eibesfeldt and Wickler, 1968.)

variability. Lorenz (1953) in consequence demonstrated that the combination of the intention movements of flight and attack in three different intensities results in nine very characteristic types of expression. The diversity of human expression might well come about by a similar combination of various basic patterns. There are strong suggestions that this is so, although the matter has still to be explored in more detail (see Chapter 8 by van Hooff, this volume). The frequent repetition of the statement that all of our expressive patterns are culturally learnt, and that except for a few reflexes of the newborn 'nothing' is inborn in man (Montagu, 1968), proves nothing: the authors promoting such views did not even consider the contrary evidence. The study of the deaf- and blind-born at least demonstrates that phylogenetic adaptation partly determines our social behaviour, though the extent of its influence is not yet known. However, the fact that at least some basic modes of communication are

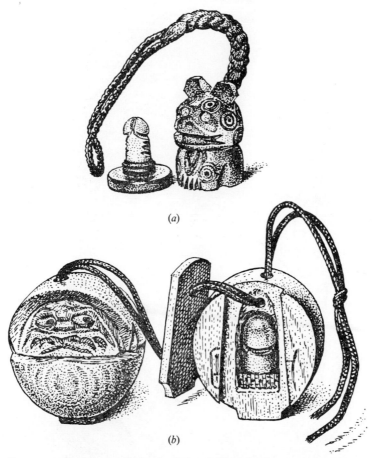

Fig. 4. Japanese amulets: (*a*) little plastic bear. The phallus can be screwed into the body of the bear; (*b*) front and rear view of an amulet showing a threat face and a penis in a shrine, normally covered by a lid. (From Eibl-Eibesfeldt, 1970*d*.)

inborn to man is of considerable theoretical importance (for a detailed discussion see Eibl-Eibesfeldt, 1970*a*).

Summing up, we can say that our conclusions are based on the motivational analyses of the observed patterns, on cross-cultural and cross-species comparisons, and finally on ontogenetic studies. The method of comparison has been discussed elsewhere (Eibl-Eibesfeldt, 1970*b*) in more detail. In principle we employ the criteria of homology used by morphologists. It

might interest the reader to see on what basis the anti-instinct argument
– to use an outdated term – is promoted. A well-known publication of
La Barre provides an example. He writes in his introduction: 'The
anthropologist is wary of those who speak of an instinctive gesture on the

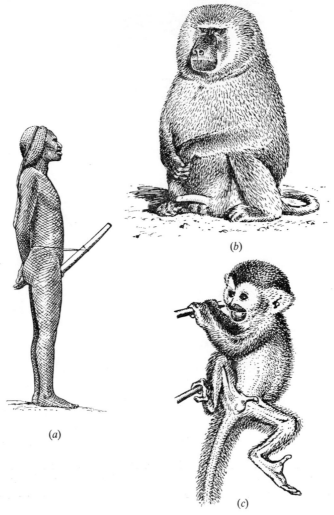

Fig. 5. (*a*) Papuan from Kogume on the Konca River; (*b*) baboon sitting guard; (*c*)
genital display in a baby squirrel monkey. (After Wickler, 1966, and Ploog *et al.*,
1963.)

part of a human being' (1947: 49), and then proceeds to enumerate
examples of supposed culturally determined expressive movements. This
is assumed to be the case wherever people react differently from culture
to culture to the same expressive movement, when the same signal is used

in different situations, or when different patterns are used to communicate the same message.

'Smiling, indeed,' he writes, 'I have found may almost be mapped after the fashion of any other culture trait; and laughter is in some senses a geographic variable. On a map of the Southwest Pacific one could perhaps even draw lines between areas of "Papuan hilarity" and others where a Cobuan, Melanesian dourness reigned. In Africa Gorer noted that laughter is used by the negro to express surprise, wonder, embarrassment and even discomfiture; it is not necessarily, or even often a sign of amusement; the significance given to "black laughter" is due to a mistake of supposing that similar symbols have identical meanings. Thus it is that even if the physiological behaviour be present, its cultural and emotional functions may differ. Indeed, even within the same culture, the laughter of adolescent girls and the laughter of corporation presidents can be functionally different things . . .' (1947: 52).

What La Barre presents as evidence is anecdotal. So far, to my knowledge, no motivational analysis of laughing exists, not even from our own culture (but see van Hooff, Chapter 8). In order to say that laughing in different cultures has a different meaning (or meanings) we need to know in detail and from many examples when people actually laugh. The statistical analysis may well reveal a common denominator.[5] Since such an analysis does not exist, statements asserting different meanings cannot be accepted. La Barre refers in his papers to many other cultural differences, for example in greeting behaviour, but again the comparison remains superficial.[6] The same applies when he discusses cultural differences in situations in which people weep. But La Barre concludes: 'So much for the expression of emotion in one culture, which is open to serious misinterpretation in another: there is no natural language of emotional gesture' (1947: 55).

This is certainly a naive and unscholarly treatment of the subject and must be criticized in face of Darwin's methodological contributions. No doubt, much of our non-verbal communication is learned, and this can be proved by ontogenetic as well as by comparative cross-cultural studies. In order to carry out such research we need *data* and not just superficial verbal descriptions. La Barre states, for example, that sticking out the tongue means different things in different cultures. To which type of tongue-showing does he refer? I have found a variety of forms of tongue-showing, each with a highly specific meaning, e.g. expressing disgust, sexual flirtation, mocking, etc. The subject is certainly not as simple as La Barre likes to present it, especially since additional signals are often added to one expressive movement and this results in a new pattern.

Careful analysis is needed and this means adequate documentation. If one wants to argue about the natural or 'instinctive' signalling code in man, one should not confuse the meaning of a movement and its pattern, as if they were the same. Both could be inborn but it is not necessarily so. The 'meaning' of an inborn motor pattern could be learned even when the motor pattern is a phylogenetically adapted one, and vice versa. A learned motor pattern might be shaped by a perceptual structure and often both are acquired from the culture.

NOTES

1 Among the various advantages of changing the film-speed is that one sees patterns, that normally escape attention. The eyebrow flash, to which we respond strongly but automatically without registering the expression consciously, was discovered when we examined slow motion films of greetings.

2 It is our intention to continue in our efforts to collect film documents, especially of populations that had had little contact with Europeans, and to build up a film encyclopaedia which publishes and archives such films. For this purpose the Max-Planck Society has established a research unit for Human Ethology in Percha, Bavaria.

3 In order not to be mistaken, I want to emphasize that I do not intend to say that this is the case and that the motor patterns are indeed learned. I only point to possibilities.

4 Whether the understanding of these signals is preprogrammed by Innate Releasing Mechanisms, is open to question. Cross-cultural understanding certainly occurs (Ekman *et al.*, 1969).

5 My cross-cultural film documents at least seem to indicate this.

6 The cross-cultural examination of greeting behaviour indeed reveals many basic similarities (Eibl-Eibesfeldt, 1970*b*).

COMMENTS

1. *Ontogeny of expressive movements*

Two extreme views are possible. One is that the facial expressions and gestures used in human non-verbal communication are 'inborn' or 'innate', which in this context means that their development does not depend on those types of experience which might be considered responsible for their cultural uniformity (see Comments after Chapter 3). The other is that they are labile, such constancy as they have being coincidental or culturally imposed. It will be clear that Eibl-Eibesfeldt takes a stand nearer the first position than the second. In the next chapter Leach expresses a different view, and comments on this issue are therefore considered there.

2. *Animal and human non-verbal communication*

Many of the themes introduced in Part B are picked up again in this chapter by Eibl-Eibesfeldt. Thus he emphasizes the use of more or less stereotyped movement patterns, often phylogenetically related to intention movements, sometimes associated with conspicuous structures, whose form and nature may have been ritualized in evolution for a signal function, and whose appearance may be associated with conflict situations. Clearly it is here that relationships between animal and human non-verbal communication should be sought.

3. *'Cultural ritualization'*

Eibl-Eibesfeldt uses this term to refer to processes leading to cultural peculiarities in the form or incidence of an expressive movement. The different senses of the term ritualization have already been referred to (see Preface): clearly the mechanisms envisaged here are quite different from those producing evolutionary ritualization.

Radcliffe-Brown's comments on the reification involved in this use of culture are mentioned on page 97. The same applies to Lorenz's usage, as quoted by Leach (Note 6 to Chapter 12).

4. *'Innate releasing mechanisms'*

This perhaps needs some explanation. As we have seen (Comments to Chapter 5), we respond selectively to the diverse types of physical energy bombarding our sense organs – some patterns of stimulation affect our behaviour, while others are 'not noticed'. Since particular types of behaviour are elicited by particular stimulus configurations, there is evidently some means whereby the latter are related to the former. This has been referred to as a 'releasing mechanism' because, according to one view of the control of behaviour no longer universally accepted, a stimulus was thought of as 'releasing' a previously inhibited response or even as 'releasing' previously stored 'psychical energy'.

Where the effectiveness of the stimulus was independent of previous experience of it, the postulated releasing mechanism was referred to as 'innate'. It can be argued that, in the experiments of Ahrens and Fantz which Eibl-Eibesfeldt cites, prior experience of the human face had not been controlled, so that the label innate is not justifiable. Furthermore, not only is the term 'releasing' related to a particular model of motivation, but the 'mechanism' is not a unitary one: it may reside partly in the sensory/perceptual apparatus and partly more centrally.

5. *Phallic displays*

The viewpoint presented by Eibl-Eibesfeldt (and earlier by Wickler) assumes that the human line once possessed a threat or boundary display in which an erect penis was conspicuous, that the display disappeared but that a fleeing or avoidance response to an erect penis persisted. It can be argued that the examples from sub-human primates cited by Wickler are heterogeneous, and that while some species certainly do use a penile display (e.g. Ploog and MacLean, 1963), in other cases which have been cited it is by no means firmly established that real displays are involved. The human amulets, etc., certainly do not themselves provide sufficient evidence for the hypothesis, and some (e.g. psychoanalysts) might argue that other explanations are possible.

12. THE INFLUENCE OF CULTURAL CONTEXT ON NON-VERBAL COMMUNICATION IN MAN

EDMUND LEACH

Faculty of Archaeology and Anthropology, Cambridge

ANIMALITY AND HUMANITY

For an earlier generation of social anthropologists the issue would have seemed very simple; the difference between man and other animals is that between Nurture and Nature, between that which is learned and that which is innate. Today the position seems much less clear cut. It has come to be realised that the presence or absence of 'culture' (i.e. a transmissible tradition of learned behaviour) does not provide any precise discrimination between the human and the non-human. In their limited way, communities of birds and beasts have their non-genetic cultural traditions just as do men. The elaborately 'ritualized'[1] systems of communication which occur among non-human animals are only in part genetically determined, and in many cases must depend to some extent on learning. For this reason some acute observers have come to believe that the difference between non-man and man is simply one of degree, that the communication systems of man are very like those of non-man, only more so. Carried to extremes this thesis leads to the position that man is no more than a naked ape.

The contention underlying this present essay is that this whole style of argument is fallacious for two very basic reasons. The first has already been indicated by Lyons (Chapter 2). Just as verbal communication in man is dependent upon context, including elements of non-verbal context, so also 'non-verbal communication' in man takes place within a context which includes language. Certainly man is an animal and, as an animal, he communicates with his fellows by signalling devices which are directly analogous to those of other animals. The eye movements discussed by Argyle provide a case in point. Such signals, in the main, are not subject to conscious control. But man as a human being is a creature that communicates by means of language. Just as his verbal performance takes place within a matrix of verbal competence, so also his non-verbal performance takes place within a wider matrix of cultural competence or cultural convention. Moreover, at the level at which a social anthropologist makes his observations, there is always a very close interpenetration of the verbal and non-verbal components of interaction; *all* customary acts which convey meaning as 'symbols' might, at a stretch, be classed as 'paralinguistic' in

Lyons' terms (i.e. those gestures, etc. which support conversation but are not integrated with its grammatical structure).

The second basic peculiarity of human non-verbal communication is that, as with verbal communication, a great deal of the discourse is not concerned at all with concrete issues which exist within the immediate context of the actors, but refers instead to abstractions and metaphysical ideas which are, as it were, 'off stage'. This lack of immediacy may be correlated with the fact that the verbally organised consciousness of man always includes an intense interest in temporal phenomena. Man, as human being, is not only aware that the living individual grows older and dies, he is also aware that the society of which the individual is a part continues. An enormous amount of human ritual – and here I am using ritual in its broad anthropological sense – is focussed around this central problem of asserting continuity in defiance of the threat of mortality.

Now this whole area of activity, which comprises very roughly everything that might be labelled religious behaviour, entails the representation, by means of metaphorical allusion, of explicitly metaphysical ideas for which the study of ethology provides no parallel. There are some forms of symbolisation which might be common both to a naked ape and to a human being. For example, sequences of human interaction which 'symbolise' relationships of dominance and submission (e.g. prostration), of affection (e.g. kissing, and other mouth contact salutations), of territorial possession (e.g. the erection of boundary markers), all have their apparent parallels in the 'ritualisations' recorded by ethologists. But gestures which stand for complex religious sequences (e.g. a Catholic's hand motion when he 'crosses himself') have no ethological counterpart. Yet it is symbolism of the latter rather than the former kind which forms the greater part of human non-verbal communication as viewed in a cultural context.

But all ritual actions which 'represent' metaphysical notions must necessarily be symbolisations of words. The verbal ideas come first, the symbolisation afterwards. So once again in searching for an answer to the question: 'What is there that is peculiar about human beings?' we are led back to a postulate about the priority of the verbal as against the non-verbal. It is a recognition of this fact which has led a long succession of European philosophers of the most diverse sorts, including Descartes, Locke, Hume, Leibniz, Vico, Rousseau and Wittgenstein to insist that the crucial distinction which separates man as human being from man as animal is his capacity for speech. For example:

'It is a very remarkable fact that (among human beings) there are none so depraved and stupid, without even excepting idiots, that they cannot arrange different words together, forming of them a statement by which they make known their thoughts; while on the other hand there is no other animal, however perfect and fortunately circumstanced it may be, which can do the same.'[2]

Or more crisply:

'Qui dit homme, dit langage, et qui dit langage, dit société' (Lévi-Strauss, 1955: 421).

This is a viewpoint which I share and which leads me to an extreme scepticism about the alleged similarities between the communicative processes of men and other animals.

SPEECH, NON-SPEECH AND TRIGGER SIGNALS

In elaborating his point that grammatical organisation is a quality of human language rather than of spoken utterance, Lyons (p. 64) introduces us to an (imaginary) community of hyper-Trappists who had acquired a non-spoken language as their 'mother-tongue'. Lyons' model has great relevance to the social anthropologist's practical experience. The field-working anthropologist regularly finds that, if he is to achieve his objective of becoming a participant observer, he must not only learn to 'speak like a native' but also must learn to 'act like a native' and that the two skills are interdependent. Until he knows the language he cannot use the appropriate gestures correctly, but until he is fully sensitive to the gestural context he cannot speak correctly. Non-verbal action of this kind, which is peripheral to ordinary speech utterance, plays a large part in human non-verbal communication. Much of it falls within the field of 'paralinguistics' (Lyons, p. 53), but other parts constitute a genuine sign language, linked with normal spoken language but distinguishable from it. For our present purposes I propose to summarise this whole area of 'meaningful action which is peripheral to speech action' as *non-speech*.

A high proportion of the statements which Lyons makes with regard to human speech are equally applicable to non-speech, using the term in this sense. The analogy between the verbal and non-verbal fields is direct. Cultural conventions constitute a 'generative grammar' which underlies the actor–observer's actual performance in the production and interpretation of symbolic gestures (Chomsky, 1966: 75). Just as one of the most characteristic features of human speech is that any speaker of a language is capable of producing and understanding an indefinitely large number of utterances that he has never previously encountered, so also, in our day-to-day interactions with our neighbours, we are constantly devising new sequences of non-verbal communicative behaviour (new *text-sentences* of non-speech) which the audience is able to understand despite the lack of previous experience.

In this respect most of the ritual behaviour discussed by social anthropologists is strikingly unlike the signalling systems which ethologists have been able to observe among non-human animals. It does not consist of isolated sets of trigger signals to which such concepts as *stimulus, response, reinforcement* can usefully be applied, but of gestures and symbols which are related to one another as a total system after the fashion of a language. Such non-speech 'languages' are both simpler and more complex than 'normal' spoken or written language. They are simpler because the syntactic rules are fewer in number and more explicit – the difference between right and wrong non-verbal behaviour is more clear-cut than the difference between right and wrong speech forms – but they are more complex because non-speech is a multi-channel phenomenon. A linguist can sensibly devote his attention to one kind of textual material at a time, either the spoken word transmitted through an auditory channel or the written word transmitted through a visual channel. But the analyst of non-speech must take account of the fact that the receiver of the message is likely to be subjected to communicative signals through several different sense channels simultaneously. Sound, sight, smell, taste, touch and rhythm may all be 'relevant', not only singly but in combination. Consider, for example, the complex ceremonial antics by which a wine connoisseur dining with a friend will display his expertise before he can decide whether a bottle of Château Latour 1928 is quite what he expected. It should be observed that what is 'communicated' in such an instance is not information about the wine but information about the skills of the actor!

As already suggested by Lyons, it is likely that the grammar of non-speech is often related to that of normal speech through a process of medium transfer and condensation, analogous to that by which vocalised speech sounds are felt to be adequately represented by patterns drawn on paper. The choreographic representation of dance movements is another example of such medium transfer, but the case of orchestral music has more general application.

What I have in mind is not just the fact that music, like speech, can be recorded in writing, but that a musician in reading the score transcribes what he reads into muscular movements. For example, qualities of loudness and pace variation, which are 'paralinguistic' in the context of speech, acquire central grammatical importance in the context of music. But, in this context, these qualities are translated by the conductor of an orchestra into a complex combination of movements of the arms, head and eyes. No two conductors code their signals in the same way, yet trained musicians seem to have little difficulty in understanding what is being 'said'. It may be inferred that the 'meaning' is somehow incorporated in the overall structure (syntax) of the conductor's actions rather than in individual signals.

The achievement of the Gardners in teaching the chimpanzee Washoe to encode a variety of English verbal concepts as gesture signs (see Thorpe, Chapter 2, this volume) is relevant here. Most commentators on their reports have been impressed by the apparently close approximation of the structure of Washoe's signalling system to that of a rudimentary human language, but it deserves note that most of the inventiveness lies with the Gardners and the inventors of ASL rather than with Washoe. Washoe was able to make speech-like communicative gestures only because human beings had previously accomplished the feat of transcribing English words into arbitrarily selected muscular movements. Two points are particularly relevant: (i) most of the signs employed in ASL are in no way 'natural symbols' either for human beings or for chimpanzees and (ii) although the Gardners in fact used an existing code (ASL), they were free to modify this code or invent a new one as they chose; they were not restricted by having to use a 'language' which they knew already. Normal Man in normal Society is constantly exhibiting resourceful inventiveness of this kind and it is for this reason that all generalisations about the 'meaning' of isolated human gestures divorced from the cultural context are highly suspect.

Lyons' discussion of speech has already given due emphasis to the importance of context and it has also made the point that, in practical cases, it is a matter of arbitrary judgement to decide just where to draw the boundary between what is linguistic and what is penumbral and 'paralinguistic'. The same difficulty applies to non-speech and this poses very real practical problems for field-working anthropologists. Just as *voice-quality* (Lyons, p. 51) is peripheral to spoken language, so also complex qualities of an aesthetic nature are peripheral to the grammar of non-speech, but there is seldom any obvious criterion by which the observer of a sequence of symbolic actions may decide just what should be considered relevant. All details of the text are likely to be in some degree *indexical* (i.e. indicative of their source, cf. Lyons, p. 71) but it may be hard to distinguish between intended and accidental communication. A lady who adorns herself in order to make a good impression upon her guests may indeed create an impression but of quite a different kind from what was intended. This again makes the point that the meaning of conventional symbols is highly dependent upon context. The comments: 'Mrs Smith looked very smart tonight' and 'Mrs Smith looked quite ridiculous tonight' could easily refer to the same lady observed on the same occasion by two different observers having only slightly different cultural expectations.

CULTURALLY STANDARDISED AND IDIOSYNCRATIC SIGNALS

The general problem here is one which presents formidable difficulties for the social anthropologist but can be evaded by ethologists and statistically oriented psychologists and sociologists. For an observer of non-human animal behaviour the norm is identical with the statistical average. He can plot on a graph the way that individual observations deviate from this norm and he may be able to make important inferences from such deviations, but he cannot take the step (which is fundamental for both the social anthropologist and the linguist) of asking the actor to give his own opinion about what is 'normal'. Human actors in ritualised situations behave in a variety of idiosyncratic ways; the same is true of non-human animals. Furthermore, in the human case, as in the animal case, it is possible to arrive at a statistical norm, e.g. 'On Easter Sunday 1970 $x\%$ of the women entering St Peter's, Rome, had their heads covered'. But in the human case (and *not* in the animal case) a third factor has to be taken into account, namely the consciously expressed normative rule, e.g. the cultural convention which says that Catholic women *ought* to cover their heads when entering a Church.

Lyons has emphasised that academic linguistics is largely preoccupied with the study of the grammar and phonology of *language systems* as distinct from the study of actual *language behaviour*; in the same way a great deal of social anthropology is concerned with the study of *cultural systems* considered as bodies of normative rules and conventions, like the grammar of a language or the rules of a game of chess rather than with empirical instances of cultural behaviour. Nevertheless, the problem of relating what is conventional – traditional – normative – to what is normal in a statistical sense is of crucial importance.

Orchestral music, which I have already proposed as an example of non-speech, illustrates some of the issues very well. Music, like speech, is peculiarly human in that an indefinitely large number of different sound patterns can be generated from the basis of a limited number of para-syntactic rules. Likewise, just as there are an enormous number of radically different human languages, so also there are an enormous number of different cultural conventions about what constitutes music (as distinct from random noise). But music differs from speech in many important ways. In particular, in speech there is ordinarily just a dyadic relationship – a message is transmitted from a speaker to a listener – but with music there is a third party; the composer is usually distinct from either the performer or the listener. In that case, although there is a sense in which the listener 'receives a message' (provided he understands the 'language' in which the music has been composed, he will not feel that he has just been subjected to a cacophony) it is difficult to decide just where this 'message' originates: does it come from the performer or from the composer, or is it in some degree self-generated by the listener? Perhaps all three. Two listeners hearing the same speaker can reasonably be expected to pick up the same message; two listeners hearing the same piece of music almost certainly will not do so.

This kind of problem is common to nearly all traditionally established ritual sequences in any cultural context. The normative cultural rule provided by 'tradition' is off-stage, like an author–composer–founding ancestor. It provides the rubric of what ought to be done but cannot control what is actually done. What is actually done will differ in detail from any previous performance of 'the same' ritual. In so far as participant observers pick up messages there is likely to be wide discrepancy; no two observers are likely to share exactly the same experience. My previous example of the over-dressed lady is a case in point. In such cases we shall not arrive at the 'truth' by averaging out any series of contradictory interpretations.

In this kind of situation we might hope to be able to distinguish at least three 'levels' of message:

(i) The message that is immediately intended by the actors: what the actors *think* they are doing. As for example when participants in a sacrificial ritual inform the anthropological observer that 'We are making a gift to the gods' or when my imaginary wine connoisseur claims to be making an objective assessment of the wine.

(ii) The intuitive and idiosyncratic interpretations of participant observers; e.g. children who look upon a sacrifice as an occasion for a good meal, cynics who look upon the wine connoisseur's performance as being the mark of a pompous ass.

(iii) The formal interpretation which depends upon the anthropologist's ability to make a syntactic analysis of the overall structure of the proceedings, so that e.g. animal sacrifices come to be interpreted in the light of 'general theory' about initiation rituals and the wine connoisseur's behaviour is placed in comparative context, as with the Englishwoman's non-rational choices regarding scent and cosmetics.

I think that all that I can usefully say on this point is that in the normal course of their professional activities social anthropologists are usually concerned with the third of these alternatives, while the other two levels are treated as raw data for analysis. To borrow from the linguist's vocabulary, the anthropologist's concern is to delineate a framework of cultural *competence* in terms of which the individual's symbolic actions can be seen to make sense. We can only interpret individual *performance* in the light of what we have already inferred about competence, but in order to make out original inferences about competence we have to abstract a standardised pattern which is not necessarily immediately apparent in the data which are directly accessible to observation. In this regard everything which Lyons writes about 'treating language as a model' (p. 57), 'idealisation' (p. 58), 'regularisation' and 'normalisation' (p. 59) has direct relevance for the social anthropologist's treatment of non-speech.

When social anthropologists argue that 'culture consists of messages' they presuppose that a preliminary sorting out of the 'raw data' has already been made. They are referring to customary behaviour rather than to idiosyncratic behaviour. But although the message bearing part of culture must necessarily conform to some degree of standardisation, this standardisation may not be immediately obvious. It may need considerable idealisation on the part of both actor and observer before syntactic order can be imposed upon what might otherwise seem to be just a random sequence of individual events. The ritual events which an anthropologist sees performed before his eyes are often barely recognisable as enactments of the drama which has been described to him in words only a few hours previously. Likewise, gestures of salutation, prayer (e.g. grace at meals), respect and so on *as directly observed*, are often so perfunctory as to be quite indecipherable without reference to some more elaborate ideal model. When the problems of interpreting human rituals are viewed in this light, their nature, and complexity, begins to become apparent.

INTENDED AND UNINTENDED SIGNALS

In his opening contribution MacKay (p. 23) has remarked that 'in trying to understand what is going on in non-verbal communication we are usually (for good or ill!) deprived of the possibility of asking the animal concerned what it thinks it is doing, or trying or intending to do'. In the human case this principle does not apply. The actor's own views about what is going on provide a central part of the evidence available for analysis. Admittedly some communicative acts among men belong to a class of unconscious or partially unconscious signals over which the actor has very little control. But such behaviour falls quite outside the purview of social anthropology and is beyond the range of what I have called non-speech.[3] In contrast non-speech, like speech proper, is usually intended by the 'speaker' to convey some sort of information. Just how far this intention can be made explicit or how far it corresponds to any 'real' consequences may be open to doubt (see e.g. above, p. 321).

The examples provided elsewhere in this book by Argyle and Lyons suffice to demonstrate that there is practically no individual human action or gesture of any kind, conscious or unconscious, which is not capable of conveying information to an interested observer. My own concern is not with the whole range of such potential signals but only with those which are patterned in accordance with cultural convention. As an anthropologist I am concerned with customary actions rather than with purely random idiosyncratic action. Nevertheless, in practical performance, customary acts are the actions of individuals who imagine that they know what they are doing. The status of such beliefs deserves consideration.

The problem at issue is one of venerable antiquity among anthropologists. At one time it was common to explain away the oddities of ethnographic fact by assuming that primitive man, unlike his civilised contemporaries, was incapable of making intentional choices. The primitive was 'a slave to custom', his non-rational actions were a kind of conditioned reflex to a cultural stimulus. This view of traditional custom as a *determinant* of ritual behaviour is no longer acceptable to anthropologists but specialists in other fields quite frequently express views which depend upon this kind of proposition,[4] so that the argument needs to be gone over.

A distinction which has been of great importance in the history of both anthropology and linguistics is that by which Sir James Frazer distinguished 'homoeopathic' and 'contageous' magic and which Jakobson has discussed under the heading of 'the metaphoric and metonymic poles of language (and other semiotic systems)'.[5]

Very roughly the point is that the relationship between a symbol and what is symbolised may be of two kinds: *either* the symbolism may depend on *similarity*, which corresponds in music to a shift of register, as when the piccolo plays 'the same note' as the bassoon but several octaves higher, *or* the symbolism may depend upon *contiguity*, which corresponds in music to the use of a melodic motif to stand for the whole composition, e.g. 'signature tunes' in modern radio. I will here refer to similarity-symbolism as *metaphoric* and contiguity-symbolism as *metonymic* (cf. Leach, 1970).

The relation between these two poles can be illustrated by a well known example from the Second World War. Churchill's V Sign with the two fingers of the right hand upraised was *metonymic* in its explicit form, 'V for Victory'; it owed much of its evocative force to its (?intentional) *metaphoric* allusion to an obscene gesture. The V was transformed *metaphorically* into the Morse code form: ···− and thence by further *metaphorical* transformation into the musical

In this last shape it became the signature tune for BBC news broadcasts

to Germany. In Germany, of course, it was not recognised as 'V for Victory' but as the opening bars of Beethoven's Fifth Symphony.

'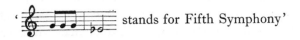 stands for Fifth Symphony'

is a *metonymic* relationship.

The particular point about the metaphoric/metonymic distinction which I want to emphasise here is that where we are concerned with customary acts performed in sequence, interpretation of the symbolism will be more arbitrary or less arbitrary according as to whether emphasis is given to the metaphoric or to the metonymic pole. Thus, in the above example, although, even in wartime, V can stand for other things besides victory and although

can suggest to the musician other compositions besides the Fifth Symphony, the possible choices within these metonymic registers are very limited. In contrast the variety of 'similarities' which can be established by metaphoric transformation is entirely without limit.

Now it is a fact that most consciously performed rituals in human society consist of *sequences* of quite 'simple' symbolic actions which the principal actors will describe as such. The *performer's* account always lays emphasis on what happens after what. That is to say he sees the linking associations between his own actions as metonymic: his first action is part of a larger whole and 'stands for' it. In such circumstances individuals who have a shared cultural background have little latitude of choice. Some performers will produce more condensed metonymic sequences than others but they are not likely to leave the track altogether. Given the opening bars of the Fifth Symphony as above, and told to whistle what follows, some performers will get on better than others but no one who knows the music is likely to end up singing *Three Blind Mice* or *Rule Britannia*. But a non-participant observer of a ritual performance has no corresponding commitment to stick to the metonymic tracks. On the contrary he is likely to inject into his private interpretations all sorts of complex nuances which originate in multiple metaphoric allusions superimposed one upon another. In this case different observers will extract different 'meanings' from what they observe according to their different prior assumptions. A Jesuit, a lay Catholic, a psycho-analyst and a social anthropologist interpreting the same performance of a Catholic Mass would each find quite different things to say about it, and each would be equally justified by his own criteria.

Another way of putting the same point is to say that when the meaning of symbolic actions is interpreted by individuals acting in their *private* capacity they are free to exercise their imaginations and they are likely to discern all kinds of metaphoric messages in the symbolism. On the other hand when the same symbolic actions are interpreted in accordance with *publicly* accepted dogma the emphasis tends to be upon metonymic rather than metaphoric associations and to show a much narrower range of variation. Furthermore, where the interpretation is metaphoric the actor (sender) of a symbolic message has no observer (receiver), but where the interpretation is metonymic (e.g. 'V stands for Victory') the symbolism is both public and intentional.

This argument may be represented schematically by Table 1.

Table 1

Context of action	The ritual as performed	
	if interpreted in the Private Arena	if interpreted in the Public Arena
Actor (sender)	Emphasis on idiosyncratic actions generated by private whim	Emphasis on culturally standardised actions performed according to a recognised precedent
Message bearing quality of symbolic action	Unintentional on part of sender	Intentional on part of sender
	Mainly metaphoric	Mainly metonymic
Observer (receiver)	Interpretation based on metaphoric intuitions. . . e.g. 'Symbol *x* means *y*'	Interpretation based on 'knowing what it is all about'. A shared competence in the language code as between sender and receiver. High evaluation of the story sequence in the ritual 'symbol *x* is a part of *y*'.

The corollary is perhaps obvious but, in the present context, important. If I wish to use non-speech to convey information to my neighbours I cannot rely on their private intuitions to pick up my intended messages in any consistent way, I shall have to use a publicly accepted code; I shall then adopt stereotyped patterns of behaviour which have become culturally standardised and which will be recognised as 'saying' something about my public role. In other words, if I *intend* to communicate I can only do so through the medium of ceremonial, and the main public emphasis in such ceremonial is nearly always sequential and metonymic rather then metaphoric.

The converse theorem would seem to be that non-verbal communicative acts which vary cross-culturally have no fixed meanings except when they are incorporated in formal ritual sequences. This is potentially a finding of some importance though in practice it may be exceedingly difficult to determine whether a particular type of signal is or is not open to cultural modification.

THE ELEMENTS OF SYMBOLIC CODING

The next question I want to consider is whether there is anything that is *generally* true about the symbolic codes embedded in public ceremonial. When we assert that these codes are culturally standardised, are we to suppose that each code is unique and peculiar to just one culture, or are there universals in non-verbal (public) languages just as there are universals in spoken languages?

Although an element of ceremony may be incorporated into even the simplest every-day action, we can all agree that some sequences of human behaviour are much more ceremonial than others. This is true of all societies. There are always some occasions, part secular, part religious, which are marked by specially elaborate 'ritual' – weddings, funerals, initiation rites, magical performances, public invocations of the deity, and so on.

Anthropological studies of ritual are ordinarily concentrated upon situations of this sort and in the normal way the proceedings are of great complexity. They involve many individuals acting over prolonged periods of time, both singly and in combination. It would be quite beyond the scope of this book to attempt any general analysis of non-verbal communication on this scale. But complex rituals may be segmented into more rudimentary elements. On every grand ritual occasion the individual actors put on a performance which serves to emphasise their mutual (formal) statuses in relation to one another and to third parties. Such performances consist of sequences of symbolic actions which in themselves are of short duration and relatively uncomplicated. The actions make use of objects as symbols. What kinds of action? What kinds of thing?

A total inventory of human culture would show that, in the context of a ritual sequence, almost anything *may* be treated as symbolically significant, but, cross-culturally, certain kinds of symbol crop up much more frequently than others. This is because man is himself a cultural universal, and actions which consist of the use of the body, or of parts of the body, or of things which are attached to or removed from the body, have a similar universality. This explains why the following 'things' are among the most widely distributed major constituents of human ritual sequences.[6]

1. *Adornments of the body*

This is an immense category. It includes markings on the body – tattoos, scars, paint – costumes in infinite variety, hair styles, physical mutilations (e.g. circumcision, knocking out teeth, shaving, etc.).

2. *Actual parts of the body*

Head, penis, vagina, breasts, hair, finger-nails, foreskin, hand, fingers, heart, etc. The common characteristic of most of the symbols of this class is that they can be cut off and separated from the body and then used metonymically on a 'part stands for whole' basis.

3. *Gross differences of posture*

Standing, sitting, lying prostrate, kneeling, etc.

4. *Relative position (as between two individuals)*

    (*a*) *level*: above or below
    (*b*) *precedence*: in front or behind
    (*c*) *confrontation*: face to face or back to back
    (*d*) *proximity*: close to or far away
    (*e*) *laterality*: on the right or on the left.

5. *Limb movements (gestures)*

In some sophisticated coding systems such as those employed in south Indian temple dancing a very great variety of limb movements are given symbolic significance, and it is fairly common for buttocks, belly, hips, etc. to be so used. But the most widely occurring range of culturally standardised gestures entails use of the arms, hands and head. (It should be noted that the limbs, as classified in the English language, are not necessarily independent symbolic units; for example, a gesture of obeisance may start as a forward bowing of the head alone, but when given emphasis this may extend to a forward bending of the whole upper half of the body. So also the arm and the hand may sometimes function as two separate signal elements within the same code and sometimes as one combined unit.)

6. *Pace*

Movement/non-movement; fast/slow.

7. *Nutrients (Relations between the individual and the food and drink which he consumes)*

This too is an immense category. Most of the generalisations which can be made about the symbolic uses of adornments to the body also apply to symbolic uses of nutrients of the body.

8. *Bodily excretions of all kinds*

Faeces, urine, semen, menstrual blood, spittle, tears, etc.

Finally, though perhaps belonging to rather a different class of sign elements there are:

9. *Facial expressions and disguises in stereotyped mask form*

Here it is necessary to exercise extreme caution in our judgement as to what should be deemed to constitute 'uniformity'. Earlier parts of this essay taken in conjunction with Lyons' Chapter have indicated the major difficulties. In the first place gross symbols of the kind I have listed are nearly always used in complex combinations, very seldom in isolation; secondly the context of use includes spoken language; thirdly, in non-speech, as in speech, the semantic value of text elements is heavily influenced by context.

SOCIAL ANTHROPOLOGY AND ETHOLOGY

At this point I have to make a digression so as to clarify my own position with regard to issues that are in dispute.

In the preliminary discussions from which most of the contributions to this symposium have been derived there was a persistent and very fundamental contrast in viewpoint between two groups. On the one hand there were those who were inclined to see the problems of non-verbal communication as focussing around the relationship between signal and response and whose ultimate concern – at least in the non-human case – was with 'the evolution and maintenance of phylogenetically-evolved rituals'; on the other there were those who stressed the importance of the linguistic analogy for all interpretations of human material.

The former group started out with the assumption that the communicative components of non-verbal rituals are compulsive behaviours of a highly repetitive kind which have been evolved either ontogenetically in the individual or phylogenetically in the species from actions of a non-communicative kind, and they kept harking back to the problem of how communicative values may have become grafted on to such actions. In this regard they were inclined to see the significance of human culture as historical rather than linguistic. Members of this group tended to underestimate the extreme variability of human culture and to assume that, at least in some statistical sense, like elements of action would elicit like responses regardless of cultural context.

The alternative view, that culturally formalised symbolic routines are best regarded as non-speech, carries with it the assumption that human beings have an innate, but very general, tendency to organise all their expressive behaviour grammatically in the manner of a language system. The critical differences between the two orientations are exhibited in full in Chomsky's review of B. F. Skinner's *Verbal Behavior* (Chomsky, 1959; Fodor and Katz, 1964).

Now since natural languages do not, in general, associate particular sounds or combinations of sounds with particular meanings, the linguistic analogy does *not* suggest that there will be any consistent relationship between non-verbal signal and response when such signals are observed in differing cultural environments, and it leads to the further conclusion that any claims to have demonstrated such consistencies in experimental situations must be considered highly suspect.

Man, admittedly, is a single species and, despite all differences of culture, the things that man is capable of doing with his limbs and his facial expression are everywhere limited by the same sets of muscles. This leads to the expectation that, in so far as the minimal segments of non-speech consist of movements of parts of the body, the same units of expressive gesture will crop up again and again in all kinds of cultural situation. This expectation is fully borne out by the ethnographic facts. But what can we infer from this? Does a raised eyebrow 'mean' the same thing in the streets of Pimlico as it does in the jungles of Peru?

The answer to that question will no doubt partly depend upon what we mean by meaning and, in this area, the gulf between the behaviourists and the linguists is often very wide. But it is twenty years since Tinbergen (1951: 24) wrote that 'the causation of behaviour is immensely more complex than was assumed in the generalisations of the past . . . the facts at our disposal are very fragmentary indeed' and the same note of extreme caution is still called for today.

According to the linguistic analogy a single gesture element such as raising the eyebrows is comparable to a single phoneme such as $/p/$ or $/k/$. These are elements which occur in all natural human languages but do not in themselves have any meaning. Consequently we are not led to suppose that any particular facial gesture – such as raising an eyebrow – will have a uniform cross-cultural significance.

The signal–response analogy leads to different expectations. In the human animal, as in other animals, particular emotional states produce predictable physiological reactions. Furthermore, conditions such as flushing of the face, tensing of the muscles, weeping, laughing, sweating, sexual tumescence and so on can be induced involuntarily by the experimental application of external stimuli. That being so we might expect that voluntary imitations of these reactions, exhibited in culturally defined contexts, would carry much the same meaning throughout the species.

Here we run into a direct conflict of evidence. Some ethologists, influenced by their expectation that like gestures exhibited in like situations will elicit like responses regardless of cultural context, have claimed to demonstrate this kind of consistency by the use of ciné film (see Eibl-Eibesfeldt, Chapter 11). In sharp contrast the majority of social anthropologists, relying on their intimate ethnographic knowledge, maintain that, even at the level of close person-to-person relationships, there is remarkably little cross-cultural standardisation of signal and response.

The evolutionary signal–response analogy has two branches. On the one hand there is an ontogenetic hypothesis which assumes that reactions which are innate in young children reappear as culturally stereotyped symbolic actions in adults, e.g. the tears shed by mourning women at a funeral are in some sense 'the same' as the tears of small children which are an automatic reflex of frustrated rage. On the other hand there is a phylogenetic hypothesis which assumes that some symbolic gestures of man are so deeply embedded in his genetic inheritance that direct comparison with the behaviour of quite different species of animal is scientifically legitimate.

To those, like myself, who put weight on the linguistic analogy (because we consider that language is a unique peculiarity of human beings), it is the second argument which seems most open to question.

Pair-bonding and hair grooming

Let me take a case in point. The ritualisation of head hair seems to be a human universal. The variety of things that can be done to the hair is almost endless: it can be cut off, shaved, worn long, worn short, curled, frizzed, drawn out straight, tied in knots, decorated with flowers, covered up, dyed, worn as a wig, used to decorate armour, offered as a sacrifice, preserved as a memento, etc. etc. etc. The emergence of the trade of barber/hairdresser as a separate profession is a marker of 'advanced' civilisation: in the modern Western world this has gone so far that the manufacture of razor blades, hair cosmetics, and the ancillary equipment of hairdressing saloons have become major industries. The phenomenon has naturally enough aroused the interest of psychologists, sociologists, social anthropologists, students of comparative religion and many others.

The literature on the subject is vast but inconclusive. Even the most rudimentary generalisations, e.g. that hair rituals always have a sexual component, have been challenged (Hallpike, 1969).

But hair grooming is not an exclusive peculiarity of man; it is a social activity of other species also. Some ethologists (e.g. Morris, 1967) consider that this similarity is highly significant and postulate a deep-rooted association between grooming behaviour and pair-bonding.

It is not at all difficult to select out from the vast variety of human hair rituals particular incidents which 'look like' those which can be observed in other species. Malinowski (1932, Plate 70) shows just such a picture in which small boys are busy delousing each others' hair in postures which closely resemble those of monkeys in a zoo. Since Malinowski also reports that in the Trobriands delousing is a recognised preliminary to sexual intercourse the pair-bonding hypothesis seems to be fully supported. My difficulty is to know what kind of generality is claimed for observations of this sort: in other places, other customs.

Delousing is a practical hygienic procedure which presupposes the existence of lice: no lice, no delousing rituals. But, quite apart from that, even where the practice is common the rules differ. For example, Firth (1936: 183, 504), writing about the Polynesian island of Tikopia, gives detailed information about who may or may not serve as a delouser. The rules are complicated but the principal point is that sexual partners, either actual or potential, may not act in this way. So in this case the pair-bonding thesis would not hold. There is no point in pursuing the matter here, but my own view would be that because of the high frequency of cultural variability of the kind I have indicated, cross-species ethological comparisons between man and other animals are nearly always thoroughly misleading.

The ontogenetic hypothesis is much more open, and just how far we accept it may depend upon where we decide to draw the boundary between non-speech (symbolic action which is culturally determined) and the 'paralinguistics of non-speech' (cf. Lyons' discussion of the boundary between linguistics and paralinguistics for the case of normal speech). At first sight the argument looks simple enough:

small children weep when they are frustrated;

adults weep when they are sad;

death evokes sadness;

therefore culturally sanctioned weeping in the context of mourning ceremonial is a 'ritualisation' of infant weeping.

Closer examination of the evidence raises doubts. The 'lamentation' which is often, but not always, felt to be appropriate to mourning situations can take many forms. Sometimes the expectation is that the mourner should quietly dab her tears (the pattern is usually sex-linked); elsewhere the expectation is a pronounced wailing and it is the noise rather than the tears which have symbolic value. Moreover it is culture rather than 'natural impulse' which decides who shall mourn and who shall not. In the Trobriands, mentioned above, children do not mourn for their mother; in many parts of the world the wailers are not relatives at all but are hired by the hour. Certainly weeping is weeping and, in some sense, it is the weeping of children which sets the model; but the use of weeping as a cultural symbol may be shifted from this source by several transfers of metaphoric register, and by that time the 'meaning' of the behaviour is anybody's guess.

SURFACE PATTERN AND DEEP STRUCTURE

The whole of the foregoing argument rests on certain basic assumptions that are implicit in the methodology of structuralism, the essentials of which have been admirably described in Lane (1970). Lane's presentation includes Fig. 1.

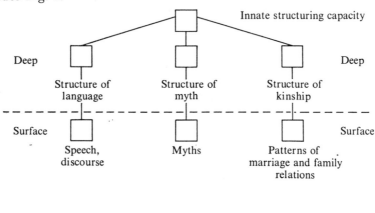

Fig. 1. Diagram to illustrate basic structuralist hypothesis (modified from Lane, 1970: 15).

At the heart of the argument is the thesis that (a) the phenomenon of medium transfer by which we are able to express the spoken word and music in written signs, and (b) the integrative capacity whereby simultaneous signals received through the senses of seeing, hearing, tasting, touching, smelling, etc. are felt to constitute a single rather than a multiple experience. Both imply that, at some level of the mind, we are endowed with an innate structuring capacity which is most easily conceived of in algebraic terms – the 'algebra' being the *structure* which is common to all the diverse cultural manifestations in which the operations of the mind may be observed.

The integrative capacity (*b*) is presumably shared with other animals but the medium transfer capacity (*a*), particularly with regard to its 'linguistic' attributes (see the discussion of *non-speech* above, pp. 316–17), appears to be a human peculiarity.

A consequential general point about structuralist method calls for emphasis. At this stage of physiological understanding, when we are still quite incapable of relating mental phenomena to the electro-chemistry of brain tissue, any inferences which are made about the 'algebraic structures of the mind' are mere guesses about the mechanism of a black box which is inaccessible to inspection, and in the case of animal studies it is perhaps best not to attempt such guesses at all. But because of the nature and importance of human language and the human capacity to answer back, hypotheses about 'how the human mind really works' become essential. Even so, structuralists fully appreciate that the sorts of guesses which they have so far been able to put forward are quite rudimentary.

In practice, most structuralist analysis invokes the kind of binary algebra which would be appropriate to the understanding of the workings of a brain designed like a digital computer. This should not be taken to imply that structuralists imagine that human brains really work just like a digital computer. It is rather that since many of the products of human brains can be shown to have characteristics which appear also in the output of man-made computers it seems reasonable to attribute computer-like qualities to human brains. This in no way precludes the probability that human brains have many other important qualities which we have not yet been able to discern.

One inference which is a direct consequence of any attempt to apply structuralist theory to the interpretation of culturally modulated symbolism is that, as in linguistics proper, we are likely to get a clearer grasp of the nature of non-speech universals if we move from the superficial level of manifest semantics to the deeper level of structural contrast. This point may be illustrated by the following example which brings out the close interdependence between verbal and non-verbal meanings.

SIGNALS OF AFFIRMATION AND DENIAL

The sign languages of nearly all human cultures include one or more methods of denoting affirmation and denial. Very frequently these signals entail some movement of the head or of the hands or of both together. Since there are only about half a dozen readily distinguishable head movements which can be used for such purposes it is hardly surprising that the codes employed often look very similar. The same signal components recur repeatedly – forward nod of the head, backward nod of the head, sideways nod of the head, partial rotation of the head. But the association of movement with meaning seems to be quite random. There is no statistical probability that the normal English convention by which a forward nod means assent and a rotational oscillation means denial will be met with in other cultures. On the other hand we can assert quite positively that wherever there is some conventional head movement which means 'yes' there will be some other conventional head movement which means 'no'. The same is true of hand movements, shoulder movements, facial grimaces and so on. Empirical data of this sort suggest that the binary 'concept' – yes/no – 'exists' as a basic structural opposition at a deep level of the 'grammar' of the human mind and that this opposition can be expressed either in speech or in any number of different kinds of non-speech. The coding of head movements is one such variety of non-speech, the coding of hand movements is another.

The 'yes'/'no' opposition of speech codes and the contrasted signals of non-speech codes are related to each other as metaphors; the structuralist hypothesis is that they are all alike transforms of a common algebraic opposition which must be assumed to occur as 'an abstract element of the human mind' in a deep level structure. The relation between this abstraction and the physical organisation of human brain tissue is a matter for later investigation.

Incidentally it is of interest that the internationally accepted symbolism of mathematics represents this binary opposition either as $0/1$ or as $-/+$. The second of these couplets is really very similar to the first since it consists of a base couplet $-/-$ with the addition of a vertical stroke to the second half of the couplet. But if a student of primitive art, who could free himself from our assumptions about the notations of arithmetic, were to encounter paired symbols $0/1$ he would immediately conclude that the opposition represented 'vagina'/'penis' and was metonymic of female/ male. A structuralist might argue that this fits very well with his assumptions about the deep structure algebra of human thought.

As another example of such 'consistency' I would point out that this same hypothesis that non-speech grammars share elements of universality with speech grammars would help to explain the fact (above, p. 318) that musical conductors of different nationality, using superficially different sets of signals, are able to convey information to members of an orchestra whom they have never previously met. The hand, head and eye movements used by different conductors are not standardised but they can be broken down into binary oppositions. The signals are certainly much more complicated and flexible than those of the Morse code, but it seems likely that they are of essentially the same type. The structuralist hypothesis is that, broadly speaking, this is true of all varieties of culturally modulated symbolism.

THE SIGNAL MARKERS OF SOCIAL STATUS AND SOCIAL RELATIONSHIP

On the face of it one implication of this whole style of argument is that the cultural selection of symbols must be completely random. How does this tie in with my claim (pp. 326–8) to have listed a whole series of categories of 'most commonly used symbols'? Quite simply it does not. In practice, the bias of choice in symbolic forms is very marked, sexual symbolism which relies on male/female opposition (e.g. penis/vagina representations) being far more common than any other. This is true at the cultural as well as at the idiosyncratic level. But the extent and cause of such biasses remain hotly disputed. Here I will confine myself to the logic rather than the psychology of the elements of non-speech.

The discussion which follows falls into four sections:

(i)   markers of interpersonal domination;

(ii) markers of sexual discrimination and social age;

(iii) markers associated with the temporal distinction between life and death;

(iv) markers associated with food choice.

I would add however that, in human contexts, these fields of interest are never fully distinct, and symbols appropriate to one register are readily shifted by metaphor into another.

1. *Markers of interpersonal domination (social hierarchy)*

(a) *'Level' as a social symbol*

A human society is 'a network of person-to-person relationships'. Every such relationship is either one of equality or of inequality; inequality implies dominance. Society as a whole is a structure of power and these power relationships are invariably exhibited in culturally stereotyped ways.

Viewed as a power structure, every political system incorporates elements of hierarchy, i.e. relations of 'superiority–inferiority'. The very words we use in English indicate how deeply ingrained is the idea that this kind of social relationship can be represented in 'the logic of the concrete' by differences of relative level, above–below. This idea certainly comes fairly close to being a human universal. The political leader is 'above' and 'in front of' those whom he 'leads' just as the head is above the feet and in front of the tail. This applies equally to the priest at his altar, the king on his throne or the college don at his 'high table', all of whom are *raised on a dais above* the commonality. The anthropologist on his travels would be very surprised to find such spatial valuations systematically reversed. On the other hand it is by no means always the case that 'leaders' are expected to walk in front of their 'followers'; if hierarchy is transformed into sequence the rank order can just as easily run 'early to late' as 'late to early'.

One of the complications here is that a time dimension often plays a part in conscious human symbolisation but verbal descriptions of time are usually linked by analogy with space. Notice, for example, how the English word 'before' means 'in front of' when applied to space but 'earlier than' when applied to time; yet even so the future is often said to 'lie before us'. Spatial–temporal ambiguities of this sort are very general throughout human culture.

Another complication is that although the positional symbolisation is in itself 'non-verbal', the logic of the positional code is always linked with a set of verbal concepts, in such a way that the binary opposition $+/-$ may very easily be reversed to take the form $-/+$. Thus, even where the linguistic relation above/below is compatible with *normal* symbolic performance, there are likely to be special circumstances in which the symbolic relation becomes inverted, thus expressing the *abnormality*.

(b) Inversion and 'abnormality'

Ethnography contains innumerable instances in which a reversal of sequence or an inversion of normality is made to carry social significance. The reversals can take all sorts of forms and are not confined to vertical level or temporal order. The substitution of *above* for *below* or *before* for *after* may alternate with substitution in such pairs as inside/outside, left/right, west/east, sea/land and so on. Here are two examples from our own culture.

(i) In days when the monarchy was a locus of greater power than is now the case, those who had caused the King offence were punished first by being incarcerated in dungeons below ground and then by being hung on gibbets with their bodies suspended above the heads of passers-by. The inner logic of such customs includes such reversals as the following: if living Kings are 'above' normal commoners, they are still more above criminals. But if dead Kings are buried below us in the ground, then dead criminals should be suspended above us in the air. Such arguments do not explain *why* offenders were alternately imprisoned below ground in dungeons and then suspended above ground on gibbets, but they do show that such apparently contrasted customs are mutually consistent.

(ii) In certain of the Shetland Islands at the present day fishing is so much the normal sphere of men's lives that women are subject to a complete taboo against having anything to do with fishing boats and their gear. Nevertheless as a marker of abnormal occasions, as at a wedding, the women will dress up as fishermen while the men engage in mock household activities which they would normally shun altogether.

My argument is that these large scale inversions which are characteristic of 'customs' or of ritual performances in which scores of different individuals participate, make sense to the actors because they incorporate, at a structural level, quite simple oppositions of the kind represented by the words yes/no in ordinary speech.

(c) Right and left

One very common coding which has attracted an extensive and sometimes polemical literature is that of laterality – very often we find that right is right (i.e. good, correct), while left is sinister, gauche and otherwise undesirable. But the exceptions are very numerous – left-wing political parties are not necessarily evil by definition, at any rate in their own estimation, and an Englishman driving his motor vehicle on the left is statistically no more murderous than his French equivalent driving on the right. Similar difficulties arise if we try to generalise about the meaning of different bodily postures.

(d) Posture

In the case of kneeling and prostration the above–below criterion nearly always applies. Both attitudes are almost invariably deferential, and prostration then denotes a greater degree of submissiveness than kneeling. On the other hand, the postures of sitting and standing are thoroughly ambiguous. Sometimes a superior sits in the presence of inferiors who stand, sometimes the inferiors must crouch on the floor, and sometimes it is the leader who stands while the followers sit. Analogies with all such behaviours can be found in the non-human animal world, but they do not explain the wide variations of human custom.

2. Markers of sexual discrimination and social age

(a) Social age, clothing and food

Man as an animal is either male or female and he gets older every day, but for man as a socialised human being things are much more complicated. Social age and even social sex are matters of culture not of nature and need to be manifested by publicly recognised signs. Small children are usually treated as sexless and then move through society in a series of stages –

infant to child, child to bachelor, bachelor to married man, married man to widower; with comparable stages for women. The changes are sudden, not gradual, and the different levels of social maturity are only loosely associated with biological age. Most of the transitions are marked by formal initiation (i.e. by rituals entailing sequences of symbolic actions) as with marriage in the English case.

Almost invariably ritualised changes of social age (for either sex) entail the adoption of body adornments, and changes in the conventions as to what is 'proper' food. Members of the opposite sex are thus discriminated as to their sexual availability by the way they dress or the body marks they carry, or the food they eat and how they eat it.

There are no easily discernible general rules about the relationship of dress to social age, each culture has its own patterns. It is not the case for example that the dress adopted at social puberty is necessarily modelled on the natural physiological changes which occur at biological puberty. Nevertheless the fact that hair dressing features prominently in nearly all social age discriminations must surely be related to the fact that the natural growth of body hair alters in the course of biological maturity.

(b) Hair

The coding of hair style symbolism shows many variations but there is an underlying approximation towards uniformity. Long hair is often used as a symbol of physical and political potency and contrasted with short hair which represents restraint and social subservience of all kinds, e.g. the celibacy of tonsured and shaven headed monks in both the Christian and the Buddhist churches and the very frequent use throughout the world of head-shaving as a sign of mourning. This pattern has been shown to accord rather nicely with a psycho-analytic thesis that the individual human psyche has a universal natural tendency to associate head hair with sexuality (cf. Berg, 1951; Leach, 1958; but contrast Hallpike, 1969).

This particular example is of some interest because it suggests the kind of connection which may exist more generally between animal trigger–response mechanisms on the one hand and culturally determined symbolic coding on the other. The relationship is not causal in any strict sense but psychological predisposition may generate certain probabilities at the cultural level.

(c) The problem of taboo

The symbolisation of sexual discrimination is everywhere complicated by notions of 'taboo' and 'sin' and some discussion of these issues will be appropriate at this point. Lorenz (1966) takes it for granted that culturally evolved rituals in man are comparable to phylogenetically evolved rituals in other animals. My own postulate would be the exact contrary of this. I assume that man has a variety of phylogenetically evolved predispositions some of which would be classed by an ethologist as ritual behaviour; but, as often as not, culture in man operates in such a way as to repress or invert these 'natural' ritualistic tendencies.

It is man's peculiarity that he is constantly imposing on himself prohibitory rules (taboos) which are then modified by further rules of special circumstances which allow the taboos to be transgressed. Thus there is a 'normal' taboo against killing other human beings, but when a human being is classified as an *enemy* it may become a duty to kill him; likewise there is a 'normal' taboo against having sex relations with members of the opposite sex, but rules of marriage and other conventions of sexual behaviour create special categories of persons with whom sex relations are not only permitted but obligatory.

On these grounds Bataille (1962) has argued that sexual behaviour in man, which he terms 'eroticism', is quite distinct from sexual behaviour in other animals. Eroticism, in this sense, always takes place within a social context in which social rules of taboo serve to establish a critical distinction between modesty and nakedness. These rules are somehow entangled with that other human peculiarity – the awareness of time through the recognition of death. The ritual attitudes which human beings everywhere adopt towards their dead seem to be psychologically linked at a very deep level with that persisting sociological mystery – the origin of the incest taboo.

We must suppose that this 'inner experience of eroticism' (and its inversions) is somehow reflected in cultural behaviour but the mode of this manifestation is not understood.

3. *Markers associated with the temporal distinction between life and death*

(a) *Rites de passage*

This last observation, that sexual taboo and concern with death are culturally very closely connected, is relevant for our understanding of a very important class of human rituals which anthropologists refer to as *rites de passage*, a term which they use as a general category to cover all rites relating to temporal transition either of the individual or of the society as such. The following generalisations concerning *rites de passage* have been current among social anthropologists for about sixty years (cf. Van Gennep, 1909, 1960).

When an individual undergoes a transition from one social status to another the symbolism of the ritual drama is repeatedly modelled on the image of death and rebirth. The individual 'dies' in his old role and is 'reborn' in his new role. The imagery of 'dying' in the old role is very commonly represented as a sloughing off of an old costume or an old skin, and quite often the rite requires the actual physical removal of some part of the initiate's body, e.g. circumcision, head-shaving, removal of teeth. This death/rebirth symbolism is 'logical' in its nature, it entails a structural opposition of the $+/-$ type that has been considered in previous examples. The view usually adopted by social anthropologists is that this logical coherence is sufficient in itself to 'explain' the pattern of symbolism that is adopted. This does not preclude the possibility that other kinds of 'explanation' are appropriate for other levels of analysis. In actual fact there are a great variety of such explanations. For example, the orthodox psycho-analytic interpretation is that all such performances represent a sublimated projection of 'castration'. On the other hand Bettelheim (1955), who is also a psycho-analyst, lays stress on the blood-letting aspect of many initiation rites and claims that the value attached to such 'symbolic wounds' lies in an association between the blood of the ritual wound and the menstrual blood of women.

But as I have already suggested, the metaphoric, as distinct from the structural, possibilities are without limit.

From a cross-cultural point of view the two *rites de passage* which are most consistently given ritual emphasis are those of marriage and death. It is a striking fact that the rituals concerned are often quite closely related and treated as a binary pair. One apparently universal element in the coding of the relevant symbols is that a married woman adorns her body differently from a marriageable woman and that, in a culture which tolerates the re-marriage of widows, the adornments appropriate to a woman entering marriage (e.g. the white veiled dress of a European bride) are similar but 'opposite' to those appropriate to a woman leaving marriage (e.g. the black veiled dress of a European widow).

In any particular instance it is quite easy to show special reasons of an historical or religious kind which might have led to this kind of symbolic reversal but the fact that these reversals are widespread seems to imply that the ultimate 'cause' must lie at a deep level of the subconscious.

MARKERS ASSOCIATED WITH FOOD CHOICE

Natural man is omnivorous. He can digest insects, reptiles, fish, meat, vegetables and fruit in wide profusion. Were it not for self-imposed inhibitions about what is or is not proper food, man-in-the-wild need seldom go hungry. Yet I have myself seen European soldiers starved to death in a jungle context where food existed in profusion on all sides. As between one society and another the cultural rules which determine food choice may vary enormously, but it is invariably the case that there are such rules and that they are treated as socially very important. It is normal rather than otherwise for these rules to be backed up by powerful religious sanctions.

The rules make a number of different kinds of discrimination. First they select between food and not-food. The ordinary individual may only become aware that these rules exist when he visits a foreign country where the rules are different. Thus it is a shock to the ordinary Englishman to discover that in France horsemeat, snails and froglegs all rate as food; at home this possibility would probably never have occurred to him. Secondly, the rules (religious taboos) specify that certain kinds of recognised food must not be eaten. The prohibition may be permanent and apply to all members of a cult group, e.g. no Jew may eat pork, no vegetarian may eat meat, or it may be temporary and refer to a transitional status of the individual concerned, e.g. menstruating women, pregnant women and widows are very commonly subjected to food taboos, and the treatment of the sick in all societies (including our own) nearly always includes non-rational dietary restrictions of one kind or another.

The other side of the same coin is that various kinds of food are treated as 'special' and reserved either for particular occasions or for particular privileged individuals. What is 'special' is not merely the food as such but the way it is prepared. This total set of discriminations provides us with a paradigm of our entire social world.

For some years past Professor Lévi-Strauss has been engaged on a large-scale demonstration that such differences in styles of food preparation and of 'table manners' embody elements of a universal semantic code. The food we eat and the way we prepare it demonstrate to ourselves and to the world at large just where we stand *vis-à-vis* nature and the rest of society. The reader of this essay must turn to other sources for fuller details of this surprising but fascinating line of enquiry – but it is clearly very relevant to the general subject matter of this book. Lévi-Strauss is interested in the social aspects of food symbolism but there must be a level at which such symbolism relates to innate motivations in the human individual.[7]

Freudian psycho-analysts claim that during the early stages of the formation of a child's personality we may distinguish oral, anal, and genital phases of interest. The ethnographic evidence is quite compatible with this formulation. In Lévi-Strauss' handling of the data the schemata, in terms of which categories of food choice are organised, appear to be markedly binary so that the structural association between the symbolism of food, of sex, of death, of time and of non-verbal noise becomes quite explicit.

The kinds of evidence which Lévi-Strauss adduces in support of his theories are extremely complicated and it is very difficult to see how his argument could ever be 'validated' in a way which would seem convincing to a sceptic deeply imbued with the canons of experimental science. But the whole point of the structuralist thesis is that when we are investigating the cultural products of human beings who are free to express their own opinions about what they are doing, the canons of 'objectivity', which are taken for granted by experimental scientists, cease to be applicable. I do not pretend that the structuralists have been able to *prove* their thesis. But even so the possibility that a wide variety of symbols systems in human culture are associated as transforms of the same structured patterns in the human mind (*l'esprit humain*) deserves serious consideration.

CONCLUSION

Some readers may have noticed that, taking my essay as a whole, the argument is inconsistent. At the beginning I went to great pains to emphasise the analogy between speech and non-speech yet in the latter part of this essay I have been writing as if it were already an established fact that collective 'rituals' of the kind that are usually described by anthropologists always build up their messages from repetitive sequences of symbols, rather than by free composition, in response to generative and transformational rules – as happens in the construction of a spoken utterance.

That I should have come to argue in this way only serves to emphasise the limitations of conventional anthropological procedures. Until very recently ethnographers working in field situations have not even considered the possibility that the customs they were studying might contain the kind of inbuilt flexibility that is implied by the linguistic analogy. Yet the dual facts that both food and sex behaviour in man are flexible rather than genetically determined and that food and sex both figure very prominently as symbolic counters in human rituals is highly suggestive. It provides a direct analogy with speech and invites us to consider the possibility that there is a generative-transformational-syntactic level at which cultural symbols may be investigated quite apart from the level of metaphor versus metonymy which has been the normal fashion among anthropologists ever since the days of Sir James Frazer (cf. Jakobson and Halle, 1956: 81).

If we are introspective about our own behaviour, we get the impression that each individual is constantly inventing new variations on the private symbolic games which he plays with his closest associates, especially with members of his immediate family, and here generative rules of symbol formation and 'sentence elaboration' certainly seem to be at work. Furthermore we all like to believe that changes in customary (public) rituals are brought about by the influence of the private arena upon the public arena.

But this brings us back to the crux of the problem as it appears to the social anthropologist. How far *is* the private individual 'free' to create his own rituals? How far, and in what ways, is he constrained (*a*) by the circumstance that he has been reared in a particular cultural community and (*b*) by the fact that he is, by his genetic constitution, a member of a particular zoological species?

Spoken language is a cultural phenomenon, yet a capacity to acquire a spoken language is evidently a genetic phenomenon. Spoken language, in its superficial manifestations, is infinitely varied but it contains at a deep structure level some elements of species-wide uniformity, and it is a plausible hypothesis that, in some as yet unexplained sense, these uniformities may be genetically transmitted. This *could* be equally true of non-verbal symbolic 'language' for here too, although the superficial manifestations are infinitely varied, there seem to be uniformities of a structural kind.

At the manifest level, the one element that is common to all human communication systems is man himself. All human beings are subject to the same physical limitations; as individuals they are capable of performing only a very limited range of actions. It is true that through the elaboration of his technology Modern Man has extended the structures of his non-verbal communication network far beyond the natural limits imposed by physiology but nevertheless it is right that we should be reminded that at the centre of every such network is man himself, an animal with animal limitations.

That being so it is a plausible hypothesis that if we concentrate our attention on those aspects of human ritual which employ, as signalling devices, parts and motions and attachments of the human body itself, then we may after all be able to demonstrate that there are structural universals implicit in the signal codes.

In this essay I have not been able to pursue these possibilities very far, but I have perhaps said enough to show where it might be profitable to look.

NOTES

1 As mentioned in the preface, it became apparent in the earlier meeting on 'ritualization' that the term 'ritual' is used in different senses by anthropologists and zoologists.

2 Descartes, 1955, vol. 1, pp. 116–17 cited in Chomsky, 1966: 4. For general discussions of the topic see Apel (1963), Hampshire (1969), de Mauro (1969).

3 The American anthropologist Edward T. Hall would disagree. Under the label *proxemics* he discusses the human use of space as a culturally determined variable and therefore a proper topic for anthropological discussion. Cultures certainly

vary greatly in their topographical and temporal presuppositions but it does not seem to me that Hall (or anyone else) has, as yet, devised a technique which enables such variables to be seriously discussed. See in particular Hall (1963, 1966) and Argyle (Chapter 9, this volume).

4 A recent example is Lorenz' assertion that: 'The role played by genetic inheritance in the evolution and maintenance of phylogenetically evolved rituals (in animals) is of course taken over by tradition in cultural ritualization (in man) (Lorenz, 1966: 280).

5 Jakobson and Halle, 1956: 76–82. The bipolar structure has been labelled in a great variety of ways by different authors: metaphor/metonym, similarity/contiguity (Jakobson and Halle, 1956); paradigmatic/syntagmatic (Lévi-Strauss, 1962); disjunction/conjunction (Parsons and Bales, 1955: 119 ff.); selective integration/progressional integration (Ruesch and Bateson, 1951: 183 ff.); identification/displacement (Freud, 1932), etc.

6 I know of no statistical study of the distribution of such elements. Other very widely used constituents derive from facts of nature – e.g. colour contrast – rather than facts of human physiology, but the distinction is not clear cut. For example the symbolic polarisation of blood and milk (or semen) which is geographically widespread has as one of its aspects the colour contrast red (or black) and white. This particular agglomeration of symbolic values has been extensively investigated by Turner for the special case of the Ndembu (Turner, 1966). Still another very widely distributed group of ritual symbols consists of nonverbal noise. This latter topic has been extensively investigated by a variety of authors, e.g. Lévi-Strauss (1964); Needham (1967). The opposition noise/silence or loud/soft is very often made to symbolise the oppositions potent/impotent, divine/human. It is at least possible that such symbolism has its roots in deepseated phylogenetic responses. After all the equation noise = fear is animal rather than human. I must emphasise again that I am here concerned with symbols which are used in a manifest way in the context of rituals which are specified by cultural convention. I am not discussing signalling mechanisms.

7 This is one of the central themes of Lévi-Strauss' *magnum opus 'Mythologiques'* of which the first three volumes are entitled *The Raw and the Cooked, From Honey to Ashes,* and *The Origin of Table Manners.* A short key section of the argument has appeared in English as *The Culinary Triangle.* See Lévi-Strauss, 1964 (1970), 1966.

COMMENTS

1. *Ontogeny of expressive movements*

In Chapter 11, Eibl-Eibesfeldt argues that many human facial expressions and gestures have considerable cross-cultural uniformity, though there may be differences of degree between cultures in their precise form, in the context in which they are used, and so on. Here Leach takes a different view. While acknowledging that some elements of expressive movements crop up in different cultures, he claims that these are comparable to the elements of human languages, and 'mean' different things according to the other elements with which they are associated and the culture in which they occur. Leach specifies the bias which leads him to this opinion – the view that human non-verbal communication is organised grammatically in some way like human speech. On this view, just because natural languages do not associate particular sounds with particular meanings, we must not expect relationships between non-verbal signals and responses – indeed, 'any claims to have demonstrated such consistencies in experimental situations must be considered highly suspect'. This issue gave rise to considerable discussion and correspondence amongst the group, during the course of which some shifting of emphasis occurred. To some extent, the issues can be resolved (see also Argyle, Chapter 9).

(i) Certain movement patterns used as signals, while intrinsically improbable, are common to all cultures so far investigated, e.g. smiling, crying, laughing. In many cases studies of the deaf- and blind-born render the view that cross-cultural similarities depend on similar conditions during development in all cultures extremely improbable.

(ii) Certain syndromes of behaviour are qualitatively (but probably not quantitatively) similar between cultures. For instance, anger involving a particular facial expression plus clenching of the fists, stamping of the feet, etc.; greeting; flirting, and so on.

(iii) In some of these cases it is possible to make hypotheses about origins from pre-human ancestors. This applies both to expressive movements (e.g. smiling and laughing, see van Hooff, Chapter 8, and Blurton Jones, Chapter 10) and whole syndromes (e.g. greeting, anger, see pp. 299–304).

(iv) In most of these cases not only the motor patterns themselves but also the motivational (= causal) basis of the movements and their interpretation by the reactor are similar between cultures; however these aspects are more labile than the movement pattern. There may be differences between cultures in that a movement pattern may be used in a wider range of emotional states, or suppressed (e.g. greeting – see p. 299), and there may be special uses of the movement (e.g. laughter in mourning; see also p. 311).

(v) Similarities between cultures occur especially with signals which concern personal or emotional characteristics.

(vi) Similarities between cultures do not necessarily imply that the movements are genetically determined independently of experience. The course of development could be due to elements of experience common to all human beings or to properties of all human beings which are themselves unrelated to communication but affect its development. To take an extreme example, all human beings are subjected to gravity, and gravity plays a major part in the development of posture. Rapidly enlarging objects are frightening to all races, and this could affect the development of both threat and, in the opposite manner, submissive movements. Intention movements of locomotion, striking, self protection and so on are similar amongst men, and this could account for similarities in signal movements derived from them. Oppositions (e.g. dominance *v.* subordinance) and conjunctions (e.g. male and female) common to all men could also affect the development of signal movements.

(vii) In many cases there are marked differences between cultures even in relatively simple expressive movements, though these can sometimes be seen as a gradual shift rather than a radical difference in convention (see also Gombrich, 1966). The diversity to be found in kissing, greeting, dominance/subordinance and so on between cultures cannot be lightly dismissed (La Barre, 1964), and attempts to specify evolutionary origins in such cases can be pretty tenuous or even absurd. Differences between cultures are especially likely in movements which take the place of words (e.g. yes/no), symbolise aspects of the culture which depend on verbal language, or have cultural implications at more than one level of complexity.

(viii) Even apparently marked differences between cultures may be the product of differing expressions of cultural universals at a deeper level, and it is these which are of primary interest to many anthropologists.

2. *Non-verbal signals and language*

The close inter-relations between such non-verbal signals and language in human cultures have recently been stressed from a slightly different viewpoint by Firth (1970) in a paper based on a talk given to this group. Referring to gestures of respect, he writes that they

'are signals, in that they release other actions by both parties. But they are not simply signals to advise other parties of appropriate action, as they might be in the animal world; they are symbolic of ideational and emotional patterns . . . The relation of these acts to language is complex. They can be described in language, they are initiated by verbal as well as non-verbal behaviour, and many of the clues to their meaning are given by language. To some extent such postures and gestures are a sufficient alternative to language, but in themselves they have a significance, a propriety, a restorative effect, a kind of creative force which words alone cannot give.'

3. *Anthropological methodology*

Leach emphasised that anthropologists are necessarily concerned with cultural systems rather than empirical instances of cultural behaviour, and compares the linguist's 'idealisation' or 'normalisation' of speech with comparable processes by the anthropologist. But it is not clear how such processes are carried out. How does the anthropologist decide what is relevant and what is noise? Cullen presumed that evidence must come from an assessment of the acceptable variation in the gesture or ritual, relevance and meaning being related to invariance. But is this in fact the way in which such issues are settled?

4. *Anthropology and ethology*

Leach criticises the phyletic leaps made from animal to man by some recent popularisers of ethology, like Ardrey, Lorenz and Morris. It would of course be quite wrong to stigmatise all students of animal behaviour on the basis of an acquaintanceship with some of its popularisers.

5. *Cultural symbols and symbolic coding*

(*a*) While Leach is at times at pains to emphasise cultural diversity in signal movements, at others he emphasises 'an underlying approximation towards uniformity'. In the latter cases it is, of course, relatively easy to point to exceptions.

(*b*) There is a considerable gap between showing that certain facets of human behaviour are compatible with particular symbolic interpretations, and showing that they demand them. For instance MacKay pointed out that there are many simple, practical and adequate explanations for the seating of monarchs, the incarceration of prisoners in dungeons and the hanging up of their bodies. 'All sorts of other considerations are relevant to the up–down dimension, such as who could find it easier to drop things on whom, or throw things at whom, or notice an intruder or miscreant, or . . .' The step from the sorts of data Leach produced to 'structure' common to all the diverse cultural manifestations seemed to some of the group a large one. It is little more than a truism that people attempt to classify objects and events in the physical world: it is hardly surprising that many of their classifications are binary. While it is easy to accept some of the symbolic equivalences suggested by this sort of analysis, an attempt to generalise too widely seems to bring one back to saying little more than 'humans classify'.

Similar criticisms have been made by Leach himself of the work of Lévi-Strauss (Leach, 1970), on whose approach his own analysis is based.

13. NON-VERBAL COMMUNICATION IN THE MENTALLY ILL

E. C. GRANT

*Basic Medical Sciences Group, Chelsea College (University of London), Manresa Road, London S.W.*3

The meaning of communication has been discussed a number of times in this volume and it has been suggested that intention to communicate is an essential aspect. However, in this chapter I wish to consider all behaviour that may transfer information from one person to another and would define my area of interest in the following terms. First, that there should be an output in the form of some recognisable act or behaviour pattern shown by one individual, and secondly a demonstration that this output can be received and reacted to by another individual. It then remains to be shown experimentally (*a*) whether the output has any constancy within or between individuals with regard to either the external situation in which it is displayed or the internal state and ongoing behaviour of the individual; and (*b*) whether the responses of the receivers indicate that the information transferred is consistent.

This description is, of course, very general, denying as it does any need for either the transmitter or receiver to intend to take part in a communicatory interaction and also denying the need for any accurate or even consistent transfer of information. Ekman, who has made considerable study of non-verbal communication, in discussing his use of the word 'communication' (Ekman, 1965) likewise stated specifically that 'there is no implication that the person . . . intended to communicate, nor any assumption that the communication is necessarily accurate'. However, he was concerned that the same information should be received by all or at least a statistically high proportion of receivers. The broader approach proposed above allows for specific reactions by individuals and so, for example, can take into account the fact that the same bit of abnormal behaviour might convey different information (mean something different) to different psychiatrists depending on their training, and at the same time might evoke yet another response from an untrained person.

Mental illness leads axiomatically to changes in behaviour since the only way of defining many kinds of mental illness is by reference to the abnormality of behaviour. Faced by an individual displaying abnormal behaviour the normal person will respond in one way or another. Commonly in the past, and still quite frequently at the present time, the response evoked is aversive, tending to increase the social distance between the individuals, or forcibly restrictive, trying to reduce the abnormal output. The abnormal behaviour is thus an output which is reacted to by other individuals, and the illness itself is communicatory. The diagnostic act, the labelling of the illness as schizophrenia, depression, psychopathy or whatever, is one kind of response to this communication.

All medical consultation is, of course, a communicatory pattern of behaviour between doctor and patient. In physical illness this communication is likely to be predominantly verbal, together with more or less co-operative, task-orientated behaviour as the doctor investigates the complaint of the patient. In mental illness it is more likely to have considerable non-verbal aspects. To begin with these non-verbal patterns may well not be directed at the doctor but, in our society at the present time, are likely to lead to a meeting with the doctor. For example, the abnormal behaviour could be total withdrawal and could go on for some time until a relative or friend informs a doctor or social worker and a medical consultation eventually takes place. With behaviour patterns such as drug dependency or some of the more bizarre behaviour of acute schizophrenic illness there may be complex interaction with other authorities first.

Attempted suicide is the non-verbal pattern which leads most directly and most frequently to medical aid, to such an extent that intention to communicate this need has been seriously implicated in its aetiology and some authorities, e.g. Cook and Stengel (1958), have stated explicitly that 'an attempted suicide is an appeal for help'.

Similar 'meaning', defined in terms of the response it evokes, can be ascribed to much mentally abnormal behaviour if the responses are recorded in the appropriate professional group; and it is expressed in phrases like 'attention seeking' and 'the cry for help' which then tend to be elevated into explanatory concepts.

The responses discussed above are to the total gestalt of abnormal behaviour, the display characteristics of the individual, what Goffman (1959) would call 'presence'. This total impression is still important at the immediate recognition level, and even in practical recording at the clinical level non-verbal cues are liable to be assessed at an all-or-none level and characterised, for instance in ward reports or clinical notes, as disturbed, confused, aggressive, etc. This is not surprising since even in normal interaction non-verbal behaviour is responded to, in the main, at a subconscious level and, if questioned about another individual, we can only respond in general terms such as 'he seemed a happy individual', or 'I couldn't get on with him'. In recent years there has been an upsurge of interest in the problem of a more precise statement of the cues that are being used to make these subjective assessments, and especially in the problem of quantification of those cues. In psychiatry this wish for objectivity and quantification led, in the first instance, to the development of rating scales. To begin with these scales simply attempted to subdivide the original assessments so that aggressive might be rated as: 1, very aggressive; 2, aggressive; and 3, not aggressive; but, as they became more sophisticated, more and more descriptive detail was added to the classification until it became clear that the necessary basic data required a full descriptive analysis of the behaviour pattern employed in communicating the overall impression.

Two developments outside psychiatry were going on at the same time. First there was a rekindling of interest in facial expression as part of the communicatory system of man. This subject had had a varied history in scientific investigation. Darwin gave it prominence in 1872 in his *Expression of the Emotions*. In the 1930s facial expression was again investigated in some detail and a number of studies were published, mainly by American psychologists. In general, these studies came to the rather surprising conclusion that human facial expressions carried no consistent meaning to observers and that no stable expression could be evoked by a given situation.

There is little doubt now that the negative and contradictory results obtained at this period were due to inappropriate experimental techniques. The commonest of these techniques was to ask an actor to display a named emotion, to photograph him, and then to ask the experimental subjects to name the emotion displayed in the photograph. This introduced artificial and static components into an essentially spontaneous and dynamic system. Even so, had they been willing to limit the investigation to the simplest emotions, e.g. pleasure, anger, they might have obtained positive results, and at the present time Ekman in America is using the 'judgement of photograph' technique to good purpose.

More recently descriptive analytical techniques have been developed for use in the study of complex ongoing behaviour in animals. Originally developed by European biologists for the study of 'instinctive' behaviour in the invertebrates and lower vertebrates, these ethological techniques are used with considerable effect for the more labile behaviour of mammals. These three areas of interest, the psychiatrists' requirement for precise statements about non-verbal behaviour, the interest in the face as a major organ of communication, and the development of recognised techniques of observation and description, now overlap and interrelate – and although very little has yet been done, some information is becoming available on non-verbal communication in the mentally ill.

Before describing some results of such studies it is necessary to limit the scope of the discussion by indicating the kinds of illness involved. These illnesses are those which, although possibly having some biochemical or physiological basis, are essentially psychological in character and are defined in terms of their effect on the thought processes and behaviour of the individual, the psychoses and neuroses. The so-called physically based illnesses, for example sub-normality and those where the malfunction in behaviour can be directly attributed to a gross neurological lesion, will not be considered. Even this distinction is, however, difficult to keep consistent: epilepsy for instance is clearly physically based but involves many psychological factors, and the disturbances in behaviour can only be directly related to the lesion during actual epileptic attacks.

DESCRIPTION AND ANALYSIS

Behaviour, especially communicative social behaviour, is not a continuous process but is made up of a series of discrete patterns. For example, if we look at an individual, we can say that he *is* smiling or he *is not* smiling and, although the smile might vary in its morphological development, this difference can generally be easily determined: the periods 'beginning to smile' and 'stopping smiling' are brief and transitory. The isolation and description of such units is an essential first stage in a descriptive analysis of behaviour and check lists of these units in man, with adequate descriptive backing, are beginning to appear. A number of examples are given in other chapters in this section and over a hundred units associated with the face have been described by this author (Grant, 1969).

In a long term study of the mentally ill only a very few units have been seen that are not also seen in the normal population (Grant, 1968, and unpublished). These new patterns have on further observation proved to be of two kinds. First, some are physical responses to side effects of medication, for example, the mimicking of Parkinsonian movements caused by high doses of the phenothiazines, or tongue and mouth movements due to the drying-of-the-mouth effect of many of the centrally acting drugs. Secondly, highly idiosyncratic behaviour patterns are occasionally developed by particular individuals. These can act as communicating patterns and will be discussed again later.

On the other hand only one behaviour pattern recorded in normal individuals was not recorded in a particular group of mentally ill. This was shrugging of the shoulders, and was never shown by individuals classified by medical staff as schizophrenic. This particular piece of behaviour is, of course, highly formalised with quite complex meanings varying from 'I could not care less' to 'What are we going to do about the situation': like the much simpler shaking and nodding of the head it could be thought of as a verbal equivalent, and its non-occurrence in schizophrenic individuals perhaps related to their general difficulty in verbal communication. Non-occurrence of an event in a finite series of observations is in any case not definite evidence of its non-occurrence in the population. In general, we can say that the normal vocabulary of non-verbal communication as used in face-to-face social meetings is also available to the mentally ill.

Evidence concerning the meaning of the behavioural units can be obtained from a variety of sources. First, from a description and assessment of the situation of occurrence: we can see, for example, that smiling increases when friends meet after a parting. Secondly, from an investigation of the inter-relationships of the units in the ongoing behaviour of the individual: if a frown is statistically associated with a physical blow we can suggest that frowning is an indication of a possibility of attack. In short, it is an aggressive signal. A third source of data is the response sequences between individuals: for example, if we can show that turning the head away and closing the eyes is a frequent response to attacking behaviour by another individual, we can suggest that these elements indicate submissiveness.

Using such methods it has been shown that human non-verbal behaviour has a logical structure and can, and does, carry meaningful information in a social situation. Using the same methods on two groups of mentally ill patients, one diagnosed as chronic schizophrenic, and the other a group of acutely ill individuals not including schizophrenics, Grant (1968) showed that the internal logic of the behaviour, and of the way and situations in which it was used, did not differ between these two groups. Nor did either group differ in these ways from normal individuals.

So it would appear that from the point of view of non-verbal communication the mentally ill individual uses the same vocabulary as the normal individual, and uses that vocabulary with the same meaning as the normal individual. On the other hand ability to recognise and respond to other peoples' non-verbal communication may well be impaired (Eysenk, 1954). Ekman, using his judgement of photographs technique, has shown that schizophrenics are poor in this ability and that individuals with other illnesses find difficulty in particular areas of emotions (pers. comm.).

There remain, however, very distinct differences between the normal and the mentally ill individual in the use of these behaviour patterns. In brief, these differences are seen in the bias of the behaviour, in rigidification and in a restriction of repertoire in the individual case.

The behaviour of the mentally ill individual is biassed towards one particular kind of behaviour. In schizophrenia this bias is towards a group of behaviour patterns associated with flight and withdrawal and away from those patterns associated with assertion. Table 1 shows the occurrence of a selection of units in schizophrenics and normals. In neurotic and psychopathic illnesses the bias is more individual, any one person tending to react in a particular way: one person might generally react assertively, another submissively, and so on.

Table 1. *Number of elements of behaviour recorded from normal and psychotic individuals. (Total elements recorded was 5000 for each group)*

	Behaviour element	Students	Psychotics
	Look at	1053	297
F.	Mouth corners back	91	147
F.	Lips in	43	57
F.	Look down	207	417
A.	Frown	92	55
A.	Thrust head	150	1
A.	Gesture	657	81
	Smile	112	141

F: elements associated with flight
A: elements associated with assertion

The apparent restriction of repertoire in the mentally ill individual may be partly an extension of the bias described above. If the behaviour is biassed towards flight and away from aggression, then it is unlikely that the more extreme units of aggression will be seen. This would not appear to be the only restriction however, and individuals show a smaller range of behaviour patterns than would a normal individual over the same period.

Evidence that the behaviour of the ill person is more rigid and less labile than normal comes from a sequence analysis of the behaviour. Although such an analysis gives the same overall result in both cases, the statistical significance of the links between the behaviour elements is higher in the mentally ill and higher still in the more severe illnesses. This greater rigidity is most evident in emotional types of behaviour. Observation makes it clear that one aspect of the lability of the normal person is his ability to generalise in situations where he is emotionally involved. If, for example, sexual problems are being discussed the normal person will generalise when the discussion enters an area that has relevance to that person. He will present the problem as an impersonal one. This lowers the emotional content of the situation and allows him to change his behaviour. The ill person is unable to do this; he becomes more and more involved and the behaviour patterns relevant to the situation go on recurring.

There is another area of facial communication where mental illness leads to breakdown and malfunction. This is the area of verbal control, the systems of gaze direction and the associated eyebrow-raising and head-tilting. In normal conversation alteration of gaze direction relates directly to the flow of conversation and helps control the interaction between the speakers. This area has been considerably investigated by a number of authors (see Chapter 9). In the mentally ill most interest has been centred on a particular illness of childhood, 'infantile autism', which is marked by an avoidance of social contact. Hutt and Ounsted (1966) and Wolff and Chess (1964) have shown that part of these children's difficulties is that their tolerance of eye-to-eye contact is exceptionally low. Many schizophrenics also show marked gaze avoidance (Argyle, 1967; Lefcourt, Rotenburg, Buckspan and Steffy, 1967), as do depressed patients (Ekman and Friesen, 1968). These, of course, are major effects on gaze. Just as important may be the smaller breakdowns of the intricate visual control of speech. The tendency to interrupt but at the same time to show longer silences than usual, seen in a number of patients, may be due to such breakdown.

TERRITORY AND DOMINANCE

Both dominance hierarchies and territorial behaviour have been described in mental patients. These social structures, as described in hospitalised schizophrenic patients (Esser, 1970; Grant, 1965), depend on non-verbal behaviour. Grant, by counting in small groups interactions in which one individual showed an aggressive behaviour pattern and was responded to by withdrawal or a submissive behaviour pattern, demonstrated stable ranks over considerable periods of time. These varied from straight line ranks to, more typically, pyramidal ranks with several non-interacting subordinates at the bottom of the pyramid. He also showed that these ranks were the same as ranking obtained by verbal counts. The more dominant individuals could give a verbal order to a subordinate and it would be carried out. After a prolonged study of a chronic ward in a New York hospital Esser described similar ranking, but also described a type of territorial behaviour within it (Esser, 1970). This behaviour consisted of the defence of particular small areas, for example a special chair or part of a radiator, by particular individuals. These individuals were not necessarily nor even commonly dominant members of the group, but within these areas held their own and refused to move if approached by a dominant. Freeman, Cameron and McGhie (1958) described similar behaviour but in defence of objects such as satchels and handbags, as well as places. They talk about a sense of ownership and relate this behaviour to their basic idea of breakdown of ego boundary, suggesting that it is a desperate attempt on the part of the individual to maintain some consciousness of self.

All of these studies were concerned with severely disabled, chronically hospitalised patients whose verbal ability was reduced to a minimum. Similar behaviour has been observed in less seriously ill patients, but in these cases behaviour with a structure as complex as this depends much more on verbal behaviour: the non-verbal component, although still important, is more subtle and less easy to describe.

The observation, over several months, of a therapeutic group meeting twice a week, gave evidence of dominance relationships within this group. The composition of this group varied slightly over this time but for the first few weeks it was clear that the mood of the meeting was determined by one particular individual. The same individual also controlled, to a large extent, who was to talk, mainly by indicating verbally that 'Mrs __ has something to say today'. She could also effectively inhibit talk in the whole group by a distinct postural display. This consisted of sitting back in her chair, folding her arms over her stomach, looking down at the floor with a smooth set face and her mouth closed and small. When the composition of the group changed there appeared to be a struggle for dominance between this individual and a new member. This struggle was mainly verbal in the first place with each trying to lead the discussion. Eventually direct aggression broke out with frowning, thrusting movements of the head and beating movements of the hands. At the same time the group split physically, some members sitting at one side of the room with one of the contestants while the rest sat on the other side with the other. Although over the weeks this overt aggression died down, the split remained for several more weeks and was still present when observations ended.

The potentiality of forming these social structures is presumably present in the normal individual, their overt expression in the mentally ill being partly due to the constricted social environment of the hospital and partly to the ease of observation that such constriction entails. The main differences are the simpler structure that the mentally ill can cope with and the greater involvement of non-verbal behaviour as verbal ability is reduced.

The development of clear rank orders in groups of schizophrenics enables us to describe certain other non-verbal behaviour patterns in fairly well defined situations and gives us clues to the information content of such behaviour. One very clear point that emerges from such observations is that the flight responses of the patients are extreme in comparison to the normal. Behaviour patterns such as hunching, drawing the shoulders up and forward and tucking the chin in, and crouching, which are only seen in face of physical danger in the normal population, are shown in response to low level aggressive acts such as frowning or even to the simple approach of one of the dominant members of the group. Similarly, sleeping postures and postures in which cat-naps occur also appear in response to aggression, this time normally indirect aggression: for example, if a dominant complains to the nurse about the behaviour of a patient, that patient may show a sleeping posture.

This ease of arousal of flight makes ambivalence a common aspect of the behaviour of these schizophrenic patients. Any positive approach whatever, even as simple as joining the others at the table, comes into immediate conflict with the wish for social withdrawal and leads to an ambivalent state. The simplest response in an ambivalent situation is probably alternating ambulatory movements towards and away from the ambivalent stimulus. This simple response is rarely seen in the normal adult, although I have seen it in the young in response to two positive stimuli in opposite directions, food on the table and a favourite toy on the floor. In the schizophrenic it is quite common. The following is an example from direct observation in a ward.

A subordinate individual, E, wanted a cigarette: she knew that one of the dominants, A_1, had some and also that a nurse had some. Another dominant, A_2, was talking to the nurse. E walked towards A_1, stopped about three yards away, turned and walked towards the nurse and A_2, stopped again, turned towards A_1 and continued this for several repetitions before eventually circling A_2 and asking the nurse.

Such behaviour can become more stereotyped and reduced to small circling movements at the balance point. Rhythmic to–fro rocking movements of the body are also shown in simple ambivalent situations. A common example in the schizophrenic groups often occurred when a dominant came and sat down near a subordinate. The subordinate would begin to rock and then, on one of the forward movements, would rise from the chair and walk away. Rocking also occurs in patients (and, of course, children) who are simply distressed, and where there is little evidence of ambivalence. The common description of this as a comfort movement may well be more relevant in these situations.

Another response shown in ambivalent situations is stroking or scratching of the body surface and particularly of the scalp. These 'grooming acts' are similar to the displacement activities recorded in many animal species. They are common in normal behaviour as well as in the mentally ill.

In discussing these responses to ambivalence I have indicated that they occur commonly in situations where a simple ambivalence would seem to be present. However, they can also be recorded as occurring when there is little evidence of any causal factors. They can all become chronically repetitive parts of the individual's behaviour and may be, as one patient told me, 'just habits'. However, even in these cases where the act is an almost permanent part of the individual's behaviour, there will be an increase in amplitude or frequency in directly ambivalent situations.

A number of patients show unique individual patterns of behaviour that appear to have no equivalent in the rest of the population. These individual acts also have been recorded in ambivalent situations: there might be an increase in amplitude or frequency of the behaviour, or it might simply be that the ambivalent situation initiates a bout of this activity. It is interesting that, just as elements that can be seen, albeit infrequently, as parts of the normal response to ambivalence, readily develop into stereotypes, so also can bizarre individual patterns of behaviour, occurring as stereotyped movements, occur as ambivalent responses in the appropriate situations.

It is clear that research into non-verbal communication in the mentally ill is still in a very early stage, as indeed it is in man as a whole. There is also little doubt that the major channel of communication in man is the verbal channel and most information transfer between doctor and patient in the psychiatric situation will continue to use this channel. Nevertheless, non-verbal behaviour is an important part of the communication between individuals and a more precise, overt knowledge of it could be of immense clinical importance, perhaps especially in those patients whose illness makes verbal communication difficult.

COMMENTS

This paper was not available for discussion by the group. As Grant himself points out, the definition of communication used is extremely broad, and much of the behaviour described would be referred to as 'symptomatic' or 'indexical' by other contributors to this volume. Nevertheless there are clear parallels with some of the movements of animals which have become adapted in evolution for intraspecific signalling (see Part B, especially Cullen in Chapter 4 and Andrew in Chapter 7). For instance many of them appear in conflict situations, and many can be described as 'intention movements' or 'displacement behaviour'. The clinician's task is to interpret these signs even though they have not become specialised as signals.

Another recent article stressing the value of the methods devised by ethologists for studying animals in providing insight into the nature of autism in children is by Tinbergen and Tinbergen (1972).

14. PLAYS AND PLAYERS

JONATHAN MILLER

In this essay I intend to deal exclusively with the non-verbal aspects of those types of performance which are specifically associated with speech. This is because, along with many of the other contributors to this volume, I assume that verbal communication takes precedence in human discourse and that non-verbal behaviour achieves most of its communicative significance in the context of syntactically organised utterances. For this reason I intend to overlook the *deliberately* non-verbal types of theatrical performance – mime, ballet and ritual dance – on the understanding that these displays are highly specialised acts comprising a deliberate abstention from spoken language. In ballet and mime, speech is conspicuous by its absence, and the audience is required to adjust its expectations in such a way that it does not experience the performance as merely impoverished versions of spoken language.

The essential feature of Western theatre therefore is a script or verbal recipe, in minimal accordance with which the actors must reproduce the propositional assertions created by the author. In other words, the central ingredient of a play consists of a series of utterances which express, as Wittgenstein points out, *the way things are*. These expressions are designed to allow the various characters to mention or refer to some state of affairs, either in the outside world, or in their own minds.

Of course this script may be extremely attenuated – in the case of certain plays by Beckett it is reduced to alarming thinness – but such as it is, the verbal recipe is the essential core giving meaning to all the silences which intervene. The gestures and grimaces which succeed one another during these silences only acquire full significance in the context of the verbal expressions which they either precede or follow. They represent changes of mood following something that *has* been said or else states of vigilance and expectation with regard to something that *might* be said. In the absence of an utterance which asserts something or other, these non-verbal passages fall off into a state of frustrated restlessness.

The different signals which jointly comprise such *interstitial communiqués* are similar in some ways to the non-verbal acoustic cues which precede the making of either a successful or an unsuccessful telephone call. A ringing tone shows that a putatively successful bid for attention has been made, but if the tone continues for a certain time without being interrupted one begins to doubt that the person is in. A high-pitched buzz shows, but does not assert, that although the bid has been correctly formulated at the caller's end, it has failed to even compete for attention at the other end. A sequence of familiar double tones, also high-pitched, signifies that the call is correct in all respects and would have gained attention but for the fact that the subscriber at the other end is already engaged in conversation. And so on. Now although these signals have an agreed significance, they only acquire it within the context of an intended exchange of verbal utterances. None of these cues would have any meaning outside the context of a possible conversation. Within this context, however, they help to eliminate certain ambiguities which arise in the course of trying to set up such an exchange. It is easy to imagine the doubts which would be created if there was a silence every time one dialled, since there would be no way for the caller to distinguish between a mis-dial, a line out of order, an engaged number or no one at home. In the same way the gaps which intervene between the utterances of a theatrical scene are filled with non-verbal signals, one of whose functions is to inform the various participants of the chances of their respective forthcoming utterances receiving attention. One character, say, will clear his throat, thereby inducing his putative communicant to register his willingness to receive the message by turning his head with an expectant raise of the eyebrows. Neither the cough nor the raising of the eyebrows have much significance on their own, but in the context of a conversation *about to* occur they considerably facilitate the chances of it bearing fruit. If character *A* did *not* clear his throat before speaking and if his partner was otherwise engaged at the time, the significance of the first few sentences might be lost on thin air. The cough is an attempt to ensure that what follows will be attended to right from the outset. On the other hand, if character *B* did not respond to the cough by turning his head and raising his eyebrows, character *A* would remain in doubt as to the success of his cough in gaining attention.

Actors who wish to reproduce an accurate version of two people trying to hold a conversation with one another must therefore take care to include imitations of these interstitial communiqués. One of the most vivid illustrations of the difference between professional and amateur actors is the failure of the amateur to come up with convincing versions of these non-verbal 'attention' signals. Not that they are as simple as I have made out. Unlike the phone example, the interstitial signals which are used for registering the various states of listeners' attention are not compiled in accordance to a strict convention. A man may bid for attention in any one of a hundred non-verbal ways; likewise a putative listener may signify his willingness to attend by choosing from a large repertoire of facial or vocal acts. Moreover each one of these acts may also indicate the *quality* of the attention which is being given. By rolling the eyes upwards and closing the lids for a moment, the listener may announce that while he is prepared to listen he is doing so with great reluctance and that he will attend to the forthcoming message in a mood of sceptical resignation. The speaker who appreciates this preliminary response to what he is about to say will often be forced to re-phrase his utterance, to make it more convincing. Not to mention the immediate modulation of the non-verbal accompaniments of the speech itself, with which the speaker will invest his remarks in the knowledge that he is about to deliver his message into a hostile ear.

This last consideration brings us to a realisation of the fact that while speech assumes priority in the flux of messages that pass between members of a given community, it is only when they are fully investigated with non-verbal signals that spoken utterances can succeed in rising above the bare grammatical meaning of their constituent expressions. When for instance, at the start of *The Seagull*, Masha announces that she is 'in mourning for her life' there is a sense in which it could be said that anyone who speaks English is bound to understand what Masha means by that particular expression. But at the same time there is clearly another level at which the mere knowledge of English grammar and vocabulary does *not* exhaust the significance of what the character is saying. The actress who is charged with the task of uttering this statement must therefore use it in such a way that it will 'mean' something over and above the significance which may be derived from a mere paraphrase of its component words. In order to do this she has at her disposal the following parameters within which variations may be introduced to modulate the full meaning.

(*a*) *Vocal.* (i) Overall volume.
            (ii) Differential volume within the phrase.
            (iii) Pitch and tone variants.
            (iv) Use of accents.
(*b*) *Facial.* Smiles, sneers, set of jaws. Elevation of eyebrows, etc.
(*c*) *Posture.* Set of head on shoulders. Spinal posture. Stance.
(*d*) *Manual gestures.*

By simultaneously modulating all these parameters the bare grammatical phrase 'I am in mourning for my life' can finally be made to render a plenum of meaning which has coherent significance within the on-going life of the character portrayed.

However this does not mean that the full 'meaning' of the utterance is finally specified when all the non-verbal variables are fixed. Even when it is fleshed out with inflections and gestures the phrase cannot be relied upon to deliver its 'full' complement of intended meaning. For the phrase, as conceived, *has* no fixed complement of antecedent meaning; but is instead a shot in the direction of some altogether hypothetical limit of 'meaning'. Added to which, the non-verbal cues which are used to inflect any given phrase towards this hypothetical limit, are so vague and equivocal that no two listeners will ever quite agree as to the precise significance of the accents, tones, emphases, expressions and gestures which jointly make up its concrete performance.

Of course in real life, whilst speaking spontaneously, no one is ever aware of the distinction that exists between the bare grammatical meaning or reference of a phrase and the swarm of non-verbal elements that help to inflect that phrase towards the meaning which the speaker intends it to convey. The phrase and its performance are conceived as one, and it would, I think, be a serious mistake to imagine the production of speech in the same terms as the manufacture of a motor car on an assembly line; with a chassis of syntax rolling up to a section of the brain which then clothes it in a coachwork of non-verbal signs.

When we come to dramatic performance, however, the metaphor of the automobile assembly line becomes awkwardly relevant, for the script exists as a grammatical entity long before it assumes life as an actual utterance, so that when the actor undertakes to perform the lines he is confronted by the peculiar task of investing them with a suitable accompaniment of non-verbal cues; and to do so moreover, as if the whole act had taken place simultaneously. This would offer no particular problems if one could read into the lines a set of specifications that automatically insisted upon the way in which they should be spoken; but as everyone knows it is possible to perform the lines assigned to Hamlet, say, in a hundred different ways, all of which are at least *compatible* with the basic semantic references of the script. Of course it would be foolish to expect each phrase taken in isolation to specify its own 'correct' performance: it is essential to consider all the speeches as a whole and from this conspectus to arrive at some instructive generalisation as to the appropriate mode of uttering any one of the constituent lines. But no two actors will arrive at the same generalisation and although the resulting performances may each, respectively, be compatible with the text, the different characters who utter this text may be quite incompatible with one another. Which only goes to highlight what Professor Strawson pointed out twenty years ago, that it is *people* who mean and not *expressions*; so that it is quite possible for two people to mean different things *by* the same expression.

The task of rehearsal therefore consists of an effort to discover a 'person' who could mean something by the sum of the expressions which comprise his part.

What is surprising perhaps is that playwrights seem to take so little trouble to specify, by means of collateral evidence, what sort of person *they* have in mind. Presumably they *do* have someone in mind, for strings of dialogue do not normally suggest themselves to the mind's ear without being set in the mouths of someone who 'means' something by them. When a writer creates speeches he must, to some extent, conceive them in the mouths of a more or less distinct individual person; which means in turn that he cannot really *avoid* entertaining an image of how the characters are standing when they speak, what they are doing with their faces, hands, tones and inflections, through the collective medium of which the character tells us what he means by the expressions he is using. And yet, more often than not, the only thing that survives in the written script is the lines themselves – like the skeleton of a prehistoric creature – leaving the actors and directors free to reconstruct the soft parts of the role, in a way that may be at odds with the creature that once lived as a whole within the writer's imagination. It has always seemed rather puzzling to me that authors who are so jealous of their conceptions should leave so much to chance in their subsequent re-creation; and I can only suggest that their almost consistent failure in this respect arises from the lack of a satisfactory notation within which the non-verbal parameters of human expression might be registered. Whatever the reason, the brute fact stands that the scripts of plays impose upon the actor the peculiar task of manufacturing a personal identity that can significantly include the lines written opposite his name in the script.

Now the way in which this synthesis is carried out depends to a very large extent upon the significance which any given actor attaches to the notion of 'a person'. In other words, whether he explicitly realises it or not, the actor compiles his performance of the lines in accordance with a philosophical assumption as to the correct definition of an individual. For in undertaking to portray the character whose name appears opposite certain lines in the text, he is accepting the tacit assumption that there is a continuous identity which spans the interval between successive exits and entrances of the character thus named. Now although he may not openly confront the logical problems that are raised by such an assumption, the actor must make a decision, at some mental level or other, as to the criteria which convince him that he is entering as the same person as whom he exited half an hour ago. More often than not he will go and sit in the green room and read a newspaper until the cue for his next entrance; which means, in effect, that he has to re-assemble the impersonation which he shed on stepping into the wings. Now the point is this, in the act of assembling himself for re-entry, he must refer to certain criteria that will define for him, no matter how vaguely, what it *means* to come back as the same person who previously exited. He must know, in some way or other, the sort of features that would count either for or against this putative continuity. It is my contention that there are two criterial extremes which might underlie this decision. At one end there is the assumption that personal identity is a self-evident sense of subjective continuity, and that a person is a sum of private experiences to which he and no one else has access. At the other extreme lies the belief that a person is nothing more than a physique, whose uniqueness and identifiability are associated with a set of behavioural dispositions which mark them off from any other claimant. Gilbert Ryle has characterised the last position in *The Concept of Mind* (1949), and many actors implicitly endorse it by approaching their parts as if identity *was* defined by behavioural features alone. Within the profession such actors are often referred to as 'physical' performers, and while their colleagues may marvel at the sparkling idiosyncrasy of their performances they often express anxiety at the so-called lack of feeling involved. However, since no other person has access to anyone else's state of mind, the accusation of lack of feeling must depend upon the observer having detected a gap in the outward display of non-verbal behaviour. In talking to actors about this, however, I have never succeeded in getting them to describe what these missing cues consist of. They can never say how to distinguish between a portrayal which is subjectively well understood but badly executed, and one in which there is no subjective sympathy but a great deal of outward accomplishment. 'I can't put my finger on it' they will say, 'I just feel it in my bones. There's no heart there' and so on.

At the other extreme there are actors who actually refuse to 'perform' during rehearsal and who prefer to mumble their lines without expression until, by some collateral route, they have arrived at a sympathy for the *state of mind* which would have led the character to say the lines assigned to him. Then, often quite suddenly, the performance will materialise, fully inflected, the whole utterance fledged with gestures and expressions.

It would be a mistake to leave the impression that actors segregate themselves sharply into one or other of these two extremes and that you could, by questioning members of the profession, get them to admit their respective allegiance to the alternative criteria of a person. An otherwise behavioural actor will often resort to a subjective justification for certain types of expression he is using, whilst an introspective performer will frequently admit that he has introduced a certain idiosyncratic inflection or gesture because he likes doing it, or simply because it helps to point up the distinctiveness of the personality he is portraying.

Besides, a considerable percentage of the ongoing stream of non-verbal behaviour takes place without it necessarily signifying *anything*; and while, without it, one would be tempted to deem a given performance as expressionless, in its presence one would almost certainly be at a loss to say *what* it actually expressed. This holds even in the case of a meticulously subjective performance. Stopped in mid-stream and asked to justify a given twist of the lips or drop in tone, the actor may not be able to say *why* he did it and in many cases he may be unaware of the fact that he has done it at all. In fact one can often recognise the really accomplished performance by the degree to which the performer is able to realise a proportion of *content-less* expression, in terms of which a sense of human vitality is conveyed without its laying an obligation upon the audience to try and decode its semantic significance. And in those performances where the actor seems hell bent on inflicting a clearly visualised meaning into every gesture we say that he is overacting.

Given the fact then that a significant proportion of facial movement and vocal intonation carries no intentional meaning anyway and that the rest is so variable in its significance, one can see how easy it is for an audience to read a performance as if it were to some extent a Rorshach blot. Even a group of perceptive critics can witness the same performance and disagree widely amongst themselves as to the meaning of what was being acted. This may assume absurd proportions, as for example on the occasion when I heard a critic give high praise for certain pauses which I, as the director of the show, was privately deploring, in the knowledge that the actor was merely forgetting his lines. Conversely in a production of *King Lear* for which I was also responsible, the actor playing the lead was widely condemned for failing to convey, presumably by non-verbal signs, the majesty which is conventionally expected of the role. Now it so happens that, for what it was worth, we had deliberately erased from the performance all gestures and facial mannerisms that might have conveyed a sense of majesty. What I would still like to know is whether the critics, on witnessing this performance, detected *formes frustes* of majestic expression and were then criticising the failure of the actor to develop these to their appropriate level (through lack of technical ability); or whether they acknowledged that a decision had been made to eliminate such expressions altogether and were condemning *this decision*.

This problem recurs with monotonous regularity without anyone ever having published the principles according to which dramatic perception is organised. What I do know is that circumstantial evidence, often quite irrelevant to the performance in question, plays a crucial part in the final judgement that is passed. For example a celebrated actor is almost invariably credited with total deliberation, with the often absurd result that both critics and audiences are led to over-interpret his gestures and expressions and read meanings into movements that had no significance whatsoever.

In the same way, if a well known comedian takes on a serious role, audiences, who have grown accustomed to the comic use to which he has previously put his face and voice, will inadvertently read humorous or even satirical intentions into his performance. The point is, to go one step further than Strawson's analysis, not only is it true to assert that *people* mean and not expressions, but it is often audiences who decide what people mean; using, in order to arrive at this decision, evidence which is not actually given in the performance itself.

For this reason I often think it is important to talk to critics before they see a play, in spite of the fact that convention insists upon the notion that the performance should stand or fall on its own merits and that critical vision is blurred by such tendentious information. For, as I see it, the dogma of critical impartiality rests upon a psychological fallacy, according to which it is assumed that perception is derived solely from what is given to the senses. Whereas modern cognitive psychology has shown quite clearly that perception is a conjectural enterprise, in which antecedent instructions are essential pre-requisites for distinguishing the figure from its ground. Far from being an unprincipled act of collusion, preliminary advice from the director may be an important contribution towards ensuring that the critic properly adjusts his perceptual horizon for the best chance of his seeing the distinctive inflection that has now been given to an otherwise familiar text. Otherwise theatrical innovation is repeatedly seen as a failed example of what the critic expects rather than as an achieved instance of something altogether new.

This argument assumes importance in the context of Western theatre due to the fact that we follow a more or less naturalistic tradition. That is to say the face, hands and voice are not used to inflect meaning in accordance to a rigidly formalised code, but are, as in real life, moved in an irregular and often unreliable relationship to the underlying expressive intentions of the utterer. Which leaves of course a large leeway for equivocal guesses. In more traditional forms of theatre the variations of non-verbal behaviour are artificially constrained within a closed series of mutually distinct expressive morphemes and in order to reinforce the formal clarity of this paradigmatic system, every effort is made to minimise the visibility (or audibility) of intervening states which might otherwise blur the issue. For example, lest irrelevant physiognomic perceptions be derived from small variations of expression that intervene between the agreed paradigms of facial movement, the features are carefully painted in order to give exclusive priority to the elements that *do* have significance within the code. Or to put it in the words of Erving Goffman 'Messages that are not part of the officially credited flow are modulated so as not to interfere seriously with the accredited messages' (Goffman, 1970). Whereas in a dramatic tradition that reflects a culture where peculiar emphasis is put upon the uniqueness of individuals, physiognomic attention spills over and applies itself to those very features which lie outside the accredited system.

During this century, however, there have been several backlashes against the sort of dramatic tradition which confers such loving emphasis upon idiosyncratic personal expression. Playwrights like Brecht, for example, have explicitly repudiated the sort of emotional sympathy which such naturalistic portrayal almost inevitably elicits from a Western audience, and by reverting to a codified and largely demonstrative type of non-verbal expression have forced their actors to *quote* the characters they represent instead of actually *inhabiting* them. It is therefore not an accident that when Brecht wrote theoretical commentaries upon his method he reverted frequently to the artistic credentials of the Chinese theatre.

'The Chinese artists' performance often strikes the Western actor as cold. That does not mean that the Chinese theatre rejects all representation of feelings. The performer portrays incidents of the utmost passion, but without his delivery becoming heated. At those points where the character portrayed is deeply excited the performer takes a lock of hair between his lips and chews it. But this is like a ritual, there is nothing eruptive about it. It is quite clearly someone else's repetition of the incident: a representation, even though an artistic one. The performer shows that this man is not in control of himself, and he points to the outward signs . . . Among all the possible signs certain particular ones are picked out with careful and deliberate consideration.' (Willett, 1964: 93.)

It is self-evident, I think, that an audience which was otherwise accustomed to inferring the character's state of mind from the continuous flux of facial and vocal expression would be made very uneasy by being made to focus its attention upon a discontinuous succession of discretely coded gestures.

In fact until the audience has successfully and unself-consciously incorporated the paradigm which confers distinctive meaning upon the various gestures and inflections that are raised to the status of signs under its aegis, they will witness the new performance with reference to the old semantic paradigms and will therefore almost inevitably condemn what they see and hear as nothing more than a stilted version of the familiar idiom.

The same problem arises in the visual arts of course, when a new paradigm of representation is introduced. Accustomed to the psychological regime that gives significance to the so-called naturalistic representation of faces and objects, the naive spectator confronting cubism for the first time will dismiss what he sees as a clumsy travesty of the pictorial reality he is used to. What he has failed to understand though, is that the self-evident immediacy of the idiom which he favours is itself organised in obedience to a representational grammar which has to be learnt in the first place. The difficulty being that once a grammatical paradigm has been incorporated, it works without its user being directly aware of the fact that he is referring to it; so that he is deluded into thinking that he is experiencing a direct and unmediated meaning.

A further example taken from linguistics may help to make this point even clearer. A native English listener understands everything that is said to him in this tongue; and although it is possible to write out the rules that define the meaningful use of English, it is part of the definition of knowing English that any utterance phrased in it can be understood without explicit reference to this set of rules.

In other words although it is true to say that a native listener does not know two things – English grammar and English language – his single competence in English depends on the fact that he has, at some time or other, incorporated a set of rules which defines the way in which meaning is assigned to the linguistic moves initiated under its legislation. We are all familiar with the semantic inaudibility of an unfamiliar language and no one expects to overcome this without consulting the rules of the new grammar; and furthermore no one except the most rabid linguistic chau-vinist conceives this new system as an incompetent foreign bosh shot at English. In the same way, unfamiliar modes of dramatic non-verbal expres-sion are organised with reference to a grammar that has to be learned before it can be forgotten.

It remains to consider some other influences upon the use of non-verbal signs in the theatre. I would like to speculate about the influence of fiction on the one hand and that of cinema and TV on the other.

Take fiction first. Although it would be difficult to verify, it seems probable that the rise of the art of fiction caused a subtle change in the emphasis that was given to naturalistic facial expression on the stage. For whilst, as we have already noted, the text of a play is limited to bare dialogue, coupled perhaps with enough stage directions to show how the action goes forward, the novelist is free to suspend both action and dialogue and tell the reader not just about the mood of the characters but about the way in which these moods were betrayed or concealed in terms of gesture, expression and intonation. An expression, for example, which might have been misinterpreted when described from outside only could be properly placed by saying what the character actually meant by it. Here for instance is Tolstoi explaining just such a situation in *Anna Karenina*.

'The Englishman puckered his lips, intending to indicate a smile that anyone should wish to check on his skill at saddling.'

This short sentence contains an enormous amount of complicated psychological information. For in addition to telling us that Vronsky's groom was proud of his skill and that he tended to show his amusement when someone called it into question, Tolstoi has also indicated that a servant is not altogether free to show his amusement blatantly, but must confine his expression to a form that is sufficiently equivocal for an aristocratic employer not to take immediate offence at it. A less careful writer might simply have said 'The Englishman smiled contemptuously that anyone should cast doubts upon his skill at saddling', which would leave the reader, knowing Vronsky's temperament, to wonder why he had not reprimanded the man for dumb insolence – an offence which would, in the eyes of a cavalry officer, stand high amongst the punishable sins. The idiom of the novel allows the writer to convey all these nuances of dissimulation and to show the complex relationship that exists between a state of mind and the expressions which might or might not bear witness to it. It is my contention that a readership which had been conditioned to such descriptive subtlety would approach theatrical performance with an eye and an ear curiously sensitised to the tell-tale variations of non-verbal expression. They would become intolerant of gross formality and by a process of natural selection would encourage the growth of physiognomic naturalism in the theatre. As I say, it would be hard to verify this assertion, and it could always be said that naturalism arose simultaneously in the novel and the drama from a common origin in social and political individualism. Nevertheless, art forms do influence one another, and it is, I think, a plausible hypothesis to suggest that this influence played a part in the development of twentieth-century dramatic style.

Finally I would like to consider the effects of certain artificial media – principally film and TV. And in spite of Marshall McLuhan's assertions to the contrary most of the claims that might be made for film hold for TV too since both are variations of moving photography.

Perhaps the most dramatic effect of film arises from the fact that the audience is no longer obliged to view the action from a fixed distance, but can, through the medium of the close-up, pay special attention to the subtle and above all continuous flux of facial expression. And although the size constancy phenomenon ensures that the spectator is not actually conscious of the vast bulk of a twenty foot face on the screen, he cannot fail to take advantage of the enormous magnification of expressive detail which is thus offered. The psychological consequences of this now almost hackneyed innovation must have been enormous, for it introduces the audience to an almost entirely new vision of the relationship that exists between mental states and the expressions by which they are registered. In the days before the movie, the large fixed distance between the performer and the auditorium made it very unlikely that a spectator would see, let alone pay much attention to, the movements of the face that intervened between relatively large scale gestures. In fact for someone sitting in the gallery, there is nothing much between a broad smile and a deep scowl. With the result that non-verbal expressions almost inevitably assume a discontinuous, quasi-atomic form. Recognising which, the actor would quite naturally put particular emphasis on those expressions which he knew could be seen and let all others go by the board.

Under such a regime of limited visibility actors and audience unconsciously co-operate to maintain the 'atomic' idiom of emotional representation. The introduction of the close-up, however, swings a high-power microscopic objective on to the scene and by magnifying the muscular activity that intervenes between the traditionally significant postures of the face succeeds in dis-establishing their mutually exclusive physiognomic integrity. Once the audience had been exposed to this transformation they would return to the theatre with eyes skinned for small scale variations of expression; and although they would still have difficulty in seeing them all, there is no doubt that, having had their attention re-trained, they would still succeed in detecting much more than they had before. Conversely the actors would soon become aware of the fact that the vision of the audience had become more sophisticated and they would soon feel obliged to satisfy this sophistication by modulating their performance in the direction of greater naturalistic subtlety.

If the close-up on its own could produce these effects, it is easy to imagine the added power that such an innovation would acquire by being integrated with the techniques of montage and editing. Eisenstein for example showed very neatly how a relatively equivocal facial expression shown in close-up could acquire strong emotional significance if it was juxtaposed in sequence with a strongly evocative frame. A fixed gaze surmounted by a frown might be taken to show horror, love or mere concentration respectively, all according to whether it was followed by a corpse, a baby or a milking machine. The point is that by putting the face to work in a dialectical context like this, Eisenstein and the directors who followed him raised the audience's interest in the subtleties of facial expression. And within the acting profession the artificial large scale gesture lost its respectability and performers began to judge their colleagues by the extent to which they could minimise their expressions and still convey an emotional meaning.

The film I think has also brought about a radical change in the way in which the voice is used. By hanging the microphone just above the frame line it is possible for the actor to drop his volume and to signal minute changes of mood with relatively minor alterations of inflection. Just as in real life in fact. And while it is not possible to reproduce such quiet modulations in the theatre, stage actors have now become very adept at giving a scaled-up version of such acoustic naturalism; with the result that dramatists are encouraged to write their dialogue to fit and to produce plays which could not have been spoken by traditional performers. In his book on Shakespeare, the critic John Wain (1968) has pointed out that the prosodic technique of blank verse was partly introduced to overcome the acoustic difficulties of the Elizabethan theatre, and that the resounding periods of Shakespearean diction were a primitive form of public address system designed to compel attention in an open air auditorium. It is interesting to note however that the declamatory style outlasted the acoustic obstacles it was designed to overcome and that artificial, demonstrative verse speaking continued to dominate the production of Shakespeare's plays until the later years of this century. And although experiments in naturalistic diction were being made in the nineteen-twenties, it is only in the last twenty years that it is widely possible to hear Shakespeare's speeches uttered in something like the tones of voice that we are accustomed to in everyday life. The remarkable thing is that Shakespeare's text is so semantically elastic that it can be naturalistically inflected without apparently seeming to undergo poetic deterioration.

This does not mean however that the effort to detumesce classical diction has not met serious critical opposition. Older members of the theatrical audience still object to the vernacularisation of their Shakespeare and expect the dignity of the Bard to be upheld in terms of grand gesture and orotund vocal delivery. According to them, anything else cheapens the great poet's imagination and because they have not yet familiarised themselves with the new code, they find it impossible to discern the moral seriousness which the new techniques try to embody. And of course by the time they *have* accustomed themselves to the new idioms of non-verbal expression, the stage will have moved on again, possibly, as I have already indicated, in the direction of *heightened* artificiality. Directors like Peter Brook are already experimenting with the hieratic techniques of Persian theatre and we can, I think, expect to see in the next ten years some extremely complicated negotiations with the various resources of non-verbal sign behaviour. Awkwardly enough, however, such dramatic experiments now tend to succeed one another with such speed that an innovation does not have time to catch on before it is overthrown by a new one. With the immediate result that no single innovation has time to establish itself within a community of understanding. Now this may or may not be a bad thing. On the one hand there is the obvious danger that the mass audience will become altogether alienated from a theatre which gives it no time to catch on, with the obvious consequence that the drama will die for lack of support. But on the other hand it is possible that the dramatic, like the visual arts, will gain attention not by virtue of the reality which is depicted by it, but as a structure of formal relationships which is of interest as a thing in itself. In the same way as a painting by Mondrian achieves acceptance through the contemplation of its own internal relationships, dramatic performances in the future could be sponsored by audiences who may be satisfied by the formal patterning of facial and manual gestures. In other words, the progressive 'dehumanization of Art' as predicted many years ago by the Spanish critic Ortega y Gasset (1969).

As it happens though I believe that this tendency towards purely aesthetic dramatic structuralism will be comparatively short-lived and while it may survive as a thin parallel stream, notably in ballet, it is extremely unlikely that the theatre will ever succeed in exorcising from its developing practices the attempt to depict meaningful states of mind. Which means moreover that propositional dialogue will continue to assume its peculiar precedence in all-out dramatic performances and that non-verbal behaviour will remain, as it has always been, a subtle and all pervading context, through the medium of which human beings will attempt to make clear to one another what it is they *mean*.

COMMENTS

This essay was not available for discussion by the group. See Comments to Chapter 15.

15. ACTION AND EXPRESSION IN WESTERN ART

E. H. GOMBRICH

I. THE PROBLEM

'It only lacks the voice.' This traditional formula in praise of works of art would suggest that painting and sculpture might present ready made material for the student of 'non-verbal communication'. But, alas, painting not only lacks speech, it also lacks most of the resources on which human beings and animals rely in their contacts and interactions. The most essential of these, of course, is movement. Art can represent neither the nod nor the headshake. The sudden blush or the frequency of eye contact are equally outside its range. Indeed if one seriously considers the elements of non-verbal communication as they are discussed in this volume, one may well wonder how art could ever have acquired the reputation of rendering human emotions.

The answer I shall propose in this paper will inevitably correspond to the one I have suggested elsewhere (Gombrich, 1960, 1965, 1968). I have argued that the creation of images which satisfy certain specific demands of verisimilitude is achieved in a secular process of trial and error. 'Making' – I suggested – 'comes before matching'. Art does not start out by observing reality and trying to match it, it starts out by constructing 'minimum models' which are gradually modified in the light of the beholder's reaction till they 'match' the impression that is desired. In this process resources which art is lacking have to be compensated for by other means till the image satisfies the requirements made on it. Seen in this light we need not doubt that works of art have in fact satisfied succeeding generations who approached them with different demands, but with a desire for the convincing rendering of human expression. Indeed the literature on art testifies to the triumph of painting and sculpture over the limitations of their media. Works of art have been traditionally praised precisely for 'only lacking the voice', in other words, for embodying everything of real life except speech.

To dismiss this reaction simply as a conventional exaggeration would exclude the student of non-verbal communication from the realm of art. It seems more fruitful and also more cautious to accept this reaction as a testimony to the combined power of convention and conditioning in creating a semblance of reality, a model to which men have responded as if it were identical with a life situation.

If art is thus seen as an experiment in 'doing without', an exercise in reduction (conventionally referred to as 'abstraction'), it is clear that the history of styles cannot be seen simply as a slow approximation to one particular solution. Different styles concentrate on different compensatory moves, largely determined by the function the image is expected to perform in a given civilization.

The most obvious way of compensating for the absence of speech is of course by the addition of writing, and this method was used with varying intensity in ancient Egypt, in archaic Greece, in medieval art where scrolls come out of the mouth of figures to show what they are saying, and once more in the modern comic with its 'balloons'. We may leave these on one side as evading the problem of 'non-verbal' communication.

2. LEGIBLE INTERACTION

There are at least two requirements for a 'still' to be legible in terms of expressive movements. The movements must result in configurations that can be easily understood and must stand in contexts which are sufficiently unambiguous to be interpreted (Gombrich, 1964).

An example (Plate 1 a), which dates from the 3rd millennium B.C., may illustrate these postulates all the better because the interaction depicted is not cultural but 'natural' and concerns 'non-verbal communication' among animals. The relief from the tomb of Ti (IVth Dynasty) shows a man carrying a calf on his shoulders which turns its head at a group of cows, one of which lifts her head towards it and appears to low. The movements represented are not only easily legible as significant deviations from the normal or expected postures of the animals, they also leave us in no doubt as to their expressive significance. We know that the man is taking the calf from the cow and thus we almost hear it low. If we asked somewhat pedantically how we can be sure, we would have to say that the man could not very well walk backwards in order to take the calf to the cow – quite apart from the fact that we would not know why he should do so, while we know very well why he takes the calf away.

Many of the resources of art for the depiction of expressive interaction are shown in this little group. It reminds us from the start that the transition from physical interaction to anticipatory movements is a gradual one, there is no intention on the part of the cow and the calf to express their reaction to the separation, only movements attempting to overcome it. It is up to us whether we interpret these movements as purposive or expressive.

3. ACTION AND EXPRESSION ON THE STAGE AND IN ART

This insight is an old one. It was developed with much subtlety by the eighteenth century actor J. J. Engel, to whose *Ideen zu einer Mimik* (1785–6) Karl Buehler (1933) has paid a justified tribute. Speaking of positive or negative reactions to an 'external object', Engel observes that they are both marked by an 'oblique position of the body' (Engel, 1785–6, letter XIV, my translation).

'When desire approaches the object either to possess it or to attack it, the head and the chest, that is the upper part of the body, is shifted forward, not only because this will enable the legs to catch up more quickly, but also because these parts are most easily set in motion and man thus strives first to satisfy his urge through them. Where disgust or fear makes him shrink back from the object the upper part of the body bends backwards before the legs have started moving . . . A second observation which will always be confirmed where a vivid desire is at work is the following: it always tends towards the object or away from the object in a straight line . . .'

Engel describes the human parallel to our animal example – the interaction between the child standing on tiptoes stretching its arms towards the mother and the mother bending down and extending her hands encouragingly towards her darling. He then proceeds to analyse reactions to more distant objects, the posture of the listener who tries to overhear a conversation, the gaping onlooker. In all these movements of orientation the most active part of the body is turned towards the object of attention.

Engel's analysis of the conflicting pulls of contrary drives strikingly foreshadows the descriptions of modern ethologists (1785–6, letter XXVI, my translation).

'When Hamlet follows the ghost of his father his longing for the desired discovery of a dreadful family secret has the overwhelming preponderance; but this longing is weakened by his fear of an unknown being from a strange world, and increasingly weakened the nearer the prince comes to the ghost and the further he moves away from his companions. Hence his movement should only be lively when he breaks away from his companions with a threat. When he begins to walk, it should be without hurry or heat though still with firmness and determination, gradually his step should become more cautious, more soft and should bestride less space, the whole movement should be more inhibited and the body increasingly pulled back into a vertical position.'

Nor is Engel unaware of the problem of the role of convention in expressive movements which must indeed obtrude itself on his kind of analysis. Having asserted that a lowering of posture is used by all nations as an expression of reverence, he discusses anthropological evidence from Tahiti which appears to contradict this claim with honesty and circumspection (Engel, 1785–6, letter XX).

As an actor, Engel also has many things to say about human reactions to internal states, to imagined situations and objects, the clenched fist of the revengeful rival who anticipates his coming fight, the movement of horror or of love in the actor's monologue. But he is also very much aware of the role which speech plays in explaining and communicating these reactions and sceptical even about the chances of pure mime. About painting, he explicitly forbears to speak (1785–6: 131).

Another eighteenth-century critic, who wrote before Engel, had actually attempted to bridge the gap between the movements of life and the 'stills' of painting by suggesting how the past and the future could somehow be made visible in the rendering of a transitory moment. He was Lord Shaftesbury who, in his *Characteristicks* (1714), discussed the ways a painter could represent the story of The Judgement of Hercules, in which the hero is confronted by Pleasure and Virtue and decides for the latter (1714, Ch. I, Section II).

'For in this momentary Turn of Action, Hercules remaining still in a situation expressive of Suspence and Doubt, wou'd discover neverthe-less that the Strength of this inward Conflict was over, and that Victory began now to declare her-self in favour of Virtue. This Transition, which seems at first so mysterious a Performance, will be easily compre-hended, if one considers, That the Body which moves much slower than the Mind, is easily out-strip'd by this latter; and that the Mind on a sudden turning itself some new way, the nearer situated and more sprightly parts of the Body (such as the Eyes and Muscles about the Mouth and Forehead) taking the alarm, and moving in an instant, may leave the heavier and more distant parts to adjust themselves and change their Attitude some moments after.'

Whatever we may think of this *a priori* analysis, it reminds us of the relevance of our two initial postulates, the need for legibility and for clear contextual clues. In art the two are obviously interdependent. The clearer the situational clues (as in our animal examples) the less may there be need for perfect legibility, and vice versa.

4. SYMBOLIC AND EXPRESSIVE GESTURES

It is well known that 'primitive' or 'conceptual' styles of representation show much concern for clarity and legibility. But it will be found that some of the devices adopted towards this very end may interfere with the lucid rendering of expressive movement. The conventions of ancient Egyptian art strenuously excluded the rendering of foreshortening (Schäfer, 1930). Every human figure had to be shown in a clear silhouette which looks somewhat distorted to us, precisely because every part of the body is so turned as to present its most lucid, 'conceptual' shape. It is partly for anatomical reasons that these needs interfered less with the rendering of animals and their movements and thus allowed the artist of our example to depict his little tragedy so movingly. Where human figures are shown interacting in violent motion, as in representations of teamwork, or of fighting, the needs of legibility sometimes lead to a wrenching and twisting of the body which somewhat hampers the convincing rendering of expressive gestures.

In the representation of social interaction recourse had therefore to be taken to social symbolism. Notoriously the important personage in Egyptian art is represented larger than are those on the lower rungs of the hierarchy. He is frequently marked with a sceptre or other insignia while those he commands or supervises are represented in more submissive postures. Moreover it is well known that all civilizations have developed standardized symbolic gestures which approximate the vocabulary of a gesture language.* I have argued elsewhere (1966) that these ritualized gestures of prayer, of greeting, of mourning at funeral rites, of teaching or triumph are among the first to be represented in art. They are much more easily fitted into the conventions of a conceptual style, such as the Egyptian, than are the spontaneous movements of human interaction. These 'performative' gestures are self-explanatory actions which set up a clear context and are not concerned with the passage of time (Groenewegen-Frankfort, 1951). The King stands before his God, the Noble receives tributes, the dead are bewailed: all these are types of juxtaposition which lend themselves to unambiguous representation even within a style which excludes the realistic approach to the human body in action.

It is well-known that it is to Greek art that we must look for the conquest of appearances. I have suggested in *Art and Illusion* that the striving for this mastery was determined by the function of art within Greek civilization, where it required the illustration or even the dramatic evocation of mythological stories as told by the epic poets. Be that as it may, Greek art certainly developed devices which compensate for the absence of movement not by symbolic expression but by the creation of images of maximal instability. Bodies are made to take up positions which we know from experience to be incapable of being maintained, muscles are tautened like a drawn bow, garments begin to flutter in the wind to indicate speed and transitoriness. By itself such a style need not be relevant to our topic, for theoretically the interaction of figures could remain on a purely physical level, fighting groups grappling and parrying blows, or athletes wrestling. But even in such situations it is no more possible than it was in our initial animal example to separate action from communication. We see the victim of aggression trying to ward off the coming thrust with a gesture of self-

* For a bibliography of symbolic gestures see Gombrich (1966), to which should be added Brandt (1965).

protection that also suggests pleading, we see the victor in an attitude of domination that suggests triumph or pride (Fig. 1). A study of these and similar motifs in ancient art reveals the need for a compromise between the conflicting demands of maximal legibility and maximal movement. The attitude of both aggressor and victim must be transitory, but lucid, and those solutions which best do justice to these demands will tend to be adopted as a formula on which only slight variations need be played. Thus

Fig. 1. Greek Fighting Group. Halicarnassus Mausoleum, British Museum, (after Reinach, *Répertoire de Reliefs*).

the moment in which 'non-verbal communication' between human beings was first specifically observed and rendered in art can never be determined with any degree of precision (Deonna, 1914; Neumann, 1965). What matters is the degree of empathy expected and aroused.

If it really became the task of Greek art to turn the beholder into an eye-witness of events he knew from Homer and other poets, it is clearly not fruitful to look beyond this demand for a hard and fast distinction between physical and psychological interaction.

5. NARRATIVE AND INTERPRETATION

We do not know how the Egyptians viewed the separation of the calf from its mother, but it is obvious that I have sentimentalized the scene in calling it a little tragedy. No empathy is likely to have been expected on the part of the beholder; even in human scenes there is little evidence of such a demand. It is this appeal to our responses which distinguishes Greek narrative art of the sixth and fifth century B.C. from most earlier styles. The fright of Eurystheus who (on a Hydria in Paris) has crept into a vat and lifts his hands in horror as Hercules brings Cerberus; the joy of Theseus' sailors (on the François Vase in Florence) gesticulating and throwing up their arms in pleasure as they land on Delos (Plate 2*a*); the sorrow of Ajax (on a cup in Vienna) who hides his head in grief when he loses his case for the arms of Achilles: the wrapt attention of the Thracians (on a wine bowl in Berlin) who hear Orpheus sing (Plate 2*b*); the Satyr (on a jug in Oxford) who dances with joy as he finds a nymph asleep; all these are examples taken from Greek vases (Buschor, 1921; Pfuhl, 1923) leading up to the period in the early fourth century when Xenophon in the *Memorabilia* represented Socrates discussing the subject of expression with the painter Parrhasios and the sculptor Cleiton. In both these little dialogues the artists must have their attention drawn to the possibility of representing not only the actions of the body but through the body the 'workings of the soul'.

'How could one imitate that which has neither shape nor colour . . . and is not even visible?' asks the puzzled Parrhasios, and is told of the effect of emotions on people's looks: 'Nobility and dignity, self-abasement and servility, prudence and understanding, insolence and vulgarity, are reflected in the face and in the attitudes of the body whether still or in motion'. (*Memorabilia*, III, x, 1–5. Ed. Marchant, 1923.)

The injunction has been repeated in countless variations throughout the literature of art which is based on the classical tradition. Not only are particular works of painting or sculpture praised for their mastery in conveying the character and emotions of the figures portrayed, many treatises on art since the Renaissance (e.g. by Alberti, Leonardo, Lomazzo, Le Brun) contain sections in which the outward symptoms of the emotions or 'passions' are described and analysed. Interesting as these discussions are for the history of our studies (Montagu, 1960), it must be admitted that most of them bypass the crucial difference between art and life which was our starting point. Expressive movements are movements and once we lack the explanatory sequence to tell us how this configuration started and where it leads to, ambiguity will increase to an unexpected extent, unless, of course, the absence of movement is compensated for by situational cues. Eurystheus in his vat may be extending his hands because he cannot wait to stroke Cerberus, the sailor who throws up his arms for joy on landing in Delos may have been hit by an arrow, Ajax may be hiding his head to conceal not his sorrow but his laughter in having brought off a splendid trick, the Thracians may be bored by Orpheus' songs and even the Satyr may jump and clap his hands in order to wake the sleeping nymph. There is a humorous book called *Captions Outrageous* (by Bob Reisner and Hal Kapplow) attempting such re-interpretations of famous masterpieces with more or less wit. Psychologists, moreover, know from the varying readings of the *Thematic Apperception Test* (Murray, 1943) how great is the spread of possible interpretations of any picture unless a firm lead is given by the context or caption. Experiments have shown that if we isolate an individual figure from the snapshot of an emotional scene it will only exceptionally allow us to guess the elements of the situation (Montagu, 1960). Even facial expression when isolated from casual snapshots turns out to be highly ambiguous. The contorted face of a wrestler may look in isolation like the expression of laughter, while a man opening his mouth to eat may appear to be yawning.*

Thus art stands in need of very clear and unambiguous cues to the situation in which the movement occurs. In particular we have to know whether a movement portrayed should be interpreted as predominantly utilitarian or expressive. Plate 3 *a* is easily misinterpreted as a gesture of submission, that is an expressive movement of extreme obeisance. We have to know the context, the story of the Gathering of Manna, to understand why the man is cowering on the ground and stretching out his hands – he is trying to grab as much as possible of the miraculous food that has fallen from heaven. A representation of interacting people is not necessarily self-explanatory. It must be interpreted and this interpretation implies setting the movements into an imaginary context.

In most periods of art such a context is given by situational cues which are familiar to members of the culture. The painter and sculptor makes use of a good deal of symbolic lore to mark the individual personages as kings or beggars, angels, or demons, he introduces further emblems or 'attributes' to label individuals so that no difficulty arises in recognizing Christ or the Buddha, the Nativity or the Rape of Proserpina.

Take the relief of Orpheus and Eurydice (Plate 1 *b*) after a Greek composition of the fifth century B.C. First we must recognize the protagonists by what are called their 'attributes', the singer's lyre, or the traveller's hat of Hermes, the guide of the dead. Only then can we really identify the

---

* The example is taken from a series of experiments made at Karl Buehler's seminar in Vienna in the early 1930s in which I took part as a subject. I was introduced to these studies by Ernst Kris who also organized a series of experiments on the reading of facial expression in art. I have applied these insights (1945), where the divergent interpretations of the expression of Venus in the *Primavera* are quoted, and (1969), which centres on the various interpretations of a figure in one of H. Bosch's compositions.

episode which is here represented, the fatal moment when Orpheus has disobeyed the condition imposed on him and has looked back at Eurydice who is therefore taken back to Hades by the god. Thus we may 'compare and contrast' it without irreverence with our Egyptian example (Plate 1 *a*). The Greek work does not deviate much from that 'conceptual' clarity that presents the posture of every figure at its most legible, it is in fact in the subtle deviation from this normal position that the relationship of the three actors is most delicately conveyed. These small deviations are in the direction postulated by Engel's analysis. Hermes is seen to bend back slightly as he gently takes Eurydice by the wrist to return her to the realm of Pluto. The two lovers face each other, her hand rests on the shoulders of the guide who had failed her, her head is slightly lowered as they gaze at each other in a mute farewell. There is no overt expression in their blank features, but nothing in these heads contradicts the mood we readily project into this composition, once we have grasped its import.

Such a subtle evocation must rely on the same kind of beholder who would also know how to appreciate the reworking of a familiar myth at the hands of a Sophocles or Euripides. The relief, in other words, is not really created to tell the story of Orpheus and Eurydice but to enable those who know the story from childhood to re-live it in human terms. This reliance on suggestion is characteristic of the great period of Greek art in which every resource of expressive movement was used to convey the interaction of individuals. These resources were lost or discarded as soon as art was predominantly used to drive home a message and proclaim a sacred truth.

6. THE PICTOGRAPHIC STYLE

During declining antiquity, with the rise of Imperial cults and, above all, with the development of Christian art, we can observe the re-emergence of frankly conceptual methods and a new standardization of symbolic or conceptual gestures (Brilliant, 1963). These gestures of prayer, instruction, teaching or mourning, help rapidly to set up the context and to make the scene legible. The Emperor sacrificing, the general addressing the army, the teacher instructing his pupils, the defeated submitting to the victor, all these are types of juxtaposition which lend themselves to as unambiguous a representation for those who know the conventions of gesture language, as are scenes of combat for those who do not. Such impressive legibility is demanded where the rendering of a holy writ almost forbids that free dramatic evocation that Greek art had evolved. Moreover it needs much mastery on the part of the artist and the beholder to isolate and interpret expressive movements in the context of vivid interaction. Thus late antique and early Christian art generally played safe in the illustration of narrative texts. An almost pictographic idiom was distilled from the freer tradition of classical art. The need for unambiguous messages stilled the vivid and subtle interplay of action and reaction that marked the masterpieces of the earlier style (Gombrich, 1935). Instead we are frequently shown the protagonist, Christ, a Saint, or a Prophet or even a pagan hero standing erect, with a gesture of 'speaking' or command, the centre of the scene to which all other figures must be related.

To the student of non-verbal communication this extreme 'pictographic' convention is of interest precisely because the need to turn art into a 'script for the illiterate' (Gombrich, 1950) brings out both the potentialities and the limitations of the medium and can serve as a point of reference in the consideration of other styles.

The pictographic style takes no chance with naturalism. There is no pretence, implied or overt, of presenting a snapshot of a given scene such as might have been seen and photographed by an imaginary witness. In fact the style makes it easy to show up the fallacies in this conception of art which has haunted criticism since Lessing's *Laocoon*. Neither the prayer nor the speech, the wailing or the submission is imagined to be recorded at a particular moment of time. The assembled pictographs relate to a story in the past which is now accomplished and complete. Christ stands with extended hand in front of an edifice that contains a mummy, to symbolize the Resurrection of Lazarus (Fig. 2). Moses is seen with outstretched hand holding a rod, while water gushes from the rock as in many catacomb paintings. It would be foolish to ask whether the act of striking is over or whether the artist has anticipated the effect by showing the jet of water. The juxtaposition simply conveys the story of the water miracle much as a brief narrative would. One might in fact translate the pictograph into a sentence in which the protagonist is the subject, the action the verb and the tomb or rock the object. The pictograph – to use a distinction I have found useful – represents the 'what' but not the 'how', the verb but not the adverb or any adjectival clause.

7. THE CHORUS EFFECT

There are several ways in which art can introduce these enrichments to convey not only the fact of the event but also some of its significance, and these invariably draw on the resources of 'non-verbal communication', that is on expressive as distinct from symbolic movement. Perhaps the most general method in art has been to clarify the meaning of the action by showing the reaction of onlookers. When Christ brings Lazarus to life, his two sisters prostrate themselves before Him in awe and gratitude while the crowd shows by their gestures and movements that they are witnessing a

Fig. 2. Resurrection of Lazarus. Ravenna, Museo Archeologico (after
Rohault de Fleury).

miracle. (Not to mention the bystanders holding their noses to remind us
that the corpse was already far gone! Plate 3*b*.) When Moses strikes water
from the rock the Elders who had come to witness the scene throw up their
arms in wonder and the thirsting Israelites extend their eager hands to
drink (Plate 4*a*). There are many themes of Western art which can best be
described in terms of this formula of action and reaction, the reacting

crowds providing the 'chorus' explaining the meaning of the action, and, in doing so, setting the key for the beholder's response. The student of expression can here verify some of the analysis by Shaftesbury and Engel mentioned above. The orientation of the figures towards or away from the central event expressing admiration, aggression, flight or awe. But art, like the stage, has also explored less obvious reactions in the depiction of great events – the 'autistic' gestures of the contemplative, the fearful movement of the hand to the head, the abstracted look of those immobilized by surprise and, to mention a frequent but very subtle formula, the way a bystander may turn away from the main event to look into his neighbour's eye as if to make sure that others, too, have seen the same and are equally moved.

8. EXPRESSION AND EMPHASIS

Needless to say, the schema purposive action–expressive reaction presents merely a convenient abstraction. Both the action and the reaction can be more or less communicative of psychological states, provided we have sufficient context to interpret them. This is a point where the 'language' of gestures can be compared with the language of words – every symbolic movement also has a 'tone' which conveys character and emotion; it can be tense or relaxed, urgent or calm. There are countless traditional subjects in Western art which allow us to study these possibilities of what Dante in the *Divine Comedy* so beautifully calls 'visible speech'. Describing a relief representing the Annunciation he says: 'The angel that came to earth with the decree of peace . . . appeared before us so truthfully carved in a gentle gesture that it did not appear to be a silent image. One would have sworn that he said "Ave" . . . and onto her attitude there was impressed that speech "Ecce ancilla Dei" exactly as a figure is sealed onto wax . . .' (*Purgatorio*, x, 37–45).

In what I have called the 'pictograph' mode this exchange would be expressed simply in the Angel extending his hand in a speaking gesture while the Virgin's reaction and response were confined to a lifting of her palms in a movement of surprise (Gombrich, 1950, Fig. 120). But on reading the gospels the artist would find more about that supreme moment of the Incarnation, he would read that on seeing the angel 'she was troubled at his saying and cast in her mind what manner of salutation this should be' before the final submission 'Behold the handmaid of the Lord; be it unto me according to thy word'.

Any artist who wanted to depart from the pictographic method of narrative to emulate the representation Dante had seen in his vision therefore had to feel his way like an actor trying to express a complex emotion – the way fear and wonder turns into unquestioning acceptance.

1 a   Detail of relief from the Tomb of Ti, Egypt, IVth Dynasty. (Hirmer Foto-
archiv, München.)

  b   Orpheus and Eurydice. Roman copy of a Greek relief, Naples, Museo
Nazionale. (*Photo* Alinari.)

(*facing p.* 384)

2a Theseus' sailors landing on Delos. Detail from the François Vase, Museo Archaologico, Florence.

b Orpheus and Thracians. From a Krater, Berlin, Staatliche Museum.

3a Aertgen van Leyden, *The Israelites in the Desert*. Berlin, Printroom, No. 12204.

b Giotto, *Resurrection of Lazarus*. Padua, Arena Chapel. (*Photo* Alinari.)

4a School of Raphael, *Moses Striking the Rock*. Vatican, Loggie. (*Photo* Alinari.)

b   Poussin, *Moses Striking the Rock*. Edinburgh, National Gallery of Scotland
(Duke of Sutherland Collection, on loan to the National Gallery of Scotland).

5 *a*   School of Botticelli, *Annunciation*. (Glasgow, Art Gallery and Museum.)

b   Botticelli, *Adoration of the Magi*. Florence, Uffizi. (*Photo* Alinari.)

6a Leonardo da Vinci,
drawing, study for
Adoration of the Magi.
Paris, Louvre.
(*Photo* Alinari.)

b Ghirlandajo,
Adoration of the Magi.
Florence, Ospedale
degli Innocenti.
(*Photo* Alinari.)

7 *a*   Mosaic, *St Peter's Denial*. Ravenna, S. Apollinare Nuovo. (*Photo* Alinari.)

b   Rembrandt, *St Peter's Denial*. Amsterdam, Rijksmuseum.

8a Haynes King, *Jealousy and Flirtation*. (Victoria and Albert Museum, London. Crown Copyright.)

b Orchardson, *The First Cloud*. (The Tate Gallery, London.)

We happen to know through the writings of Leonardo da Vinci that the right extent of departing from pictographic clarity towards the Greek style of dramatic evocation was a subject of debate among artists of the Renaissance. Chiding those of his fellow artists whom he regarded as mere 'face painters' – specialists in portraiture – Leonardo comes to speak of his favourite topic, the need for universality in an artist and especially the importance of observing the expression of mental states. Those Florentine artists of the *quattrocento* who had gone furthest in exploring the representation of movement encountered a certain amount of opposition in the name of 'decorum'. Alberti in the 1430s speaks in general terms of the need for restraint, since figures throwing their limbs about look like 'duellers' (Alberti, 1435, ed. Janitschek, 1877: 127). Filarete (1460, ed. Spencer, 1965: 306), paraphrasing this remark some twenty years later, identifies the target of these strictures – he says that Donatello's disputing Apostles are gesticulating like jugglers. Now Leonardo, who must have heard similar remarks passed about his own paintings, goes over to the counter-attack. Specialists in portraiture, he remarks, lack judgement in these matters because their own works are without movement and they themselves are lazy and sluggish. Thus when they see works showing more movement and greater alertness than their own, they attack them for looking as if 'possessed' or like morris dancers (Leonardo da Vinci, 1500, ed. McMahon, 1956: 58).

Admittedly, Leonardo concedes, there can also be excesses in the other direction. (Cod. Urb. fol. 33.)

> 'One must observe decorum, that is the movements must be in accord with the movements of the mind . . . thus if one has to represent a figure which should display a timid reverence it should not be represented with such audacity and presumption that the effect looks like despair . . . I have seen these days an angel who lookes as if in the annunciation he wanted to chase Our Lady out of her chamber with gestures which looked as offensive as one would make towards the vilest enemy, and Our Lady looked as if she wanted to throw herself out of the window in despair.'

A painting in Glasgow (Plate 5a) from the workshop of Botticelli almost answers to Leonardo's satirical description.

But Leonardo being Leonardo did not remain content with these polemical remarks. He went on reflecting on the problem posed by such disparate judgements about works of art and came to the conclusion that the reactions of his fellow artists were invariably connected with their own style and temperament. 'He who moves his own figures too much will think that he who moves them as they should, makes them look sleepy, and he who moves them but little will call the correct and proper movement "possessed"' (ibid.).

It is interesting to watch Leonardo himself groping for the 'correct and proper' rendering of 'timid reverence' in an early study for an Adoration of the Magi (Plate 6 a). Once more it may be instructive to recall the 'pictographic' mode of illustrating the Biblical episode on early Christian sarcophagi, where the Virgin and Child are approached by three identical figures in the recognizable garb of the Magi in the symbolic act of paying homage, their hands carrying the presents often covered by a cloth (Fig. 3). For Leonardo, of course, the symbolic act must also be expressive of

Fig. 3. Adoration of the Magi. Rome, S. Marcello (after Garrucci, vol. V, pl. 310, 3).

what goes on in the minds of the Kings who have come from afar to greet the newborn Saviour. He varies the gesture of presentation and submission from the upright and rather unmoved youngster in the left-hand corner who does not even look at the child, to the old King who humbles himself as he moves forward on his knees to extend his gift. But the right degree of emphasis is obviously only one of the needs the artist seeks to satisfy. He also takes great care that the posture and movement remain completely legible, turning the actors in such a way as to present the clearest silhouette. None of these attitudes is really a movement caught on the wing; each could be taken up and held in a *tableau vivant* and it is this among other things which Leonardo clearly wanted if the figures are not to incur the justified strictures of excessive movement.

Comparing his solution with that of his contemporaries one might imagine that the criticism that Leonardo's figures looked too lethargic might have come from Botticelli whose later style is indeed almost 'possessed' (Plate 5*b*). The opposite objection that Leonardo's own figures are 'gesticulating like mad' might have come from his other Florentine rival Ghirlandajo, whose Adoration provides a foil of stolid immobility to Leonardo's dramatic gestures (Plate 6*b*).

Leonardo's interesting observations can be generalized to apply not only to the varying standards of artists, but also to those of other critics. We all know that the Northerner will tend to find the expressive movements of the Latin nations over-emphatic and theatrical. In writing about Leonardo's *Last Supper*, Goethe (1817) had to remind his German readers of this characteristic of Italian culture, and I have found that contemporary English students can be incredulous if they are told that Leonardo may really have intended the intensity of gesticulation he used in the *Last Supper* to convey the disciples' reaction to Christ's words that one of them would betray Him. We certainly judge the emotional import of an expressive movement by comparing it with some mean, just as we do with the loudness of speech or other dimensions of emphasis.

Thus the style of movement represented in art will depend on a great many variables including the current level of emphasis, or the demand for restraint which varies in its turn not only from period to period and nation to nation but also from class to class. Few aspects of 'manners' and behaviour were more eagerly discussed in treatises on acting and on art than this question of 'decorum'.

Nobility, on the whole, implied restraint or at least a stylized type of emphasis, while the vulgar could disport themselves more freely and more spontaneously, as in pictures of carnivals, of taverns, of the barber pulling a tooth. Naturally the resources of expressiveness continued to be adapted to different ends in conformity with these different ideals. It has been claimed (Mâle, 1932) that the Church of the Counter Reformation favoured the representation of martyrdoms to rouse the beholder, compositions in which the roles of the chorus and the protagonist tend to be reversed. It is certain that the seventeenth century in Italy developed new formulae for the expression of extreme and ecstatic states.

By that time, of course, art may be said largely to have returned to a function akin to its role in classical Greek art. It was not mainly there to tell the sacred story to the illiterate but rather to evoke it in a convincing and imaginative way to those who knew it. It is characteristic of art that ultimately the display of resources may become part of a novel purpose. This is certainly true of the rendering of expressive movement. The artist's mastery in conveying human emotions was so much admired that the illustration of a sacred or secular story becomes rather the occasion for the exercise of such mastery. Just as the libretto of the average opera was chosen to allow the composer to express or depict the widest range of human passions in his music, so the subjects selected by post-Renaissance artists were frequently intended to permit a maximum of dramatic effects.

Naturally in art no less than in drama these effects in their turn were subject to the rules of 'decorum', particularly in seventeenth century France. When Poussin illustrated the story of Moses striking the rock (Plate 4*b*) – as he did three times in his life – he took great care that the thirsting Israelites in the desert were made to express their response to the miracle with nobility and restraint. His rendering of the Gathering of Manna was the subject of a famous academy discourse by the painter Le Brun who stressed the conformity of the various types to classical precedents (Testelin, 1696). The very approximation of pictorial representations to the stage, however, also produced a reaction. Epithets such as 'stagey' or 'theatrical' are not necessarily words of praise when applied to works of art, and the gradual eclipse suffered by academic art with its 'grand manner' is closely linked with the reaction against classical rhetoric and in favour of a less formal and less public display of emphatic emotion.

9. INWARDNESS AND AMBIGUITY

The tradition of Northern art less immediately affected by classical influences had developed pictorial devices earlier on which appealed to this taste for a more inward, more lyrical and less dramatic expression in art. Instead of concentrating on expansive movement, the artist relied on the characterization of physiognomies and facial expression. A type of composition appeared towards the end of the fifteenth century in which the *dramatis personae* are shown in close-ups and all the psychological inter-action must be read in the features (Ringbom, 1965). Thus the artist may concentrate on representing the expression of devotion in the heads of the three Magi who offer their gifts to the Christ child or the fierce aggression of Christ's tormentors contrasted with the Saviour's patience.

Northern art, like the drama of Shakespeare, was altogether less hemmed in by classical rhetoric and decorum and this may explain the fact that in painting too the greatest portrayer of human reactions is to be found in the Protestant North, in Rembrandt who had studied and absorbed both traditions. Once more it is instructive to compare his way of narration with the 'pictographic' method of early Christian art. One of the mosaics of S. Apollinare Nuovo (Plate 7*a*) illustrates the episode of the Gospel of the Denial of Peter (Luke 22: 54–62).

'Then took they him, and led him, and brought him into the high priest's house. And Peter followed afar off. And when they had kindled a fire in the midst of the hall, and were set down together, Peter sat down among them. But a certain maid beheld him as he sat by the fire, and earnestly looked upon him, and said, This man was also with him. And he denied him, saying, Woman, I know him not. And after a little while another saw him, and said, Thou art also of them. And Peter said, Man I am not. And about the space of one hour after another confidently affirmed, saying, Of a truth this fellow also was with him: for he is a Galilaean. And Peter said, Man, I know not what thou sayest. And immediately, while he yet spake, the cock crew. And the Lord turned, and looked upon Peter. And Peter remembered the word of the Lord, how he had said unto him, Before the cock crow, thou shalt deny me thrice. And Peter went out, and wept bitterly.'

The Ravenna mosaic represents the essential elements in the story. The maid raising her hand in a speaking gesture towards St Peter who shrinks back and vividly signals his denial. Rembrandt (Plate 7 *b*) evokes the entire scene by the camp fire, but at first glance it would seem that he is less intent on translating speech into movement. The maid holds a candle close to Peter's face to scrutinize his features but he merely lifts one hand in a movement which is much less unambiguous than that of the early Christian mosaicists. Indeed taken in isolation the figure may simply be shown to speak or even to make an inviting gesture asking one of the other figures to come forward. But the figure is not in isolation and thus Rembrandt compels us to picture the whole tragic scene in our mind, the anxious old man sadly facing the inquisitive woman and the two tough soldiers whose presence amply accounts for his denial. But what makes the picture particularly unforgettable is the barely visible figure of Christ in the dark background who has been facing his accusers and is turning round, as the Bible says, to look at His erring disciple. It is the absence of any 'theatrical', that is of any unambiguous, gesture which prevents us from reading off the story as if it were written on scrolls and involves us all the more deeply in the event. The very element of ambiguity and of mystery makes us read the drama in terms of inner emotions and once we are attuned to this reading we increasingly project more intensity into these calm gestures and expressions than we are likely to read into the extrovert gesticulations of the Latin style. The painting by Rembrandt demands quiet scrutiny and prolonged meditation. Moreover it demands much more active participation on the part of the beholder who must know the biblical story and have pondered its universal significance if he is to understand the poignancy of Peter's expression and of Christ's invisible gaze.

Speaking somewhat schematically it may be argued that from the Renaissance to the eighteenth century the function of art was conceived in the same way as it had been in ancient Greece – the artist should show his mettle by interpreting known texts. It was the 'how' and not the 'what' the connoisseur admired and pondered. He appreciated the way the painter rendered a particular episode from the Bible or from the Classics and desired to share and understand the reaction of participants through an act of imaginative empathy. It is here, of course, that Rembrandt is supreme precisely because he has discovered and developed the perfect mean between the unrealistic pictographic gesture and the indeterminate representation of an enigmatic movement.

The importance for art of mobilizing the beholder's projective activities in order to compensate for the limitations of the medium can be demonstrated in a variety of fields. The indeterminate outlines of Impressionist pictures which suggest light and movement are a case in point. Such experiments should be of interest to the psychologist of perception for what they tell us about our reactions to real-life situations. This may also apply to the study of non-verbal communication. There is no reason to think that in real-life situations the most unambiguous gesture or expression is also the most telling or moving. We learn to appreciate ambiguity, ambivalence and conflict in the reactions of our fellow human beings. It is this richness and depth of our response that a great artist such as Rembrandt knows how to evoke. Provided therefore we do not make the mistake of looking in the greatest works of dramatic narration for a realistic record of movements such as actually occur in non-verbal communication we can study these illustrations with much more profit than the limitations of the medium would allow one to accept.

This is true despite the fact that an inventory of expressive movements used in old master paintings would be likely to reveal a surprisingly limited range. The reasons for this restriction should have become clear from the preceding examples. Perhaps the most decisive of them is the need for conceptual clarity in the posture presented to the beholder, which rules out a large range of movements in which limbs would be too much fore-shortened or hidden for the movement to explain itself. Needless to say, neither this nor any other rule is absolute and subsidiary figures can often be shown in postures of greater complexity or obscurity. However, the astonishment with which the first snapshots were greeted shows that the average observer rarely notices, let alone remembers, the more transient movements which were therefore excluded from the traditional vocabulary of art. We must stress once more that in this as in other respects the realistic rendering of life situations did not arise from a simple imitation and adjustment of a conceptual or pictographic tradition.

10. ALTERNATIVE FUNCTIONS

I have emphasized the interdependence of art and function because its recognition helps us to escape from a dilemma which still haunts the history and criticism of art. Originally this history was told in terms of progress, interrupted by periods of decline. It is this conception of history we find in the authors of classical antiquity and in those from the Renaissance to the nineteenth century, who describe the gradual acquisition of mastery in the rendering of the human anatomy, of space, of light, texture and expression. To the twentieth century which has witnessed the deliberate abandonment of these skills on the part of its artists this interpretation of history has come to look naïve. No style of art is said to be better or worse than any other. We may accept this verdict within certain limits provided it does not tempt us into an untenable relativism concerning the achievement of certain things – and the rendering of non-verbal communication is such a case in point.

We have a right to speak of evolution and of progress in the mastery of certain problems and in the discovery of perfect solutions. Kenneth Clark, in a perceptive essay (1963), has singled out such a problem, the meeting and embrace as it occurs in the story of the Visitation, and has shown its progressive perfection towards what may be called a 'classic' form. We can acknowledge such perfection without forgetting the possibility of alternative solutions once a shift in the problem occurs.

Unfortunately the history of art has tended for too long to fight shy of this type of investigation. We have no systematic study of eye contacts in art (but see Nordenfalk, 1950) and even the exact development of facial expression is all but unknown. Clearly it would not be possible for this essay to reduce these large blank patches on the map of our knowledge. All that can still be done, in conclusion, is to point to their existence and to the location of some of them.

I have mentioned one at least by implication: there must be a great difference between a painting that illustrates a known story and another that wishes to *tell* a story. No history exists of this second category, the so-called anecdotal painting which flourished most in the nineteenth-century salon pictures. Indeed twentieth-century critics have covered the whole genre with such a blanket of disapproval that we are only now beginning to notice this phase in the history of art.

It is likely, however, that the student of non-verbal communication would find a good deal of interest in these systematic attempts to condense a typical dramatic scene into a picture without any more contextual aids than, at the most, a caption. Clearly many of these painters must have profited from a study of the realistic stage rather than from an observation of life, but the fact remains that they made use of a very much enriched vocabulary. The painting by Haynes King *Jealousy and Flirtation* (1874, Plate 8*a*) hardly stands in need of a caption. The flirting girl with her inviting look, her hands resting on her head, is immediately intelligible as is the awkward but pleased reaction of the young man. The expression of jealousy may be a little too obvious and genteel, though the 'autistic' gesture of the girl's left hand is expressive enough.

Or take *The First Cloud* (1887) by Orchardson (Plate 8*b*). It would be interesting to test the interpretations of this scene by subjects who do not know the caption. One could certainly think of alternative interpretations, for after all we only see the woman from the back and have to project into her movements whatever we read in the man's expression. According to Raymond Lister (1966) 'she is walking off in a huff . . . the man's eyes following her with a somewhat puzzled though obstinate expression . . . .'. At any rate his expressive posture is a novelty to art.

It would be interesting to trace the development of these novel means and in particular to examine the role which book illustration on the one hand and photography on the other played in this development. One thing seems to me sure. Given the new story-telling function of anecdotal art we should be enabled also to trace another series of progressive skills in this as in any other type of representation. In fact, if we go back to the roots of this art in the genre paintings of the Netherlands and if we stop to examine the methods used by the first deliberate story-teller, William Hogarth, we will in all likelihood find that there is a gradual process of enrichment and refinement regardless of whether we like or dislike the ultimate result.

One could think of other topics and social functions which have driven the artist towards the exploration of non-verbal communication. Advertising, for instance, frequently demands the signalling of rapturous satisfaction on the part of the child who eats his breakfast cereals, the housewife who uses a washing powder or the young man smoking a cigarette. It has equally specialized in the exploration of erotic enticement, the 'come hither look' of the pretty girl or the inviting smile of the secretary who ostensibly recommends a typewriter. Clearly the commercial artist and the commercial photographer are likely to know a great deal about the degree of realism and stylization that produces the optimum results for this purpose and also about the changing reactions of the public to certain means and methods. Finally we may point once more to the unexplored realm of the 'comics' with their own conventions of facial expressions and gestures which have penetrated into 'pop' art. Art is long and life is short.

POSTSCRIPT

In conclusion a further elucidation of the use of the term 'expression' in relation to art may be useful. The traditional usage here adopted, which applies this term to the expression of the emotions of the figures in a dramatic *illustration* (*Laocoon*, the *Pietà*), has indeed been partly superseded by the approach of twentieth-century aesthetics which so frequently regards the work of art as an expression of the *artist's* inner states (Gombrich, 1962, 1966). To these may be added the most ancient usage which relates art predominantly to the emotions it is capable of *arousing*. The interplay of these usages can best be exemplified in the history of musical theory. The Greeks (including Plato) concentrated on the *effects* of music on the emotions which ranged from magic efficacy to the creation of moods. The dramatic theory of music favoured by the revivers of opera in the Renaissance and the Baroque stressed the power of music to *depict* or paint the emotions of the noble hero or the desolate lover. It was only in the Romantic period that music was interpreted as an expression of the composer's moods and sentiments. It will be observed that this change of attitude may leave the correlation between certain types of music and certain types of emotion unaffected: the proverbial trumpet call may be seen as arousing, depicting or manifesting war-like feelings. Interest in these aspects changes with the changing social functions of music. It is the same with the visual arts. The magic function of arousal may reach far back to apotropaic images and survives in religious, erotic and commercial art. Interest in art as an expression of the artist's personality and emotion presupposes an autonomy of art only found in certain societies such as Renaissance Italy. Indeed Leonardo's observations on the link between an artist's character and his dramatic powers quoted above (Section 8) point the way to this evaluation.

COMMENTS

1. *Language, art and non-verbal communication*

In the earlier chapters we were led towards distinctions between signals which do not depend on language and whose evolutionary origins may be traced in sub-human forms, those paralinguistic features which augment our spoken language, and gestures and rituals which, while not immediately associated with verbal language, depend either on it or on 'structures' basic to it. But while such categories form convenient pigeon-holes, we must not think that every element of non-verbal communication will fit conveniently into one or the other. The difficulty is obvious enough with the stage: the actor's use of expressive movements which are themselves independent of language, as well as paralinguistic and prosodic features, must depend on an interpretation of the written verbal language of the script.

With the visual arts the relations are even more involved. The portrayal of simple human expressive movements on canvas at first sight seems to be quite independent of language, and Sebeok indeed (1968, see also pp. 91–2, this volume) classified the visual arts on that basis. But this may prove to be a superficial view. The visual arts are limited to man, so their emergence is at least correlated with that of verbal language. Furthermore, Gombrich emphasizes that the ritualized gestures of prayer, greeting, mourning and so on which, as Leach (Chapter 12) argues, depend on verbal language, are among the first to be represented in art – perhaps because they are 'more easily fitted into the conventions of a conceptual style'. Only later did the 'spontaneous movements of human interaction' come to be depicted in art. This suggests either that both the visual arts and language were independent expressions of a faculty necessary for both or, more probably, that the visual arts were dependent on a language-based culture and were concerned first with its most obvious expressions.

2. *Context*

Both Gombrich and Miller emphasize the importance of context at a number of different levels. Playwright, actor and painter are influenced by the conventions of their time both in their aims and in how they strive to achieve them: in their turn, of course, they influence those conventions. In particular, they are much influenced by other techniques in their culture – the actor by films and television, the artist by photography and the requirements of book illustration. And both play-goer and beholder interpret each facet of the work in terms of the context in which it is set – the chorus, the rest of the play or picture.

EPILOGUE

In these chapters we have attempted to convey some indication of the range of phenomena which can be described as non-verbal communication in man and animals. We have discussed some of the terminological and classificatory difficulties which arise in its study, and tried to air some of the points of current controversy. It is both fashionable and satisfying to the enquiring mind to establish unifying principles for an area of study, but it will be clear that, within this area, we are far from doing so. Since the greater the range of phenomena, the more remote must be the level of abstraction of any unifying framework, it may be as well to conclude by emphasizing that students of communication are certainly not agreed as to whether it is even worthwhile to search for such a framework. Tavolga (1968), in insisting that the methods and theory useful in the study of human language are unlikely to be useful in the study of communication in another species at a different evolutionary level, is applying Schneirla's view that there are qualitative differences in organization between animals of different phyletic level. His conclusion, though reached by a different route, is the same as Chomsky's (1967), who believes that human language, human gestures and animal communication systems can be studied together only at a remote level of abstraction. On the other hand Sebeok (1968) sees the establishment of that level of abstraction as precisely the challenge of semiotics, and points to the useful cross-phyletic comparisons which have already been made. His drive stems from the dilemma that, if all languages have a common and unique basis, the description of that basis leaves us with a sample of one: how then can we ever explain that basis in the sense of showing that it is deducible from a higher-level set of laws? Sebeok's hope is that comparison with other systems will permit discrimination between what is necessary and what is contingent in communication.

Be that as it may, such success as cross-phyletic comparisons have achieved so far has come when clearly defined questions have been asked and limited goals have been sought. Perhaps the most salutary lesson learnt by the members of the study group in the meetings which gave rise to this book was the need not only to search for generalizations, but to specify as precisely as possible the limitations of the generalizations that they made.

REFERENCES

INTRODUCTION TO PART A

Frings, H. and Frings, M. (1964). *Animal Communication*. New York; Blaisdell.

Hockett, C. F. and Altmann, S. A. (1968). A note on design features. *Animal Communication*. Ed. T. A. Sebeok, Bloomington, Indiana and London; Indiana University Press.

Ploog, D. and Melnechuk, T. (1969). Primate communication. *Neurosci. Res. Bull.* **7**, 419–510.

Tavolga, W. N. (1968). Fishes. In *Animal Communication*. Ed. T. A. Sebeok, Bloomington, Indiana and London; Indiana University Press.

1. FORMAL ANALYSIS OF COMMUNICATIVE PROCESSES

Argyle, M. (1969). *Social Interaction*, London; Methuen, New York; Atherton Press.

Fant, G. (1962). Sound spectrography. *Proc. 4th Int. Congr. of Phonetic Sci. Helsinki 1961*, 14–33; The Hague; Mouton.

Hamming, R. W. (1950). Error-detecting and error-correcting codes. *Bell Syst. Tech. J.* **29**, 147–60.

Hinde, R. A. and Stevenson, J. R. (1970). Goals and Response Control. In *Development and Evolution of Behavior*. Ed. L. R. Aronson, E. Tobach, D. S. Lehrman and J. S. Rosenblatt. San Francisco; Freeman.

Lindblom, B. (1962). Accuracy and Limitations of sonagraph measurements. *Proc. 4th Int. Congr. of Phonetic Sci. Helsinki 1961*, 190–202, Mouton; The Hague.

MacKay, D. M. (1951). Mindlike behaviour in artefacts. *Br. J. Phil. Sci.* **2**, 105–21.

MacKay, D. M. (1956). Towards an information-flow model of human behaviour. *Br. J. Psychol.* **47**, 30–43.

MacKay, D. M. (1957). Information theory and human information systems. *Impact of Science on Society*, **8**, 86–101. Reprinted (revised) in *Readings in Psychology*, 214–35. Ed. J. Cohen, Allen and Unwin, 1964.

MacKay, D. M. (1966). Cerebral organization and the conscious control of action. In *Brain and Conscious Experience*. Ed. J. C. Eccles, New York; Springer-Verlag.

MacKay, D. M. (1967). Ways of looking at perception. In *Models for the Perception of Speech and Visual Form*, 25–43. Ed. W. Wathen-Dunn, Boston; M.I.T. Press.

MacKay, D. M. (1969). *Information, Mechanism and Meaning*. Boston; M.I.T. Press.

Shannon, C. E. (1948). Mathematical theory of communication. *Bell. Syst. Tech. J.* **27**, 379–423; 623–56.

Shannon, C. E. (1951 a). In *Cybernetics*, 219. Ed. H. von Foerster, *Trans. 8th Conf. Josiah Macy Jr. Foundation*, New York.

Shannon, C. E. (1951 b). Prediction and entropy of printed English. *Bell Syst. Tech. J.* **30**, 50–64.

Thorpe, W. H. (1967). Vocal imitation and antiphonal song. *Proc. 14th Int. Ornithol. Congr.* 245–63. Ed. D. W. Snow, Oxford; Blackwell.

2. THE COMPARISON OF VOCAL COMMUNICATION IN ANIMALS AND MAN

Brown, R. (1970). The first sentences of child and chimpanzee. *Selected Psycholinguistical Papers*. New York; Macmillan.

Brown, R. and Bellugi, U. (1964). Three processes in the child's acquisition of syntax. *Harvard Educational Review*, **34**, 133–51.

Chomsky, N. (1957). *Syntactic Structures*. The Hague; Mouton.

Gardner, R. A. and Gardner, B. T. (1969). Teaching sign language to a chimpanzee. *Science, N.Y.* **165**, 664–72.

Gardner, R. A. and Gardner, B. T. (1970). Progress reports currently circulated in cyclostyled form only.

Gardner, R. A. and Gardner, B. T. (1971). In *Behaviour of Non-human Primates*, vol. 3. Ed. A. Schrier and F. Stollnitz, New York; Academic Press.

Hayes, K. G. and Hayes, C. (1952). Imitation in a home-raised chimpanzee. *J. comp. physiol. Psychol.* **45**, 450–9.

Hayes, K. G. and Hayes, C. (1955). In *The Non-human Primates and Human Evolution*. Ed. G. J. D. Wayne, Detroit; Detroit University Press.

Hebb, D. O. and Thompson, W. H. (1954). The social significance of animal studies. In *Handbook of Social Psychology*. Ed. G. Lindzey, Cambridge, Mass; Addison-Wesley.

Herrnstein, R. J. and Loveland, D. H. (1964). Complex visual concepts in the pigeon. *Science, N.Y.* **146**, 549–51.

Hockett, C. F. (1960a). Logical considerations in the study of animal communication. In *Animal Sounds and Communication*, 392–430. Ed. W. E. Lanyon and W. N. Tavolga, Washington D.C., American Inst. Biol. Sciences.

Hockett, C. F. (1960b). The origin of speech. *Sci. Amer.* **203**, 89–96.

Hockett, C. F. and Altmann, S. A. (1968). A note on design features. In *Animal Communication*, 61–72. Ed. T. A. Sebeok, Bloomington, Indiana and London; Indiana University Press.

Lehr, E. (1967). Experimentelle Untersuchungen an Affen und Halbaffen über Generalisation von Insekten – und Blütenabbildungen. *Z. Tierpsychol.* **24**, 208–44.

Mowrer, O. H. (1950). The psychology of talking birds: a contribution to language and personality theory. In *Learning Theory and Personality Dynamics*, 688–726, New York, The Ronald Press.

Pantin, C. F. A. (1968). In *The Relations Between the Sciences*. Ed. with an introduction and notes by A. M. Pantin and W. H. Thorpe, London; Cambridge University Press.

Premack, D. (1970a). A functional analysis of language. *J. exp. Analysis Behav.* **14**, 107–25.

Premack, D. (1970b). The education of Sarah. *Psychology Today*, **4**, 55–8.

Teuber, H.-L. (1967). Summary. In *Brain Mechanisms' Underlying Speech and Language*. Ed. C. H. Milliken and F. L. Darley, New York and London; Grune and Stratton.

Thorpe, W. H. (1966). Ethology and consciousness. In *Brain and Conscious Experience*, 470–505. Ed. J. C. Eccles, New York; Springer-Verlag.

Thorpe, W. H. (1968). Perceptual bases for group organisation in social vertebrates, especially birds. *Nature, Lond.* **220**, 124–8.

Thorpe, W. H. (1969). Vitalism and organicism. In *The Uniqueness of Man*, 71–99. Ed. J. D. Roslansky, Amsterdam and London; North-Holland Pub. Co.

3. HUMAN LANGUAGE

Abercrombie, D. (1967). *Elements of General Phonetics*. Edinburgh; Edinburgh University Press.

Abercrombie, D. (1968). Paralanguage. *Br. J. dis. Comm.* **3**, 55–9.

Argyle, M. (1967). *The Psychology of Interpersonal Behaviour*. Harmondsworth; Penguin.

Austin, J. (1962). *How To Do Things With Words*. Cambridge, Mass.; Harvard University Press.

Austin, W. A. (1967). Non-verbal communication. In *Language Resource Information for the Culturally Disadvantaged*. Ed. A. L. Davis, Champaign, Ill.; National Council of Teachers of English.

Bar-Hillel, Y. (1954). Indexical expressions. *Mind* **63**, 359–79. Reprinted in Bar-Hillel, 1970.

Bar-Hillel, Y. (1957). Three methodological remarks on *Foundations of Language*. *Word* **13**, 223–35. Reprinted in Bar-Hillel, 1970.

Bar-Hillel, Y. (1967). Review of Fodor and Katz, *Structure of Language*. In *Language* **43**, 526–50. Reprinted in Bar-Hillel, 1970.

Bar-Hillel, Y. (1969). Review of Lyons, *Theoretical Linguistics*. In *Semiotica* **1**, 449–59. Reprinted in Bar-Hillel, 1970.

Bar-Hillel, Y. (1970). *Aspects of Language*. Jerusalem; Magnes Press.

Brown, R. (1970). The first sentences of child and chimpanzee. In *Psycholinguistics: Selected Papers*. New York; Free Press.

Campbell, R. and Wales, R. J. (1970). The study of language acquisition. In Lyons, 1970*b*.

Chomsky, N. (1956). Three models for the description of language. *IRE Trans. Information Theory*. **IT-2**, 113–24. Reprinted in Smith, 1966.

Chomsky, N. (1957). *Syntactic Structures*. The Hague; Mouton.

Chomsky, N. (1965). *Aspects of the Theory of Syntax*. Cambridge, Mass.; M.I.T. Press.

Chomsky, N. (1968). *Language and Mind*. New York; Harcourt, Brace and World.

Crystal, D. (1969). *Prosodic Systems and Intonation in English*. London and New York; Cambridge University Press.

Crystal, D. (1971). Paralinguistics In *Current Trends in Linguistics*, vol. 12. Ed. T. A. Sebeok, The Hague; Mouton.

Gimson, A. C. (1970). *An Introduction to the Pronunciation of English*, 2nd edition. London; Arnold.

Grice, H. P. (1957). Meaning. *Phil. Rev.* **66**, 377–88.

Halliday, M. A. K. (1970). Language structure and language function. In Lyons, 1970*b*.

Halliday, M. A. K., McIntosh, A. and Stevens, P. D. (1964). *The Linguistic Sciences and Language Teaching*. London; Longmans.

Hjelmslev, L. (1958). *Prolegomena to a Theory of Language*. Bloomington, Indiana, and London; Indiana University Press.

Hockett, C. F. (1960). The origin of speech. *Sci. Amer.* **203**, 89–96.

Honikman, B. (1964). Articulatory settings. In *In Honour of Daniel Jones*. Ed. D. Abercrombie *et al.*, London; Longmans.

Householder, F. W. (1970). *Linguistic Speculations*. London and New York; Cambridge University Press.

Hymes, D. (1962). The ethnography of speaking. In *Anthropology and Human Behaviour*. Ed. T. Gladwin and W. Sturtevant, Washington; American Anthropological Association.

Hymes, D. (1970). On communicative competence. In *Directions in Sociolinguistics*. Ed. J. J. Gumperz and D. Hymes, New York; Holt Rinehart and Winston.

Labov, W. (1966). *The Social Stratification of English in New York City*. Washington, D.C.; Center for Applied Linguistics.

Ladefoged, P. (1964). *Three Areas of Experimental Phonetics*. London; Oxford University Press.

Laver, J. D. M. (1968). Voice quality and indexical information. *Br. J. Disorders Comm.* **3**, 43–54.

Laver, J. D. M. (1970). The production of speech. In Lyons, 1970*b*.

Lenneberg, E. H. (1964). *New Directions in the Study of Language*. Cambridge, Mass.; M.I.T. Press.

Lenneberg, E. H. (1967). *Biological Foundations of Language*. New York; Wiley.

Lyons, J. (1968). *Introduction to Theoretical Linguistics*. London and New York; Cambridge University Press.

Lyons, J. (1970*a*). *Chomsky*. London; Fontana. (American edition, *Noam Chomsky*. New York; Viking.)

Lyons, J. (ed.) (1970*b*). *New Horizons in Linguistics*. Harmondsworth; Penguin.

Malinowski, B. (1946). Supplement 1 to *The Meaning of Meaning*, 8th edition. Ed. C. K. Ogden and I. A. Richards, London; Routledge and Kegan Paul.

Marshall, J. C. (1970). The biology of communication in man and animals. In Lyons, 1970.

Martinet, A. (1960). *Eléments de Linguistique Générale*. Paris; Armand Colin. (English translation, *Elements of General Linguistics*. London; Faber and Faber, 1964.)

Matthews, P. H. (1967). Review of Chomsky, 1965. *J. Ling.* **3**, 119–52.

Matthews, P. H. (1970). Recent developments in morphology. In Lyons, 1970*b*.

Moravscik, J. (1969). Competence, creativity and innateness. *Phil. Forum* **1**, 407–37.

Morris, C. (1946). *Signs, Language, and Behavior*. New York; Prentice-Hall.

Peirce, C. S. (1931). *Collected Papers*. Ed. C. Hartshorne and P. Weiss, Cambridge, Mass.; Harvard University Press.

Peirce, C. S. (1940). *The Philosophy of Peirce: Selected Writings*. Ed. J. Buchler, London; Kegan Paul, Trench and Trubner.

Pike, Kenneth L. (1967). *Language in Relation to a Unified Theory of Human Behavior*, 2nd revised edition. The Hague; Mouton.

Pride, J. B. (1970). Sociolinguistics. In Lyons, 1970*b*.

Quine, W. V. (1960). *Word and Object*. Cambridge, Mass.; M.I.T. Press.

Quirk, R. (1970). *The Use of English*, 2nd edition. London; Longmans.

Robins, R. H. (1964). *General Linguistics: An Introductory Survey*. London; Longmans.

Searle, John R. (1969). *Speech Acts*. London and New York; Cambridge University Press.

Skinner, B. F. (1957). *Verbal Behavior*. New York; Appleton-Century-Crofts, and London; Methuen.

Smith, A. G. (ed.) (1966). *Communication and Culture: Readings in the Codes of Human Interaction*. New York; Holt Rinehart and Winston.

Strang, B. M. H. (1968). *Modern English Structure*, 2nd edition. London; Arnold.

Strawson, P. F. (1970). *Meaning and Truth*. Oxford; Clarendon Press.

Trager, G. L. and Smith, H. L. (1951). *An Outline of English Structure*. (Studies in Linguistics, Occasional Papers, 3.) Washington; American Council of Learned Societies and Norman, Okla.; Battenburg Press.

COMMENTS TO PART A

Altmann, S. A. (1962). A field study of the sociobiology of rhesus monkeys, *Macaca mulatta*. *Ann. N.Y. Acad. Sci.* **102**, 338–435.

Altmann, S. A. (1965). Sociobiology of rhesus monkeys. II Stochastics of social communication. *J. theor. Biol.* **8**, 490–522.

Altmann, S. A. (1967). The structure of primate social communication. In *Social Communication among Primates*. Ed. S. A. Altmann, University of Chicago Press.

Aronson, L., Tobach, E., Lehrman, D. S. and Rosenblatt, J. S. (eds.) (1970). *Development and Evolution of Behaviour*. New York and San Francisco; Freeman.

Austin, J. L. (1962). *How to Do Things with Words*. Cambridge, Mass.; Harvard University Press.

Barthes, R. (1967). *Systèmes de la Mode*. Paris; Editions du Seuil.

Bowlby, J. (1969). *Attachment and Loss,* vol. 1. London; Hogarth Press.

Bronowski, J. (1967). Human and animal languages. In *To Honor Roman Jakobson.* Mouton, Hague.

Bronowski, J. and Bellugi, U. (1970). Language, name and concept. *Science, N.Y.* **168**, 669–73.

Chomsky, N. (1959). Review of Skinner, *Verbal Behavior. Language* **35**, 26–58.

Chomsky, N. (1968). *Language and Mind.* New York; Harcourt, Brace and World.

Chomsky, N. (1971). On interpreting the world. *Cambridge Review* **92**, 77–93.

Held, R. (1967–8). Dissociation of visual functions by deprivation and rearrangement. *Psychol. Forschung* **31**, 338–48.

Hinde, R. A. (1959). Unitary Drives. *Anim. Behav.* **7**, 130–41.

Hinde, R. A. (1968). Dichotomies in the study of development. In *Genetic and Environmental Influences on Behaviour.* Ed. J. M. Thoday and A. S. Parkes, Edinburgh; Oliver and Boyd.

Hinde, R. A. (1970). *Animal Behaviour,* 2nd ed. New York; McGraw-Hill.

Hinde, R. A. and Stevenson, J. G. (1970). Goals and response control. In Aronson *et al.,* 1970.

Hockett, C. F. and Altmann, S. A. (1968). A note on design features. In Sebeok, 1968.

Hull, C. L. (1943). *Principles of Behavior.* New York; Appleton-Century-Crofts.

Lehrman, D. S. (1953). A critique of Konrad Lorenz's theory of instinctive behavior. *Quart. Rev. Biol.* **28**, 337–63.

Lehrman, D. S. (1970). Semantic and conceptual issues in the nature-nurture problem. In Aronson *et al.,* 1970.

Lenneberg, E. H. (1967). *Biological Foundations of Language.* New York; Wiley.

Lévi-Strauss, C. (1966). The culinary triangle. *Partisan Rev.* **33**, 586–95.

Lyons, J. (1963). *Structural Semantics.* Oxford; Blackwell.

Lyons, J. (1970). *Chomsky.* London; Fontana, Collins.

Marler, P. (1959). Developments in the study of animal communication. In *Darwin's Biological Work.* Ed. P. R. Bell, London; Cambridge University Press.

Matthews, G. V. T. (1968). *Bird Navigation.* London; Cambridge University Press.

Miller, N. E. (1959). Liberalization of S-R concepts. In *Psychology, a Study of a Science,* study 1, vol. 2. Ed. S. Koch, New York; McGraw-Hill.

Morris, C. (1946). *Signs, Language and Behavior.* New Jersey; Prentice-Hall.

Osgood, C. E. (1953). *Method and Theory in Experimental Psychology.* New York; Oxford University Press.

Radcliffe-Brown, A. R. (1948). *A Natural Science of Society.* Chicago; Univ. Chicago Press.

Ramsay, A. (1969). Time, space and hierarchy in zoosemiotics. In Sebeok and Ramsay, 1969.

Schneirla, T. C. (1953). The concept of levels in the study of social phenomena. In *Groups in Harmony and Tension.* Ed. M. and C. Sherif, New York; Harper.

Sebeok, T. A. (ed.) (1968). *Animal Communication.* Bloomington, Indiana and London; Indiana University Press.

Sebeok, T. A. (1969). Semiotics and ethology. In Sebeok and Ramsay, 1969.

Sebeok, T. A. and Ramsay, A. (eds.) (1969). *Approaches to Animal Communication.* Mouton, The Hague.

Struhsaker, T. T. (1967). Auditory communication among vervet monkeys (*Cercopithecus aethiops*). In Altmann, 1967.

Tavolga, W. N. (1970). Levels of interaction in animal communication. In Aronson *et al.,* 1970.

Thorpe, W. H. (1967). Animal vocalization and communication. In *Brain Mechanisms underlying Speech and Language.* Ed. C. H. Millikan and F. L. Darley, Grume and Stratton, USA.

Tinbergen, N. (1959). Comparative studies of the behaviour of gulls (Laridae): a progress report. *Behaviour* **15**, 1–70.

Tolman, E. C. (1932). *Purposive Behavior in Animals and Man.* New York; Century.

Welford, A. T. (1962). Experimental psychology and the study of social behaviour. In *Society: Problems and Methods of Study.* Ed. A. T. Welford, London; Routledge and Kegan-Paul.

INTRODUCTION TO PART B

Busnel, R. G. (1963). *Acoustic Behaviour of Animals.* Amsterdam; Elsevier.

Frings, H. and Frings, M. (1964). *Animal Communication.* New York; Blaisdell.

Sebeok, T. A. (ed.) (1968). *Animal Communication.* Bloomington; Indiana University Press.

4. SOME PRINCIPLES OF ANIMAL COMMUNICATION

Alexander, R. D. (1968). Arthropods. In Sebeok, 1968, 167–216.

Altmann, S. A. (1967). The structure of primate social communication. In *Social Communication among Primates*, 325–62. Ed. S. A. Altmann, Univ. Chicago Press.

Andrew, R. J. (1961). The displays given by passerines in courtship and reproductive fighting: a review. *Ibis* **103a**, 315–48.

Atz, J. W. (1970). The application of the idea of homology to behavior. In *Development and Evolution of Behavior*, 53–74. Ed. L. R. Aronson, E. Tobach, D. S. Lehrman and J. S. Rosenblatt, New York and San Francisco; Freeman.

Baerends, G. P. (1950). Specialisations in organs and movements with a releasing function. *Symp. Soc. Exp. Biol.* **4**, 337–60.

Bastock, M. (1967). *Courtship: a zoological study.* London; Heinemann.

Bateson, P. P. G. (1966). The characteristics and context of imprinting. *Biol. Rev.* **41**, 177–220.

Blair, W. F. (1964). Isolating mechanisms and interspecies interactions in anuran amphibians. *Quart. Rev. Biol.* **39**, 334–44.

Blest, A. D. (1957a). The function of eyespot patterns in the Lepidoptera. *Behaviour* **11**, 209–56.

Blest, A. D. (1957b). The evolution of protective displays in the Saturnoidea and Sphyngidae (Lepidoptera). *Behaviour* **11**, 257–309.

Blest, A. D. (1961). The concept of ritualisation. In *Current Problems in Animal Behaviour*, 102–24. Ed. W. H. Thorpe and O. Zangwill, London; Cambridge University Press.

Bremond, J.-C. (1967). Reconnaissance de schémas réactogènes liés à l'information contenue dans le chant territorial du rouge-gorge. *Proc. 14th Int. Ornithol. Congr.* 217–29. Ed. D. W. Snow, Oxford; Blackwell.

Bremond, J.-C. (1969). Valeur spécifique de la syntaxe dans le signal de défense territoriale de Troglodyte (*Troglodytes troglodytes*). *Behaviour* **30**, 66–75.

Chomsky, N. (1968). *Language and Mind.* New York; Harcourt, Brace and World.

Crook, J. H. (1964). The evolution of social organisation and visual communication in the weaver birds (Ploceinae). *Behaviour* Suppl. **10**, 1–178.

Crook, J. H. (1965). The adaptive significance of avian social organisations. *Symp. zool. Soc. Lond.* **14**, 181–218.

Delgado, J. M. R. (1967). Social behaviour and radio-stimulated aggression in monkeys. *J. Nerv. Mental Diseases* **144**, 383–90.

Dingle, H. (1969). A statistical and information analysis of aggressive communication in the mantis shrimp *Gonodactylus bredeni* Manning. *Anim. Behav.* **17**, 561–75.

402 References

Eisenberg, J. (1966). The social organisation of mammals. In *Kukenthal's Handbuch der Zoologie*, **8**, 1–92.

Franck, D. (1969). Genetische Grundlagen der Evolution tierischer Verhaltensweisen. *Zool. Anz.* **183**, 31–46.

Frisch, K. von (1967). *The Dance Language and Orientation of Bees*. Cambridge, Mass.; Harvard University Press.

Hailman, J. (1967). The ontogeny of an instinct. *Behaviour* Suppl. **15**, 1–159.

Haldane, J. B. S. and Spurway, H. (1954). A statistical analysis of communication in *Apis melifera* and a comparison with communication in other animals. *Insectes Sociaux* **1**, 247–83.

Hazlett, B. A. and Bossert, W. H. (1965). A statistical analysis of the aggressive communications system of some hermit crabs. *Anim. Behav.* **13**, 357–73.

Hinde, R. A. (1959). Behaviour and speciation in birds and lower vertebrates. *Biol. Rev.* **34**, 85–128.

Hinde, R. A. (1970). *Animal Behaviour*, 2nd edition, New York; McGraw-Hill.

Hunsaker, D. (1962). Ethological isolating mechanisms in the *Scleropus torquatus* group of lizards. *Evolution* **16**, 62–74.

Huxley, J. S. (1966). A discussion of ritualisation of behaviour in animals and man: Introduction. *Phil. Trans. Roy. Soc. Lond.* **251**, 247–71.

Immelmann, K. (1959). Experimentelle Untersuchungen über die biologische Bedeutung artspezifischer Merkmale beim Zebrafinken (*Taeniopygia guttata*). *Zool. Jb.* (*Syst.*) **86**, 437–592.

Lack, D. (1968). *Ecological Adaptations for Breeding in Birds*. London; Methuen.

Lehrman, D. and Friedman, M. (1969). Auditory stimulation of ovarian activity in the Ring dove (*Streptopelia risoria*). *Anim. Behav.* **17**, 494–7.

Lill, A. and Wood-Gush, D. (1965). Potential ethological isolating mechanisms and assortative mating in the domestic fowl. *Behaviour* **25**, 16–44.

Loizos, C. (1967). Play behaviour in higher primates: a review. In *Primate Ethology*, 226–82. Ed. D. Morris, London; Weidenfeld and Nicolson.

Lorenz, K. Z. (1966). *Evolution and Modification of Behavior*. London; Methuen.

McKinney, F. (1961). An analysis of the displays of the European Eider *Somateria mollissima* (Linnaeus) and the Pacific Eider *Somateria mollissima v. nigra* Bonaparte. *Behaviour* Suppl. **7**, 1–124.

Manning, A. (1965). *Drosophila* and the evolution of behaviour. In *Viewpoints in Biology*, **4**, 125–69.

Marler, P. (1957). Specific distinctiveness in the communication signals of birds. *Behaviour* **11**, 13–39.

Marler, P. (1965). Communication in monkeys and apes. In *Primate Behaviour*, 544–84. Ed. I. De Vore, New York; Holt Rinehart and Wilson.

Mayr, E. (1963). *Animal Species and Evolution*. London; Oxford University Press.

Morris, D. (1957). 'Typical Intensity' and its relation to the problem of ritualisation. *Behaviour* **11**, 1–22.

Nelson, J. B. (1969). The breeding behaviour of the Red-footed Booby *Sula sula*. *Ibis* **111**, 357–85.

Nicolai, J. (1964). Der Brutparasitismus der Viduinae als ethologisches Problem. *Z. Tierpsychol.* **21**, 129–204.

Schenkel, R. (1956). Zur Deutung der Balzleistungen einiger Phasianiden und Tetraoniden. *Orn. Beob.* **53**, 182–201.

Schneirla, T. C. and Rosenblatt, J. S. (1961). Behavioral organisation and genesis of the social bond in insects and mammals. *Amer. J. Orthopsych.* **31**, 223–53.

Sebeok, T. A. (ed.) (1968). *Animal Communication*. Bloomington, Indiana and London; Indiana University Press.

Smith, W. J. (1966). Communication and relationships in the genus *Tyrannus*. Publications of the Nuttall Ornithological Club, No. 6. Cambridge, Mass.

Smith, W. J. (1969). Messages of vertebrate communication. *Science, N.Y.* **165**, 145–50.

Stokes, A. W. (1962). The comparative ethology of Great, Blue, Marsh and Coal tits at a winter feeding station. *Behaviour* **19**, 208–18.

Stout, J. F. and Brass, M. E. (1969). Aggressive communication by *Larus glaucescens*. Part II. Visual communication. *Behaviour* **34**, 42–54.

Thielcke, G. (1969). Geographical variation in bird vocalizations. In *Bird Vocalizations*, 311–39. Ed. R. A. Hinde, London; Cambridge University Press.

Tinbergen, N. (1952). 'Derived' activities; their causation, biological significance, origin and emancipation during evolution. *Quart. Rev. Biol.* **27**, 1–32.

Tinbergen, N. (1959*a*). Comparative studies of the behaviour of gulls (Laridae): a progress report. *Behaviour* **15**, 1–70.

Tinbergen, N. (1959*b*). Einige Gedanken über Beschwichtigungsgebärden. *Z. Tierpsychol.* **16**, 651–65.

Tinbergen, N. (1965). Some recent studies of the evolution of sexual behavior. In *Sex and Behavior*, 1–33. Ed. F. A. Beach, New York; Wiley.

Waddington, C. H. (1957). *The Strategy of the Genes*. London; Allen and Unwin.

Wickler, W. (1966). Sexualdimorphismus, Paarbildung und Versteckbrüten bei Cichliden. (Pisces; Perciformes). *Zool. Jb. (Syst.)* **93**, 127–38.

Wickler, W. (1968). *Mimicry*. London; Weidenfeld and Nicolson.

Wickler, W. (1969). Zur Soziologie des Brabantbuntbarsches, *Tropheus moorei*. (Pisces, Cichlidae). *Z. Tierpsychol.* **26**, 967–87.

Wilson, E. O. (1962). Chemical communication among workers of the Fire-ant *Solenopsis saevissima* (Fr. Smith). *Anim. Behav.* **10**, 148–58.

Wortis, R. P. (1969). The transition from dependent to independent feeding in the young Ring dove. *Anim. Behav. Monogr.* **2**, 1–53.

5. THE LOWER VERTEBRATES AND THE INVERTEBRATES

Alexander, R. D. (1960). Sound communication in Orthoptera and Cicadidae. In Lanyon and Tavolga, 1960, 38–93.

Alexander, R. D. (1967). Acoustical communication in arthropods. *A. Rev. Ent.* **12**, 495–536.

Alexander, R. D. (1968). Arthropods. In Sebeok, 1968, 167–215.

Altevogt, R. (1957). Untersuchungen für Biologie, Ökologie und Physiologie indischer Winkerkrabben. *Z. Morph. Ökol. Tiere* **46**, 1–110.

Altevogt, R. (1959). Ökologische und Ethologische Studien an Europas Winkerkrabbe *Uca tangeri*. *Z. Morph. Ökol. Tiere* **48**, 132.

Boeckh, J., Kaissling, K.-E. and Schneider, D. (1965). Insect olfactory receptors. *Cold Spring Harbor Symp. Quant. Biol.* **30**, 263–80.

Bogert, C. M. (1960). The influence of sound on the behavior of amphibians and reptiles. In Lanyon and Tavolga, 1960, 137–320.

Bonner, J. T. (1959). *The Cellular Slime Moulds*. Princeton; Princeton Univ. Press.

Bossert, W. H. (1968). Temporal patterning in olfactory communication. *J. theor. Biol.* **18**, 157–70.

Böttger, K. (1962). Zur Biologie und Ethologie der einheimischen Wassermilben *Arrenurus (Megaloiracarus) globator* (Müll.), 1776, *Piona nodata nodata* (Müll.), 1776, *Eylais infundibulifera meridionalis* (Thou), 1899 (*Hydrachnellae*, Acari). *Zool. Jb. (Syst.)* **89**, 501–84.

Bristowe, W. S. (1939–41). *The Comity of Spiders*. London; The Ray Society.

Bull, H. O. (1928–39). Studies on conditioned responses in fishes. *J.M.B.A. and Dove Marine Laboratory Reports*, Parts I–IX.

Bullock, T. H. and Horridge, G. A. (1965). *Structure and Function in the Nervous Systems of Invertebrates*, 2 vols. San Francisco; Freeman.

Butler, C. G. (1967). Insect pheromones. *Biol. Rev.* **42**, 42–87.

Capranica, R. R. (1965). *The Evoked Vocal Response of the Bullfrog.* Cambridge, Mass.; M.I.T. Press.

Capranica, R. R. (1966). Vocal response of the bullfrog to natural and synthetic mating calls. *J. acoust. Soc. Am.* **40**, 1131–9.

Capranica, R. R. (1968). The vocal repertoire of the bullfrog. *Behaviour* **31**, 302–25.

Capranica, R. R. and Frishkopf, L. S. (1966). Responses of auditory units in the medulla of the cricket frog. *J. acoust. Soc. Am.* **40**, 1263.

Carlisle, D. B. and Knowles, F. (1959). *Endocrine Control in Crustaceans.* London; Cambridge University Press.

Cloudesley-Thompson, J. L. (1958). *Spiders, Scorpions, Centipedes and Mites.* London; Pergamon Press.

Cott, H. B. (1961). Scientific results of an inquiry into the ecology and economic status of the Nile crocodile in Uganda and Northern Rhodesia. *Proc. zool. Soc. Lond.* **29**, 211–356.

Crane, J. (1943). Display breeding and relationships of fiddler crabs *Brachyura* genus *Uca* in the N.E. United States. *Zoologica* **28**, 217–23.

Crane, J. (1949). Comparative biology of salticid spiders of Rancho Grande, Venezuela. IV. An analysis of display. *Zoologica* **34**, 159–214.

Crane, J. (1957). Basic patterns of display in fiddler crabs *Ocypodidae* genus *Uca.* *Zoologica* **42**, 69–82.

Crane, J. (1958). Aspects of social behaviour in the fiddler crabs, with special reference to *Uca maracoani* (Latreille). *Zoologica* **43**, 113–30.

Crisp, D. J. (1961). Territorial behaviour in a barnacle settlement. *J. exp. Biol.* **38**, 429–46.

Crisp, D. J. and Mellers, P. S. (1962). The chemical basis of gregariousness in Cirripedes. *Proc. Roy. Soc.* B **156**, 500–20.

Davenport, D. (1966). The experimental analysis of behaviour in symbioses. In *Symbiosis*, 381–429. Ed. S. M. Henry, New York and London; Academic Press.

Davenport, D. and Norris, K. S. (1958). Observations on the symbiosis of the sea anemone *Stoichactis* and the pomacentrid fish *Amphiprion percula. Biol. Bull.* **115**, 397–410.

Delco, E. A., Jr. (1960). Sound discrimination by males of two cyprinid fishes. *Tex. J. Sci.* **12**, 48–54.

Dietrich, G. (1963). *General Oceanography.* New York; Wiley.

Eibl-Eibesfeldt, I. (1967). *Grundriss der vergleichenden Verhaltensforschung.* München; Piper.

Evans, L. T. (1961). Structure as related to behavior in the organization of populations in reptiles. In *Vertebrate Speciation.* Ed. W. F. Blair, Austin, Texas; Univ. Texas Press.

Frings, H. and Frings, M. (1968). Other invertebrates. In Sebeok, 1968, 244–70.

Frisch, K. von (1941). Über einen Schreckstoff der Fischhaut und seine biologische Bedeutung. *Z. vergl. Physiol.* **29**, 46–145.

Frisch, K. von (1946). Die Tanz der Bienen. *Ostrr. Zool. Zeit.* **1**, 1–48. A further summary in the same year was given in *Experienta*, Basle, **2**, 197–404.

Frisch, K. von (1954). *The Dancing Bees.* London; Methuen.

Frisch, K. von (1967a). *The Dance Language and Orientation of Bees.* Cambridge, Mass.; Harvard University Press.

Frisch, K. von (1967b). Honey bees: Do they use the information as to direction and distance provided by their dances? *Science, N.Y.* **158**, (3804), 1072–5.

Frisch, K. von (1968). Do bees really not understand their own language? *Allg. dt. Imkerztg.* **2**(2), 35–41.

Frisch, K. von (1969). The foraging bee: How she finds and exploits sources of food. *Bee World* **50**, (4), 141–52.

Frishkopf, L. S. and Goldstein, M. H., Jr. (1963). Responses to acoustic stimuli from single units in the eighth nerve of the bullfrog. *J. acoust. Soc. Am.* **35**, 1219–28.

Froloff, J. P. (1925). Bedingte Reflexe bei Fischen. I. *Pflüg. Arch.* **208**, 261–71.

Froloff, J. P. (1928). Bedingte Reflexe bei Fischen. II. *Pflüg. Arch.* **220**, 339–49.

Galloway, T. W. (1908). A case of phosphorescence as a mating adaptation. *School Sci. and Math.* **8**, 411–15.

Galloway, T. W. and Welsh, P. S. (1911). Studies on a phosphorescent Bermudan annelid *Odontosyllis enopla* Verrill. *Trans. Am. microsc. Soc.* **30**, 13–39.

Galtsoff, P. S. (1938). Physiology of reproduction of *Ostrea virginica*. II. Stimulation of spawning in the female oyster. *Biol. Bull.* **75**, 286–307.

Gould, J. E., Henerey, M. and MacLeod, M. C. (1970). Communication of direction by the Honey bee. *Science, N.Y.* **169**, 544–54.

Harris, C. G. and Bergeijk, W. A. van (1962). Evidence that the lateral line organ responds to near-field displacements of sound sources in water. *J. acoust. Soc. Am.* **34**, 1831–41.

Hasler, A. D. and Wisby, W. J. (1951). Discrimination of stream odour by fishes and its relation to parent stream behavior. *Am. Nat.* **85**, 223–38.

Hinde, R. A. (1970). *Animal Behaviour: A Synthesis of Ethology and Comparative Psychology*, 2nd edition. New York; McGraw-Hill.

Hockett, C. F. (1960). Logical considerations in the study of animal communication. In Lanyon and Tavolga, 1960, 392–430.

Huxley, J. S. (1966). A discussion on ritualization of behaviour in animals and men. *Phil. Trans. Roy. Soc.* B **251**, 249–71.

Johnson, D. L. (1967). Communication among Honey bees with field experience. *Anim. Behav.* **15**, 487–92.

Kleerekoper, H. and Mogensen, J. A. (1959). The chemical composition of scent of fresh-water fish with special reference to amines and amino acids. *Z. vergl. Physiol.* **42**, 494–500.

Kleerekoper, H. and Mogensen, J. A. (1963). Role of olfaction in the orientation of *Petromyzon marinus*. I. Response to a single amine in prey's body odor. *Physiol. Zool.* **36**, 347–60.

Lanyon, W. E. and Tavolga, W. N. (eds.) (1960). *Animal Sounds and Communication*. Washington, D.C.; Am. Inst. Biol. Sci.

Leslie, C. J. (1951). Mating behavior of leeches. *J. Bombay Nat. Hist. Soc.* **50**, 422–3.

Lindauer, M. (1961). *Communication Amongst Social Bees*. Cambridge, Mass.; Harvard University Press.

Lindauer, M. (1967). Recent advances in bee communication and orientation. *A. Rev. Entomol.* **12**, 439–70.

Lissmann, H. W. (1951). Continuous electrical signals from the tail of a fish *Gymnarchus niloticus* Cur. *Nature, Lond.* **167**, 201–2.

Lissmann, H. W. (1958). The function and evolution of the electric organs of fish. *J. exp. Biol.* **35**, 156–91.

Lissmann, H. W. (1963). Electric location by fishes. *Sci. Am.* **208** (3), 50–9.

Lloyd, J. E. (1966). Studies on the flash communication system in *Photinus* fireflies. *Univ. Michigan Mus. Zool. Mis. Publ.* **130**, 1–95.

Lutz, B. (1947). Trends towards aquatic and direct development in frogs. *Copeia* **4**, 242–52.

Marler, P. and Hamilton, W. J. (1966). *Mechanisms of Animal Behavior*. New York; Wiley.

Michelsen, A. (1968). Frequency discrimination in the locust ear by means of four groups of receptor cells. *Nature, Lond.* **220**, 585–6.

Michener, C. D. (1969). Comparative social behavior of bees. *A. Rev. Entomol.* **14**, 299–342.

Moulton, J. M. (1969). The classification of acoustic communication behavior amongst teleost fishes. In *Approaches to Animal Communication*. Ed. T. A. Sebeok and A. Ramsay, Mouton, The Hague.

Noble, G. K. and Aronson, L. R. (1942). The sexual behavior of Anura. *Bull. Am. Mus. Nat. Hist.* **80**, 127–42.

Parry, D. A. (1965). The signals generated by an insect in a spider's web. *J. exp. Biol.* **43**, 185–92.

Ross, D. M. (1960). The association between the hermit crab *Eupagurus bernhardus* (L) and the sea anemone *Calliactis parasitica* (Couch). *Proc. zool. Soc. Lond.* **134**, 43–57.

Schneider, D. (1963). Electrophysiological investigation of insect olfaction. In *Olfaction and Taste*, 85–103. Ed. Y. Zotterman, New York; Macmillan.

Sebeok, T. A. (ed.) (1968). *Animal Communication*. Bloomington, Indiana and London; Indiana University Press.

Shaffer, B. M. (1953). Aggregation in cellular slime moulds: *in vitro* isolation of acrasin. *Nature, Lond.* **171**, 975.

Shaffer, B. M. (1956a). Properties of acrasin. *Science, N.Y.* **123**, 1172–3.

Shaffer, B. M. (1956b). Acrasin, the chemotactic agent in cellular slime moulds. *J. exp. Biol.* **33**, 645–57.

Shaffer, B. M. (1957a). Aspects of aggregation in cellular slime moulds. I. Orientation and chemotaxis. *Am. Nat.* **91**, 19–35.

Shaffer, B. M. (1957b). Properties of slime mould amoebae of significance for aggregation. *Q. J. microsc. Sci.* **98**, 377–92.

Shaffer, B. M. (1957c). Variability of behaviour of aggregating cellular slime moulds. *Q. J. microsc. Sci.* **98**, 393–405.

Smith, M. (1954). *The British Amphibians and Reptiles*, 2nd edition. London; Collins.

Stout, J. (1963). The significance of sound production during the reproductive behaviour of *Notropus analostaneus* family Cyprinidae. *Anim. Behav.* **11**, 83–92.

Tavolga, W. N. (1956). Visual, chemical and sound stimuli as cues in the sex-discriminatory behaviour of the Gobiid fish *Bathygobius soporator*. *Zoologica* **41**, 49–64.

Tavolga, W. N. (1958). The significance of underwater sounds produced by males of the Gobiid fish *Bathygobius soporator*. *Physiol. Zool.* **31**, 259–71.

Tavolga, W. N. (1960). Sound production and underwater communication in fishes. In Lanyon and Tavolga, 1960, 93–136.

Thorpe, W. H. (1949). Orientation and methods of communication in the Honey bee and its sensitivity to the polarisation of light. *Nature, Lond.* **164**, 11.

Thorpe, W. H. and Davenport, D. (eds.) (1965). Learning and associated phenomena in the invertebrates. *Anim. Behav. Supplement No. 1*.

Tinbergen, N. (1953). *Social Behaviour in Animals*. London; Methuen.

Welsh, J. H. (1931). Specific influence of the host on the light response of parasitic water mites. *Biol. Bull.* **61**, 497–9.

Wenner, A. M. (1967). Honey bees: Do they use the distance information contained in their dance maneuver? *Science, N.Y.* **155**, (3764), 847–9.

Wenner, A. M. and Johnson, D. L. (1967). Honey bees: Do they use the direction and distance information provided by their dancers? *Science, N.Y.* **158**, (3804), 1076–7.

Wenner, A. M., Wells, P. H. and Rohlf, F. J. (1967). An analysis of recruitment in Honey bees. *Physiol. Zool.* **40** (4), 317–4.

Wever, E. G., Hepp-Reymond, M.-C. and Vernon, J. A. (1966). Vocalization and hearing in the leopard lizard. *Proc. Nat. Acad. Sci.* **55**, 98–106.

Wilson, E. O. (1968). Chemical systems. In Sebeok, 1968, 75–102.

Wilson, E. O. and Bossert, W. H. (1963). Chemical communication among animals. *Recent Progr. Hormone Res.* **19**, 673–716.

6. VOCAL COMMUNICATION IN BIRDS

Ashmole, N. P. and Humbertotova, A. (1968). Prolonged parental care in Royal terns and other birds. *Auk* **85**, 90–100.

Beer, C. G. (1970). Individual recognition of voice in the social behavior of birds. *Adr. Study Behav.* **3**, 27–74.

Bertram, B. C. R. (1970). The vocal behaviour of the Indian hill mynah, *Gracula religiosa*. *Anim. Behav. Monog.* **3**, Pt. 2.

Brémond, J.-C. (1967). Reconnaissance de schémas réactogènes liés à l'information contenue dans le chant territorial du Rouge-gorge. *Proc. 14th Int. Ornith. Congr.* 217–29. Oxford; Blackwell.

Busnel, R.-G. (1968). Acoustic communication. In *Animal Communication*, 127–53. Ed. T. A. Sebeok, Bloomington, Indiana and London; Indiana University Press.

Chomsky, N. (1967). The general properties of language. In *Brain Mechanisms underlying Speech and Language*, 63–88. Ed. C. H. Millikan and F. L. Darley, New York and London; Grune and Stratton.

Chomsky, N. (1968). *Language and Mind*. New York; Harcourt, Brace and World.

Coulson, J. C. (1966a). The influence of change of mate on the breeding biology of the kittiwake. *Anim. Behav.* **14**, 189–90.

Coulson, J. C. (1966b). The significance of pair-bond and age on the breeding biology of the Kittiwake gull (*Rissa tridactyla*). *J. Anim. Ecol.* **35**, 269–79.

Falls, J. B. (1969). Functions of territorial song in the White-throated sparrow. In *Bird Vocalizations*, 207–32. Ed. R. A. Hinde, London; Cambridge Univ. Press.

Gibson, J. J. (1968). *The Senses Considered as Perceptual Systems*. London; Allen and Unwin.

Hall-Craggs, J. (1962). The development of song in the Blackbird (*Turdus merula*). *Ibis* **104**, 277–300.

Hall-Craggs, J. (1969). The aesthetic content of bird-song. In *Bird Vocalizations*, 367–81. Ed. R. A. Hinde, London; Cambridge University Press.

Hartshorne, C. (1958). The relation of bird song to music. *Ibis* **100**, 421–45.

Hinde, R. A. (1958). The nest building behaviour of domesticated canaries. *Proc. zool. Soc. Lond.* **131**, 1–48.

Hooker, T. and Hooker, B. I. (Lade) (1969). Duetting. In *Bird Vocalizations*, 185–205. Ed. R. A. Hinde, London; Cambridge University Press.

Hutchison, R. E., Stevenson, J. and Thorpe, W. H. (1968). The basis for individual recognition by voice in the Sandwich tern (*Sterna sandvicensis*). *Behaviour* **32**, 150–7.

Kandel, E., Castellucci, V., Pinsker, H. and Kuppermann, I. (1970). The role of synaptic plasticity in the short-term modification of behaviour. In *Short-term Changes in Neural Activity and Behaviour*. Ed. G. Horn and R. A. Hinde, London; Cambridge University Press.

Konishi, M. (1963). The role of auditory feedback in the vocal behaviour of the domestic fowl. *Z. Tierpsychol.* **20**, 349–67.

Lack, D. L. (1939). The behaviour of the Robin. *Proc. zool. Soc. Lond.* **109**, 169–78.

Lade, B. I. and Thorpe, W. H. (1964). Dove songs as innately coded patterns of specific behaviour. *Nature, Lond.* **202**, 366–8.

Marler, P. (1955). Characteristics of some animal calls. *Nature, Lond.* **176**, 6–7.

Marler, P. (1960). Bird songs and mate selection. In *Animal Sounds and Communication*, 348–67. Ed. W. E. Lanyon and W. N. Tavolga, Washington, D.C.; Am. Inst. Biol. Sci.

Marler, P. (1969). Animals and man: communication and its development. In *Communication*. Ed. J. D. Roslansky, North-Holland; Amsterdam.

Marler, P. (1967). Comparative study of song development in Emberizine Finches. *Proc. 14th Int. Ornithol. Congr.* 231–44. Oxford; Blackwell.

Marler, P. (1970). A comparative approach to vocal learning: song development in white-crowned sparrows. *J. comp. physiol. psychol.*, *Monogr.* **71**, 1–25.

Nicolai, J. (1959). Familientradition in der Gesangentwicklung des Gimpels (*Pyrrula pyrrhula* L.). *Ornith.* **100**, 39–46.

Nottebohm, F. (1968). Auditory experience and song development in the Chaffinch (*Fringilla coelebs*): Ontogeny of a complex motor pattern. *Ibis* **110**, 549–68.

Nottebohm, F. (1970). The ontogeny of bird song. *Science, N.Y.* **167**, 950–6.

Schwartzkopff, J. (1960). Vergleichende Physiologie des Gehörs. *Fortschr. Zool.* **12**, 206–64.

Smith, W. J. (1968). Message-meaning analysis. In *Animal Communication*, 44–60. Ed. T. A. Sebeok, Bloomington, Indiana and London; Indiana University Press.

Stevenson, J., Hutchison, R. E., Hutchison, J., Bertram, B. C. R. and Thorpe, W. H. (1970). Individual recognition by auditory cues in the Common tern (*Sterna hirundo*). *Nature, Lond.* **226**, 562–3.

Teuber, H. L. (1967). Lacunae and Research Approaches to Them. In *Brain Mechanisms underlying Speech and Language*, 204–16. Eds. C. H. Millikan and F. L. Darley, New York and London; Grune and Stratton.

Thielcke, G. (1969). Geographic variation in bird vocalization. In *Bird Vocalizations*, 311–39. Ed. R. A. Hinde, London; Cambridge University Press.

Thorpe, W. H. (1958). The learning of song-patterns by birds, with especial reference to the song of the Chaffinch, *Fringilla coelebs*. *Ibis* **100**, 535–70.

Thorpe, W. H. (1961). *Bird Song: The Biology of Vocal Communication and Expression in Birds.* London; Cambridge University Press.

Thorpe, W. H. (1963). Antiphonal singing in birds as evidence for avian auditory reaction time. *Nature, Lond.* **197**, 774–6.

Thorpe, W. H. (1966). Ritualization in the individual development of bird song. *Phil. Trans. Roy. Soc.* B **251**, 351–8.

Thorpe, W. H. (1967). Vocal imitation and antiphonal song and its implications. *Proc. 14th Int. Ornithol. Congr.* 245–63.

Thorpe, W. H. (1968). Perceptual bases for group organization in social vertebrates, especially birds. *Nature, Lond.* **220**, 124–8.

Thorpe, W. H. (1972). Duetting and antiphonal song in birds: its extent and significance. *Behaviour: Monograph Supplement* No. **18**.

Thorpe, W. H. and North, M. E. W. (1965). Origin and significance of the power of vocal imitation with special reference to the antiphonal singing of birds. *Nature, Lond.* **208**, 219–22.

Thorpe, W. H. and North, M. E. W. (1966). Vocal imitation in the Tropical Bou-Bou shrike (*Laniarius aethiopicus major*) as a means of establishing and maintaining social bonds. *Ibis* **108**, 432–5.

Tschanz, B. (1968). Trottellummen. *Z. Tierpsychol.* **4**.

White, S. J. (1971). Selective responsiveness by the Gannet to played-back calls. *Anim. Behav.* **19**.

White, S. J., White, R. E. C. and Thorpe, W. H. (1970). Acoustic basis for individual recognition in the Gannet. *Nature, Lond.* **225**, 1156–8.

References 409

7. THE INFORMATION POTENTIALLY AVAILABLE IN ANIMAL DISPLAYS

Abrahams, V. C., Hilton, S. M. and Zbrozyna, A. (1960). Active muscle vasodilation produced by stimulation of the brain stem: its significance in the defence reaction. *J. Physiol.* **154**, 491–513.

Abrahams, V. C., Hilton, S. M. and Zbrozyna, A. (1964). The role of active muscle vasodilatation in the alerting stage of the defence reaction. *J. Physiol.* **171**, 189–202.

Adams, D. B., Baccelli, G., Mancia, G. and Zanchetti, A. (1968). Cardiovascular changes during preparation for fighting behaviour in the cat. *Nature, Lond.* **220**, 1239–40.

Altmann, S. A. (1962). A field study of the sociobiology of rhesus monkeys, *Macaca mulatta*. *Ann. N.Y. Acad. Sci.* **102**, 338–435.

Andrew, R. J. (1956a). Fear responses in *Emberiza* spp. *Br. J. Anim. Behav.* **4**, 125–32.

Andrew, R. J. (1956b). Some remarks on behaviour in conflict situations, with special reference to *Emberiza* spp. *Br. J. Anim. Behav.* **4**, 41–5.

Andrew, R. J. (1961a). The displays given by passerines in courtship and reproductive fighting: a review. *Ibis*, **103a**, 315–48.

Andrew, R. J. (1961b). The motivational organisation controlling the mobbing calls of the Blackbird (*Turdus merula*): I, II, III and IV. *Behaviour* **17**, 224–46, 228–321; **18**, 25–43, 161–76.

Andrew, R. J. (1962). The situations that evoke vocalization in primates. *Ann. N.Y. Acad. Sci.* **102**, 296–315.

Andrew, R. J. (1963a). The displays of the primates. In *Genetic and Evolutionary Biology of the Primates*. Ed. J. Buethner-Janusch, New York; Academic Press.

Andrew, R. J. (1963b). The origins and evolution of the calls and facial expressions of the primates. *Behaviour* **20**, 1–109.

Andrew, R. J. (1964). Vocalization in chicks, and the concept of 'stimulus contrast'. *Anim. Behav.* **12**, 64–76.

Andrew, R. J. (1969). The effects of testosterone on avian vocalizations. In *Bird Vocalizations*. Ed. R. A. Hinde, London; Cambridge University Press.

Azrin, N. H., Hutchinson, R. R. and Hake, D. F. (1966). Extinction-induced aggression. *J. exp. Anal. Behav.* **9**, 191–204.

Baden-Powell, R. (1929). *Scouting for Boys*. London; Arthur Pearson.

Blurton Jones, N. G. (1968). Observations and experiments on causation of threat displays of the great tit (*Parus major*) *Anim. Behav. Monogr.* **1**, 2.

Brown, J. L. and Hunsperger, R. W. (1963). Neuroethology and the motivation of agonistic behaviour. *Anim. Behav.* **11**, 439–48.

Cannon, W. B. (1929). *Bodily Changes in Pain, Hunger, Fear and Rage*. New York; Appleton-Century-Crofts.

Cooper, K. E. (1966). Temperature regulation and the hypothalamus. *Br. Med. Bull.* **22**, 238–42.

Eibl-Eibesfeldt, I. (1970). *Ethology*. New York; Holt Rinehart and Winston.

Hilton, S. M. (1963). Inhibition of baroceptor reflexes on hypothalamic stimulation. *J. Physiol.* **165**, 56–7P.

Hinde, R. A. (1954). Factors governing the changes in strength of a partially inborn response, as shown by the mobbing behaviour of the chaffinch (*Fringilla coelebs*). *Proc. Roy. Soc.* B **142**, 306–31.

Hinde, R. A. (1955–6). A comparative study of the courtship of certain finches (*Frigillidae*). *Ibis* **97**, 706–45; **98**, 1–23.

Hinde, R. A. (1970). *Animal Behaviour*, 2nd edition. New York; McGraw-Hill.

Hooff, J. A. R. A. M. van (1967). The facial displays of the catarrhine monkeys and apes. In *Primate Ethology*. Ed. D. Morris, London; Weidenfeld and Nicolson.

Horn, G. (1967). Neuronal mechanisms of habituation. *Nature, Lond.* **215**, 707–11.

Kiley, M. (1969). *A comparative study of some displays in ungulates, canids and felids.* D.Phil thesis. University of Sussex.

Lacey, J. I. (1966). Somatic response patterning and stress: some revisions of activation theory. Paper delivered at a symposium on 'Issue in Stress', York University, Toronto, Canada, May 10–12, 1965.

Leyhausen, P. (1956). Verhaltensstudien bei Katzen. *Z. Tierpsychol.* **2**.

McCleary, R. A. (1966). Response-modulating functions of the limbic system. In *Progress in Physiological Psychology* I; N.Y.; Academic.

Michael, R. P. and Herbert, J. (1963). Menstrual cycle influences grooming behavior and sexual activity in the Rhesus monkey. *Science, N.Y.* **140**, 500–1.

Montagna, W. (1962). The skin of lemurs. *Ann. N.Y. Acad. Sci.* **102**, 190–209.

Montagna, W. (1965). The Skin. *Sci. Amer.* Feb., 56–66.

Moynihan, M. (1958). Notes on the behavior of some North American gulls: II. Non-aerial hostile behaviour of adults. *Behaviour* **12**, 95–182.

Moynihan, M. (1964). Some behavior patterns of Platyrrhine monkeys I. *Smithson. Misc. Coll.* **146**, 1–84.

Neal, E. (1958). *The Badger.* Harmondsworth; Penguin Books.

Pearson, M. (1970). *Causation and development of behaviour in the guinea pig.* D.Phil thesis; University of Sussex.

Roberts, W. W. (1962). Fear-like behavior elicited from dorso-medial thalamus of cat. *J. comp. physiol. Psychol.* **55**, 191–7.

Roberts, W. W., Steinberg, M. L. and Means, L. W. (1967). Hypothalamic mechanisms for sexual, aggressive and other motivational behaviors in the Opossum, *Didelphis virginiana. J. comp. physiol. Psychol.* **64**, 1–15.

Rushmer, R. F., Smith, O. A. and Lasher, E. P. (1960). Neural mechanisms of cardiac control during exertion. *Physiol. Rev.* **40**, Suppl. 4, 27–34.

Schenkel, R. (1948). Ausdruck-Studien an Wölfen. *Behaviour* **1**, 81–130.

Sebeok, T. A. (ed.) (1968). *Animal Communication.* Bloomington, Indiana and London; Indiana University Press.

Sokolov, E. N. (1960). Neuronal models and the orienting reflex. In *The Central Nervous System and Behavior.* Ed. M. A. B. Brazier, Macy Foundation, New York.

Thompson, T. I. (1963). Visual reinforcement in Siamese fighting fish. *Science, N.Y.* **141**, 55–7.

Tinbergen, N. (1952). Derived activities: their causation, biological significance, origin and emancipation during evolution. *Quart. Rev. Biol.* **27**, 1–32.

Tinbergen, N. (1959 a). Comparative studies of the behaviour of gulls (Laridae): a progress report. *Behaviour* **15**, 1–70.

Tinbergen, N. (1959 b). Einige Gedanken uber 'Beschwichtigungsgebärden'. *Z. Tierpsychol.* **16**, 651.

Tinbergen, N. (1966). Ritualization of courtship postures of *Larus ridibundus* L. *Phil. Trans. Roy. Soc.* B **251**, 457.

Wiepkema, P. R. (1961). An ethological analysis of the reproductive behaviour of the bitterling. *Arch. néerl. Zool.* **14**, 103–99.

8. A COMPARATIVE APPROACH TO THE PHYLOGENY OF LAUGHTER AND SMILING

Allin, J. T. and Banks, E. M. (1968). Behavioural biology of the Collared lemming (*Dicrostonyx groenlandicus* (Traill)). I. Agnostic behaviour. *Anim. Behav.* **16**, 245–62.

Ambrose, J. A. (1963). The age of onset of ambivalence in early infancy: indications from the study of laughing. *J. Child. Psychiat.* **4**, 167–84.

Andrew, R. J. (1963). The origin and evolution of the calls and facial expressions of primates. *Behaviour* **20**, 1–109.

Baerends, G. P. (1959). Comparative methods and the concept of homology in the study of behaviour. *Arch. néerl. Zool.* Suppl. 1, **13**, 401–17.

Bain, A. (1859). *The emotions and the will.* London; Parker and Son.

Bateson, G. (1955). A theory of play and phantasy. *Psychiat. Res. Rept.* A **2**, 39–51.

Bergson, H. (1900). *Le rire. Essai sur la signification du comique.* Paris; Alcan. 17th ed. 1918.

Bertrand, M. (1968). The behavioral repertoire of the stumptail Macaque. *Bibl. Primatol.* **11**, Basel; Karger.

Blurton Jones, N. G. (1967). An ethological study of some aspects of social behaviour of children in nursery schools. In *Primate Ethology*, 347–68. Ed. D. Morris, London; Weidenfeld and Nicolson.

Bolwig, N. (1964). Facial expression in primates with remarks on a parallel development in certain Carnivores (a preliminary report on work in progress). *Behaviour* **22**, 167–93.

Brannigan, C. and Humphries D. (1969). I see what you mean. *New Scientist* **42**, 406–8.

Buytendijk, F. J. J. (1947). *De eerste glimlach van het kind.* Nijmegen; Dekker and v.d. Vegt.

Buytendijk, F. J. J. (1948). *Algemene theorie der menselijke houding en beweging.* Utrecht; Het Spectrum.

Darwin, C. (1872). *The expression of the emotions in man and animals.* London; Murray.

Dries, W. van den (in prep.). The appeasing denials; a teleological approach to laughter and shrugging.

Dumas, G. (1937). *Nouveau traité de psychologie.* **3**, fasc. 2. *L'expression des émotions*; fasc. 3. *Le rire – les larmes.* Paris.

Eibl-Eibesfeldt, I. (1957). Ausdrucksformen der Säugetiere. *Handb. Zool. Berl.* **8**, (8) 10, 6, 1–26.

Eibl-Eibesfeldt, I. (1967). *Grundriss der vergleichenden Verhaltensforschung.* München; Piper.

Eibl-Eibesfeldt, I. (1968). Zur Ethologie des menschlichen Grussverhaltens. I. Beobachtungen an Balinesen, Papuas und Samoanern nebst vergleichende Bemerkungen. *Z. Tierpsychol.* **25**, 727–44.

Epple, G. (1967). Vergleichende Untersuchungen über Sexual- und Sozialverhalten der Krallenaffen (Hapalidae). *Folia primat.* **7**, 37–65.

Ewer, R. F. (1968). A preliminary survey of the behaviour in captivity of the Dasyurid Marsupial *Sminthopsis crassicaudata* (Gould). *Z. Tierpsychol.* **25**, 319–65.

Foley, J. P. (1935). Judgement of facial expression of emotion in the chimpanzee. *J. soc. Psychol.* **6**, 31–67.

Fox, M. W. (1970). A comparative study of the development of facial expression in Canids; wolf, coyote and foxes. *Behaviour* **36**, 49–73.

Freud, S. (1912). *Der Witz und seine Beziehung zum Unbewussten.* Wien; Denticke.

Frijda, N. H. (1956). *De betekenis van de gelaatsexpressie.* Amsterdam; Van Oorschot.

Frijda, N. H. (1965). Mimik und Pantomimik. In *Handbuch der Psychologie 5, Ausdruckspsychologie*, 352–421. Ed. R. Kirckhoff, Göttingen; Hogrefe.

Frijda, N. H. (1968). Emotion and recognition of emotion. *Contr. 3rd Feelings and Emotions Symp.* Loyola Univ.; Chicago.

Gewalt, W. (1966). Kleine Beobachtungen an seltenen Beuteltieren im Berliner Zoo. III. Tüpfelbeutelmarder (*Satanellus hallucatus albopunctatus* Schlegel 1880). *D. zool. Garten* **32**, 99–107.

Gewirtz, J. L. (1965). The course of infant smiling in four childrearing environments in Israel. In *Determinants of Infant Behaviour*, vol. 3, 205–59. Ed. B. M. Foss, London; Methuen.

Goethe, F. (1939). Über das Anstoss-Nehmen bei Vögeln. Z. *Tierpsychol.* **3**, 371–4.

Goodall, J. (1965). Chimpanzees of the Gombe Stream Reserve. In *Primate Behavior*, 425–73. Ed. I. DeVore, New York; Holt Rinehart and Winston.

Grant, E. C. (1968). An ethological description of non-verbal behaviour during interviews. *Br. J. med. Psychol.* **41**, 177–84.

Grant, E. C. (1969). Human facial expression. *Man* **4** (n.s.), 525–36.

Grzimek, B. (1941). Beobachtungen an einem kleinen Schimpansenmädchen. *Z. Tierpsychol.* **4**, 295–306.

Guilford, J. P. (1956). *Fundamental statistics in psychology and education*. New York; McGraw-Hill.

Hall, K. R. L., Boelkins, R. C. and Goswell, M. J. (1965). Behaviour of Patas monkeys, *Erythrocebus patas*, in captivity, with notes on the natural habitat. *Folia Primat.* **3**, 22–49.

Hayworth, D. (1928). The social origin and function of laughter. *Psychol. Rev.* **35**, 367–84.

Herter, K. (1957). Das Verhalten der Insektivoren. *Handb. Zool. Berlin* **8**, 10, (10), 1–50.

Hoesch, W. (1964). Beobachtungen an einem zahmen Honigdachs (*Mellivora capensis*). *D. zool. Garten* **28**, 182–8.

Hooff, J. A. R. A. M. van (1962). Facial expressions in higher primates. *Symp. zool. Soc. Lond.* **8**, 97–125.

Hooff, J. A. R. A. M. van (1967). The facial displays of the catarrhine monkeys and apes. In *Primate Ethology*, 7–68. Ed. D. Morris, London; Weidenfeld and Nicolson.

Hooff, J. A. R. A. M. van (1970). A component analysis of the structure of the social behaviour of a semi-captive chimpanzee group. *Experientia* **26**, 549–50.

Hooff, J. A. R. A. M. van (in press). A structural analysis of the social behaviour of a semi-captive group of chimpanzees. In *Expressive movement and non-verbal communication*. Ed. M. von Cranach and I. Vine, London; Academic Press.

Kant, I. (1781). *Kritik der reinen Vernunft*. Berlin; Cassiner, ed. 1913.

Kaufmann, J. H. (1965). Studies on the behavior of captive tree shrews (*Tupaia glis*). *Folia primat.* **3**, 50–74.

Kawai, I. (1966). Changes in social behavior following bilateral removal of the posterior parts of the superior temporal gyri in Japanese monkeys. *Primates* **7**, 1–20.

Koestler, A. (1949). *Insight and Outlook*. London; Macmillan.

Kohts, N. (1937). La conduite du petit chimpanzé et de l'enfant de l'homme. *J. Psychol. norm. pathol.* **34**, 494–531.

Lahiri, R. K. and Southwick, C. H. (1966). Parental care in *Macaca sylvana*. *Folia primat.* **4**, 257–64.

Laroche, J. G. and Tcheng, F. (1963). *La sourire du Nourisson*. (*La voix comme facteur déclenchant*.) Publ. Univ. de Louvain.

Lawick-Goodall, J. van (1968). The behaviour of free-living chimpanzees in the Gombe Stream Reserve. *Anim. Behav. Monogr.* **1**, 161–311.

Lersch, P. (1957). Zur Theorie des mimischen Ausdrucks. *Z. exp. angew. Psychol.* **4**, 409–19.

Lipps, Th. (1898). *Komik und Humor*. Leipzig; Voss.

Loizos, C. (1967). Play behaviour in higher primates: a review. In *Primate Ethology*, 176–218. Ed. D. Morris, London; Weidenfeld and Nicolson.

Loizos, C. (1968). An ethological study of chimpanzee play. *Proc. 2nd Int. Congr. Primatol. Atlanta*, 1968, **1**, 87–93. Basel; Karger.

McComas, H. C. (1926). The origin of laughter. *Psychol. Rev.* **33**, 45–55.

McGrew, W. C. (1969). An ethological study of agonistic behaviour in preschool children. *Proc. 2nd. Int. Congr. Primatol. Atlanta*, 1968, **1**, 149–59. Basel; Karger.

Osman Hill, W. C. and Bernstein, I. S. (1969). On the morphology, behaviour and systematic status of the Assam macaque (*Macaca assamensis* McCleiland, 1839). *Primates* **10**, 1–17.

Plessner, H. (1942). *Lachen und Weinen, eine Untersuchung nach den Grenzen menschliches Verhalten*. Arnhem; van Loghum Slaterus.

Plessner, H. (1950). *Lachen und Weinen*. Bern; Francke.

Plessner, H. (1953). *Zwischen Philosophie und Gesellschaft*. Bern; Francke.

Pohl, A. (1967). Beiträge zur Ethologie und Biologie des Sonnendachses (*Helictis personata* Gray 1831) in Gefangenschaft. *D. zool. Garten* **33**, 225–47.

Radcliffe-Brown, A. R. (1956). *Structure and Function in Primitive Society*, Glencoe, Ill.; Free Press.

Reynolds, V. and Reynolds, F. (1965). Chimpanzees of the Budongo Forest. In *Primate Behaviour*, 368–424. Ed. I. DeVore, New York; Holt Rinehart and Winston.

Schloeth, R. (1956). Zur Psychologie der Begegnung zwischen Tieren. *Behaviour* **10**, 1–80.

Spencer, H. (1870). *Principles of Psychology*. London; Williams and Norgate.

Spivak, H. (1968). *Ausdrucksformen und soziale Beziehungen in einer Dschelada-Gruppe (Theropithecus gelada) im Zoo*. Zürich; Juris-Verlag.

Washburn, R. W. (1929). A study of the smiling and laughing of infants in the first year of life. *Genet. Psychol. Monogr.* **6**, 397–535.

Yerkes, R. M. (1943). *Chimpanzees, a laboratory colony*. New Haven; Yale Univ. Press.

9. NON-VERBAL COMMUNICATION IN HUMAN SOCIAL INTERACTION

Abercrombie, K. (1968). Paralanguage. *Br. J. dis. Comm.* **3**, 55–9.

Allport, G. W. (1961). *Pattern and Growth in Personality*. New York; Holt Rinehart and Winston.

Allport, G. W. and Vernon, P. E. (1932). *Studies in Expressive Movement*. New York; Macmillan.

Argyle, M. (1969). *Social Interaction*. London; Methuen, New York; Atherton.

Argyle, M. (1971). *The Social Psychology of Work*. London; The Penguin Press.

Argyle, M., Alkema, F. and Gilmour, R. (1972). The communication of friendly and hostile attitudes by verbal and non-verbal signals. *Europ. J. soc. Psychol.* (in press).

Argyle, M. and Dean, J. (1965). Eye-contact, distance and affiliation. *Sociometry* **28**, 289–304.

Argyle, M. and Ingham, R. (1972). Gaze, mutual gaze and proximity. *Semiotica* (in press).

Argyle, M. and Kendon, A. (1967). The experimental analysis of social performance. *Adv. exp. soc. Psychol.* **3**, 55–98.

Argyle, M., Lalljee, M. G. and Cook, M. (1968). The effects of visibility on interaction in a dyad. *Hum. Relat.* **21**, 3–17.

Argyle, M. and Little, B. R. (1971). Do personality traits apply social behaviour? Paper to B.P.S., unpublished.

Argyle, M. and McHenry, R. (1970). Do spectacles really affect judgements of intelligence? *Br. J. soc. clin. Psychol.* **10**, 27–9.

Argyle, M., Salter, V., Nicholson, H., Williams, M. and Burgess, P. (1970). The communication of inferior and superior attitudes by verbal and non-verbal signals. *Br. J. soc. clin. Psychol.* **9**, 221–31.

Argyle, M. and Williams, M. (1969). Observer or observed? A reversible perspective in person perception. *Sociometry* **32**, 396–412.

Bandura, A. (1962). Social learning through imitation. In Nebraska Symposium on *Motivation*. Ed. M. R. Jones, Lincoln; Nebraska University Press.

Barker, R. G. and Wright, H. F. (1954). *Midwest and its children: the Psychological Ecology of an American Town*. Evanston, Ill.; Row, Peterson.

Berkowitz, L. (1968). Responsibility, reciprocity and social distance in help-giving: an experimental investigation of English social class differences. *J. exp. soc. psychol.* **4**, 46–63.

Birdwhistell, R. L. (1952). *Introduction to Kinesics*. Louisville University Press.

Brun, T. (1969). *The International Dictionary of Sign Language*. London; Wolfe.

Condon, W. S. and Ogston, W. D. (1966). Sound film analysis of normal and pathological behaviour patterns. *J. Nerv. ment. Dis.* **143**, 338–47.

Cook, M. (1970). Experiments on orientation and proxemics. *Hum. Relat.* **23**, 61–76.

Crystal, D. (1969). *Prosodic Systems and Intonation in English*. London; Cambridge University Press.

Darwin, C. R. (1872). *The Expression of the Emotions in Man and Animals*. London; John Murray.

Davitz, J. R. (1964). *The Communication of Emotional Meaning*. New York; McGraw-Hill.

Ekman, P. (1969a). Non-verbal leakage and clues to deception. *Psychiatry* **32**, 88–106.

Ekman, P. (1969b). Pan-cultural elements in facial displays of emotion. *Science, N.Y.* **164**, 86–8.

Ekman, P. and Friesen, W. V. (1967). Origin, usage, and coding: the basis for five categories of non-verbal behaviour. *Semiotica* (in press).

Eldred, S. H. and Price, D. B. (1958). Linguistic evaluation of feeling states in psychotherapy. *Psychiatry* **21**, 115–21.

Exline, R. V., Gray, D. and Schuette, D. (1965). Visual behavior in a dyad as affected by interview content and sex of respondent. *J. pers. soc. Psychol.* **1**, 201–9.

Exline, R. V. and Winters, L. C. (1965). Affective relations and mutual gaze in dyads. In *Affect, Cognition and Personality*. Ed. S. Tomkins and C. Izzard, New York; Springer.

Felipe, N. J. and Sommer, R. (1966). Invasions of personal space. *Social Problems* **14**, 206–14.

Garfinkel, H. (1963). Trust and stable actions. In *Motivation and Social Interaction*. Ed. O. J. Harvey, New York; Ronald.

Goffman, E. (1956). *The Presentation of Self in Everyday Life*. Edinburgh; Edinburgh University Press.

Goffman, E. (1961). *Encounters*. Indiana; Bobbs-Merrill.

Goffman, E. (1963). *Behavior in Public Places*. Glencoe, Ill.; Free Press.

Goffman, E. (1971). *Relations in Public*. Harmondsworth; Allen Lane.

Haggard, E. A. and Isaacs, K. S. (1966). Micromomentary facial expressions as indicators of ego mechanisms in psychotherapy. In *Methods of Research in Psychotherapy*. Ed. L. A. Gottschalk and A. H. Auerback, New York; Appleton Century.

Hutt, S. J. and Hutt, C. (1970). *Direct Observation and Measurement of Behaviour*. Springfield, Ill.; Thomas.

Ingham, R. (1971). *Cultural differences in social behaviour*. D.Phil thesis. Oxford University.

Jecker, J. D., Maccoby, N. and Breitrose, H. S. (1965). Improving accuracy in interpreting non-verbal cues of comprehension. *Psych. in the Schools.* **2**, 239–44.

Jourard, S. M. (1966). An exploratory study of body-accessibility. *Br. J. soc. clin. Psychol.* **5**, 221–31.

Kendon, A. (1967). Some functions of gaze direction in social interaction. *Acta psychol.* **26**, 1–47.

Kendon, A. (1970). Movement coordination in social interaction: some examples considered. *Acta psychol.* **32**, 1–25.

Kendon, A. (1971*a*). The role of visible behaviour in the organisation of social interaction. In *Symposium on Human Communication*. Ed. M. von Cranach and I. Vine, London and New York; Academic Press.

Kendon, A. (1971*b*). Some relationships between body motion and speech: an analysis of an example. In *Studies in Dyadic Communication*. Ed. A. Siegman and B. Pope, Elmsford, N.Y.; Pergamon.

Kendon, A. and Cook, M. (1969). The consistency of gaze patterns in social interaction. *Br. J. Psychol.* **60**, 481–94.

Lennard, H. L. and Bernstein, A. (1960). *The Anatomy of Psychotherapy*. Columbia University Press.

Lott, E. E., Clark, W. and Altman, I. (1969). A propositional inventory of research on interpersonal space. *Naval Medical Research Institute Research Report*.

McPhail, P. (1967). The development of social skill in adolescents. Paper to B.P.S. Oxford Department of Educational Studies.

Mahl, G. F. and Schulze, G. (1964). Psychological research in the extralinguistic area. In *Approaches to Semiotics*. Ed. T. A. Sebeok, A. S. Hages and M. C. Bateson, The Hague; Mouton.

Mehrabian, A. (1968). The inference of attitudes from the posture, orientation, and distance of a communicator. *J. Consult. Psychol.* **32**, 296–308.

Miller, G. A., Galanter, E. and Pribram, K. H. (1960). *Plans and the Structure of Behavior*. New York; Holt Rinehart and Winston.

Mischel, W. (1969). *Personality and Assessment*. New York; Wiley.

Morris, D. (1967). *The Naked Ape*. London; Cape.

Porter, E. R., Argyle, M. and Salter, V. (1970). What is signalled by proximity? *Perc. Motor Skills* **30**, 39–42.

Rosenfeld, H. M. (1967). Non-verbal reciprocation of approval: an experimental analysis, *J. exp. soc. Psychol.* **3**, 102–11.

Sahlins, M. D. (1965). On the sociology of primitive exchange. In *The Relevance of Models for Social Anthropology*. A.S.A. monographs I. London; Tavistock Publications.

Scheflen, A. E. (1965). *Stream and Structure of Communicational Behavior*. Commonwealth of Pennsylvania; Eastern Pennsylvania Psychiatric Institute.

Schutz, W. C. (1958). *FIRO: A Three Dimensional Theory of Interpersonal Behavior*. New York; Holt Rinehart and Winston.

Sissons, M. (1970). The psychology of social class. In *Money, Wealth and Class*. London; Oxford University Press.

Sommer, R. (1965). Further studies of small group ecology. *Sociometry* **28**, 337–48.

Szasz, T. S. (1961). *The Myth of Mental Illness*. London; Secker and Warburg.

Tagiuri, R. (1958). Social performance and its perception. In *Person Perception and Interpersonal Behavior*. Ed. R. Tagiuri and L. Petrullo, Stanford University Press.

Verplanck, W. S. (1955). The control of the content of conversation: reinforcement of statements of opinion. *J. abnorm. soc. Psychol.* **51**, 668–76.

Vine, I. (1971). Communication by Facial-Visual Signals. In *Social Behaviour in Animals and Man*. Ed. J. H. Crook, New York and London; Academic Press.

Watson, O. M. and Graves, T. D. (1966). Quantitative research in proxemic behavior. *Amer. Anthrop.* **68**, 971–85.

Williams, J. H. (1964). Conditioning of verbalization: a review. *Psychol. Bull.* **62**, 383–93.

10. NON-VERBAL COMMUNICATION IN CHILDREN

Ainsworth, M. D. S. (1969). Object relations, dependency, and attachment: A theoretical review of the infant–mother relationship. *Child Develop.* **40**, 969–1026.

Altmann, S. A. (1962). A field study of the sociobiology of rhesus monkeys, *Macaca mulatta. Ann. N.Y. Acad. Sci.* **102**, 338–435.

Ambrose, J. A. (1961). The development of the smiling response in early infancy. In *Determinants of Infant Behaviour*. Ed. B. M. Foss, London; Methuen.

Ambrose, J. A. (1963). The age of onset of ambivalence in early infancy: indications from the study of laughing. *J. Child Psychiat.* **4**, 167–84.

Ames, L. B. (1949). Development of interpersonal smiling responses in the preschool years. *J. genet. Psychol.* **74**, 273–91.

Anderson, J. W. (1972). Attachment behaviour out of doors. In Blurton Jones, 1972.

Andrew, R. J. (1957). The aggressive and courtship behaviour of certain emberizinae. *Behaviour* **10**, 255–308.

Andrew, R. J. (1963). The origin and evolution of the calls and facial expressions of the primates. *Behaviour* **20**, 1–110.

Berg, I. (1966). A note on observations of young children with their mothers in a child psychiatric clinic. *J. Child Psychol. Psychiat.* **7**, 69–73.

Blurton Jones, N. G. (1967). An ethological study of some aspects of social behaviour of children in nursery school. In *Primate Ethology*. Ed. D. Morris, London; Weidenfeld and Nicolson.

Blurton Jones, N. G. (1968). Observations and experiments on the causation of threat displays of the Great tit *Parus major. Anim. Behav. Monogr.* **1**, 75–158.

Blurton Jones, N. G. (in press). Criteria for describing facial expressions. *Hum. Biol.*

Blurton Jones, N. G. (1972). Categories of child–child interaction. In *Ethological Studies of Child Behaviour*. Ed. N. G. Blurton Jones, London; Cambridge University Press.

Blurton Jones, N. G. and Leach, G. M. (1972). Behaviour of children and their mothers at separation and greeting. In Blurton Jones, 1972.

Blurton Jones, N. G. and Trollope, J. (1968). Social behaviour of Stump-tailed macaques in captivity. *Primates* **9**, 365–94.

Bowlby, J. (1969). *Attachment and Loss.* **1**, Attachment, London; Hogarth Press.

Brannigan, C. and Humphries, D. (1972). Human non-verbal behaviour: A means of communication. In Blurton Jones, 1972.

Bridges, K. M. B. (1932). Emotional development in early infancy. *Child Develop.* **3**, 324–41.

Bridges, K. M. B. (1933). A study of social development in early infancy. *Child Develop.* **4**, 36–49.

Buhler, Charlotte and Hetzer, H. (1928). The first understanding of expression in the first year of life. *Z. Psychol.* **107**, 50–61.

Connolly, K. and Smith, P. (1972). Reactions of preschool children to a strange observer. In Blurton Jones, 1972.

Currie, K. H. and Brannigan, C. R. (1970). Behavioural analysis and modification with an autistic child. In *Behaviour Studies in Psychiatry*. Ed. C. Hutt and S. J. Hutt, Oxford; Pergamon.

Dawe, H. C. (1934). An analysis of 200 quarrels of preschool children. *Child Develop.* **5**, 139–57.

Ding, G. F. and Jersild, A. T. (1932). A study of the laughing and smiling of preschool children. *J. genet. Psychol.* **40**, 452–72.

Duncan, Starkey (1969). Non-verbal Communication. *Psych. Bull.* 118–37.

Enders, A. C. (1927). A study of the laughter of the pre-school child in the Merrill–Palmer Nursery School. *Papers Mich. Acad. Sci., Art & Letters*, **8**, 341–56.

Every, R. G. (1965). The teeth as weapons, their influence on behaviour. *The Lancet* **1**, 685–8.

Fairbanks, G., Wiley, J. H. and Lassman, F. M. (1949a). An acoustical study of vocal pitch in seven and eight-year-old boys. *Child Develop.* **20**, 63–9.

Fairbanks, G., Herbert, E. C. and Hammond, J. M. (1949b). An acoustical study of vocal pitch in seven and eight-year-old girls. *Child Develop.* **20**, 71–8.

Freedman, D. G. (1964). Smiling in blind infants and the issue of innate vs. acquired. *J. child Psychol. Psychiat.* **5**, 171–84.

Gardner, Beatrice T. and Wallach, Lise (1965). Shapes of figures identified as a baby's head. *Percept. Motor Skills* **20**, 135–42.

Grant, E. C. (1965). An ethological description of some schizophrenic patterns of behaviour. In *Proceedings of the Leeds Symposium on Behavioural Disorders*, May and Baker, Dagenham.

Grant, E. C. (1968). An ethological description of non-verbal behaviour during interviews. *Br. J. med. Psychol.* **41**, 177–83.

Grant, E. C. (1969). Human facial expression. *Man* **4**, 525–36.

Hattwick, M. S. (1932). A preliminary study of pitch inflection in the speech of pre-school children. *Proc. Iowa Acad. Sci.* **39**, 237–42.

Hattwick, M. S. (1933). The role of pitch level and pitch range in the singing of pre-school, first grade, and second grade children. *Child Develop.* **4**, 281–91.

Hinde, R. A. (1958). The nest-building behaviour of domesticated canaries. *Proc. zool. Soc. Lond.* **131**, 1–48.

Hooff, J. A. R. A. M. van (1962). Facial expressions in higher primates. *Symp. zool. Soc. Lond.* **8**, 97–128.

Justine, Florence (1932). A genetic study of laughter provoking stimuli. *Child Develop.* **3**, 114–36.

Kehrer, H. E. and Tente, D. (1969). Observations on displacement activities in children. *J. Child Psychol. Psychiat.* **10**, 225–32.

Kenderine, Margaret (1931). Laughter in the pre-school child. *Child Develop.* **2**, 228–30.

Koch, Helen L. (1935). An analysis of certain forms of so-called 'nervous habits' in young children. *J. genet. Psychol.* **46**, 139–70.

Landreth, C. (1941). Factors associated with crying in young children in the nursery school and the home. *Child Develop.* **12**, 81–97.

Leach, G. M. (1972). Comparison of social behaviour of anxious children and normal children in a playgroup setting. In Blurton Jones, 1972.

Lorenz, K. (1943). Die angeborenen Formen moglicher Erfahrang. *Z. Tierpsychol.* **5**, 235–409.

MacKay, D. M. (1965). Cerebral organisation and the conscious control of action. In *Semoine d'Etude sur Cerreau et expérience consciente*. Pontificiae Academiae Scientiarum Scripta Varia 30. Vatican.

McGrew, W. C. (1969). An ethological study of agonistic behaviour in preschool children. *Proc. 2nd Internat. Congr. Primatol., Behaviour* **1**, 149–59.

McGrew, W. C. (1970). Ph.D. thesis. University of Edinburgh.

McGrew, W. C. (1972). Aspects of social development in nursery school children, with emphasis on introduction to the group. In Blurton Jones, 1972.

Michael, G. and Willis, F. N. (1968). The development of gestures as a function of social class, education and sex. *Psychol. Record* **18**, 515.

Miller, N. E., Galanter, E. and Pribram, K. H. (1960). *Plans and the Structure of Behaviour*. New York; Holt Rinehart and Winston.

Mitchell, J. C. (1968). Dermatological aspects of displacement activity: attention to the body surface as a substitute for fright or flight. *Canad. med. Ass. J.* **98**, 962–4.

Morris, D. (1957). 'Typical Intensity' and its relation to the problem of ritualisation. *Behaviour* **11**, 1–12.

Osgood, C. E. (1966). Dimensionality of the semantic space for communication via facial expressions. *Scand. J. Psychol.* **7**, 1–30.

Rheingold, H. L. and Keene, G. C. (1965). Transport of the human young. In *Determinants of Infant Behaviour*, vol. 3. Ed. B. M. Foss, London; Methuen, New York; Wiley.

Richards, M. P. M. and Bernal, Judith (1972). An observational study of mother–infant interaction. In Blurton Jones, 1972.

Robson, K. S. (1967). The role of eye-to-eye contact in maternal–infant attachment. *J. Child Psychol. Psychiat.* **8**, 13–25.

Schaffer, H. R. and Emerson, P. E. (1964). The development of social attachments in infancy. *Monogr. Soc. Res. Child Dev.* **29**, No. 3, 1–77.

Smith, P. and Connolly, P. (1972). Patterns of play and social interaction in preschool children. In Blurton Jones, 1972.

Thompson, J. (1941). Development of facial expressions in blind and seeing children. *Arch. Psychol.* **37**, No. 264 (New York).

Tinbergen, N. (1951). *The Study of Instinct*. London; Oxford University Press.

Tinbergen, N. (1959). Comparative studies of the behaviour of gulls (Laridae): a progress report. *Behaviour* **15**, 1–70.

Tugendhat, B. (1960). The disturbed feeding behaviour of the three spined stickle-back. 1. Electric shock is administered in the food area. *Behaviour* **16**, 159–87.

Vine, I. (1970). Communication by facial–visual signals: a review and analysis of their role in face-to-face encounters. In *Social Behaviour in Animals and Man*. Ed. J. H. Crook, London and New York; Academic Press.

Vine, I. (in preparation). The significance of facial–visual signalling in human social development.

Washburn, Ruth W. (1929). A study of the smiling and laughing of infants in the first year of life. *Genet. Psychol. Monogr.* **6**, 397–457.

Wasz-Hockert, O., Lind, J., Vuorenkoski, V., Partanen, T. and Valanne, E. (1968). The infant cry. A spectrographic and auditory analysis. *Clinics in Developmental Medicine*, No. 29. Spastics International Medical Publications and Wm. Heinemann Ltd., London.

Wolff, P. H. (1963). Observations on the early development of smiling. In *Determinants of Infant Behaviour*, vol. 2. Ed. B. M. Foss, London; Methuen.

11. SIMILARITIES AND DIFFERENCES BETWEEN CULTURES IN EXPRESSIVE MOVEMENTS

Ahrens, R. (1953). Beitrag zur Entwicklung des Physiognomie und Mimikerkennens. *Z. Exptl Angew. Psychol.* **2**, 412–54, 599–633.

Andree, R. (1889). *Ethnographische Parallelen und Vergleiche*. Leipzig.

Bateson, G. and Mead, M. (1942). *Balinese Character*. Special Publ. of the New York Academy of Sciences, II.

Birdwhistell, R. L. (1963). The kinesis level in the investigation of the emotions. In *Expressions of the Emotions in Man*. Ed. P. H. Knapp, New York; Int. Univ. Press.

Birdwhistell, R. L. (1967). Communication without words. In *L'Aventure Humaine* (Paris). Ed. P. Alexandre, Société d'Etudes littéraires et Artistiques.

Coss, R. G. (1969). Perceptual aspects of eyespot patterns and their relevance to gaze behaviour. In *Behaviour Studies in Psychiatry*. Ed. S. H. Hutt and C. Hutt, Oxford; Pergamon.

Cranach, M. von (in press). *Über die Signalfunktion des Blickes*, Soziale Theorie und Praxis, Festschrift Baumgarten. Meinsenheim; A. Hain.

Darwin, C. (1872). *The Expression of the Emotions in Man and Animals*. London; Murray.

Eibl-Eibesfeldt, I. (1968). Zur Ethologie menschlichen Grußverhaltens I. Beobachtungen an Balinesen, Papuas und Samoanern nebst vergleichenden Bemerkungen. *Z. Tierpsychol*. **25**, 727–44.

Eibl-Eibesfeldt, I. (1970*a*). The expressive behaviour of the deaf and blind born. In *Non-verbal behaviour and Expressive Movements*. Ed. M. von Cranach and I. Vine, London; Academic Press.

Eibl-Eibesfeldt, I. (1970*b*). *Ethology, The Biology of Behaviour*. New York; Holt Rinehart and Winston.

Eibl-Eibesfeldt, I. (1970*c*). *Liebe und Hass – Zur Naturgeschichte elementarer Verhaltensweisen*. Munich; Piper.

Eibl-Eibesfeldt, I. (1970*d*). Männliche und weibliche Amulette im modernen Japan. *Homo* **20**, 175–88.

Eibl-Eibesfeldt, I. and Hass, H. (1967). Neue Wege der Humanethologie. *Homo* **18**, 13–23.

Eibl-Eibesfeldt, I. and Wickler, W. (1968). Die ethologische Deutung einiger dämonenabwehrender Figuren von Bali. *Z. Tierpsychol*. **25**, 719–26.

Ekman, P., Sorenson, E. R. and Friesen, W. V. (1969). Pan Cultural Elements in the Facial Displays of Emotion. *Science, N.Y.* **164**, 86–8.

Erikson, E. H. (1966). Ontogeny of Ritualisation in Man. *Phil. Trans. Roy. Soc.* B **251**, 337–49.

Fantz, R. L. (1967). Visual perception and experience in infancy. In *Early Behaviour*. Ed. H. W. Stevenson, New York; Wiley.

Frijda, N. H. (1965). Mimik und Pantomimik. In *Handb. d. Psychol.* **5**, 351–421. Ausdruckspsychologie.

Gardner, R. and Heider, K. G. (1968). *Gardens of War*. New York; Random House.

Hass, H. (1968). *Wir Menschen*, Wien; Molden.

Hess, E. H. (1965). Attitude and Pupil Size. *Sci. Amer.* **212**, 46–54.

Hooff, J. A. R. A. M. van (1967). The facial displays of the catarrhine monkeys and apes. In *Primate Ethology*, 7–68. Ed. D. Morris, London; Weidenfeld and Nicolson.

Koenig, O. (1970). *Kultur und Verhaltensforschung*, München; Deutscher Taschenbuch Verlag.

La Barre, W. (1947). The cultural basis of emotions and gestures. *J. Personality* **16**.

Lawick-Goodall, J. van (1968). The behaviour of free-living chimpanzees in the Gombe Stream Reserve. *Anim. Behav. Monogr.* **1**, (3), 161–311.

Lorenz, K. (1953). Die Entwicklung der vergleichenden Verhaltensforschung in den letzten 12 Jahren. *Zool. Anz. Suppl.* **16**, 36–58.

Lorenz, K. (1965). *Evolution and Modification of Behaviour*. Chicago; University of Chicago Press.

Montagu, M. R. A. (1968). *Man and Aggression*, New York; Oxford University Press.

Ploog, D. W., Blitz, J. and Ploog, F. (1963). Studies on Social and Sexual Behaviour of the Squirrel Monkey (*Saimiri sciureus*). *Folia Primat.* **1**, 29–66.

Ploog, D. W. and MacLean, P. D. (1963). Display of penile erection in squirrel monkey (*Saimiri sciureus*). *Anim. Behav.* **11**, 32–9.

Schenkel, R. (1947). Ausdrucksstudien an Wölfen. *Behaviour*, **1**, 81–129.

Sorenson, E. R. and Gajdussek, D. C. (1966). The Study of Child Behaviour and Development in Primitive Cultures. *Pediatrics, Suppl.* **37**, 149–243.

Wickler, W. (1966). Ursprung und biologische Deutung des Genitalpresentierens männlicher Primaten. *Z. Tierpsychol.* **23**, 422–37.

12. THE INFLUENCE OF CULTURAL CONTEXT ON NON-VERBAL COMMUNICATION IN MAN

Apel, K. O. (1963). Die Idee der Sprache in der Tradition des Humanismus von Dante bis Vico. *Arch. Begriffsgesch.* **8**.

Bataille, G. (1962). *Eroticism*. London; Calder.

Berg, C. (1951). *The Unconscious Significance of Hair*. London; Allen and Unwin.

Bettelheim, B. (1955). *Symbolic Wounds*. London; Thames and Hudson.

Chomsky, N. (1959). A Review of B. F. Skinner's *Verbal Behaviour*. In *Language* **35**, 26–58. Reprinted in Fodor and Katz, 1964, 547–76.

Chomsky, N. (1966). *Cartesian Linguistics*. New York; Harper and Row.

Descartes, R. (1955). *The Philosophical Works of Descartes*. Trans. E. S. Haldane and G. R. T. Ross, New York; Dover Publications.

Firth, R. (1936). *We, the Tikopia*. London; Allen and Unwin.

Firth, R. (1970). Postures and gestures of respect. In *Échanges et Communications*. Ed. J. Pouillon and P. Marande, The Hague; Mouton.

Fodor, J. A. and Katz, J. J. (ed.) (1964). *The Structure of Language: Readings in the Philosophy of Language*. Englewood Cliffs, N.J.; Prentice-Hall.

Freud, S. (1932). *The Interpretation of Dreams*. London; Allen and Unwin.

Gombrich, E. H. (1966). Ritualized gesture and expression in art. *Phil. Trans. Roy. Soc.* B **251**, 393–402.

Hall, E. T. (1963). A System for the notation of proxemic behaviour. *Am. Anthropol.* **65**, 1003–26.

Hall, E. T. (1966). *The Hidden Dimension*. New York; Doubleday.

Hallpike, C. R. (1969). Social Hair. *Man* **4**, 256–64.

Hampshire, S. (1969). Vico and the Contemporary Philosophy of Language. In Tagliacozzo and White, 1969, 475–82.

Huxley, Sir J. S. (1966). A discussion on ritualization of behaviour in animals and man. *Phil. Trans. Roy. Soc. Lond.* B **772**, vol. 251.

Jakobson, R. and Halle, M. (1956). *Fundamentals of Language*. The Hague; Mouton.

La Barre, W. (1964). Paralinguistics, kinesics and cultural anthropology. In *Approaches to Semiotics*. Ed. T. A. Sebeok, A. S. Hayes and M. C. Bateson, The Hague; Mouton.

Lane, M. (1970). 'Introduction' to M. Lane (editor) *Structuralism: A Reader*, pp. 11–42; London; Cape.

Leach, E. R. (1958). Magical Hair. *Man* (n.s.) **88**, 147–64.

Leach, E. R. (1970). *Lévi-Strauss*. London; Fontana.

Lévi-Strauss, C. (1955). *Tristes Tropiques*. Paris; Plon.

Lévi-Strauss, C. (1962). *La Pensée Sauvage*. Paris; Plon.

Lévi-Strauss, C. (1964). *Mythologiques: I: Le Cru et le Cuit*. Paris; Plon. (English edition 1970, London; Cape.)

Lévi-Strauss, C. (1966). The Culinary Triangle, *New Society*, 22 Dec., 937–40.

Lorenz, K. Z. (1966). Evolution of Ritualization. . . . In Huxley, 1966, 273–84.

Malinowski, B. (1932). *The Sexual Life of Savages in North-Western Melanesia*, 3rd edition, London; Routledge.

Mauro, T. de (1969). Giambattista Vico: From Rhetoric to Linguistic Historicism. In Tagliacozzo and White, 1969, 279–95.

Morris, D. (1967). *The Naked Ape*. London; Cape.

Needham, R. (1967). Percussion and transition. *Man* (n.s.) **2**, 606–14.

Parsons, T. and Bales, R. F. (1955). *Family, Socialization and Interaction Process*. Glencoe; The Free Press.

Ruesch, J. and Bateson, G. (1951). *Communication, the Social Matrix of Psychiatry*. New York; W. W. Norton.

Skinner, B. F. (1957). *Verbal Behavior*. New York; Appleton-Century-Crofts.

Tagliacozzo, G. and White, H. V. (ed.) (1969). *Giambattista Vico: an International Symposium*. Baltimore; Johns Hopkins Press.

Tinbergen, N. (1951). *The Study of Instinct*. London; Oxford University Press.

Turner, V. W. (1966). Colour Classification in Ndenbu Ritual. In *Anthropological Approaches to the Study of Religion*, 47–84. Ed. M. Banton, London; Tavistock.

Van Gennep, A. (1909). *Les Rites de Passage*, Paris. English translation 1960, *The Rites of Passage*. London; Routledge.

13. NON-VERBAL COMMUNICATION IN THE MENTALLY ILL

Argyle, M. (1967). *The Psychology of Interpersonal Behaviour*. Harmondsworth; Penguin Books.

Cook, N. G. and Stengel, E. (1958). *Attempted Suicide*. London; Oxford Univ. Press.

Darwin, C. (1872). *The Expression of the Emotions in Man and Animals*. London; John Murray.

Ekman, P. (1965). Communication through non-verbal behaviour: a source of information about interpersonal relationship. In *Affect, Cognition and Personality*, 725–35. Ed. S. S. Tompkins and C. E. Izard, New York; Springer Press.

Ekman, P. and Friesen, W. V. (1967). Non-verbal behaviour in research in psychotherapy. In *Research in Psychotherapy*. Ed. J. Schlein, vol. 3, pp. 179–216. Am. Psych. Assoc., Washington, D.C.

Esser, A. H. (1970). International hierarchy and power structure on a psychiatric ward: – ethological studies on dominance behaviour in a total institution. In *Behavioural Studies in Psychiatry*, 25–59. Ed. C. Hutt and J. Hutt, Oxford; Pergamon.

Eysenk, H. J. (1954). Problems der diagnostichen Untersuchung und Demonstration des Charakter-Interpretationstestes. *Z. exp. angen. Psychol.* **2**, 1–32.

Freeman, T., Cameron, J. L. and McGhie, A. (1958). *Chronic Schizophrenia*. London; Tavistock.

Goffman, E. (1959). *The Presentation of Self in Everyday Life*. New York; Doubleday.

Grant, E. C. (1965). An ethological description of some schizophrenic patterns of behaviour. *Proc. Leeds Symp. Behav. Disorders*, Dagenham; May and Baker.

Grant, E. C. (1968). An ethological description of non-verbal behaviour during interviews. *Br. J. med. Psychol.* **41**, 177–84.

Grant, E. C. (1969). Human facial expression. *Man* (n.s.) **4**, 525–36.

Hutt, C. and Ounsted, C. (1966). A behavioural and electroencephalographic study of autistic children. *J. Psychiat. Res.* **3**, 181–97.

Lefcourt, H. M., Rotenberg, F., Buckspan, B. and Steffy R. A. (1967). Visual interaction and performance of process and reactive schizophrenics as a function of the examiner's sex. *J. Pers.* **35**, 535–46.

Tinbergen, E. A. and Tinbergen, N. (1972). Early childhood autism – an ethological approach. *Z. Tierpsychol.* (in press).

Wolff, S. and Chess, A. (1964). A behavioural study of schizophrenic children. *Acta Psychiat. Scand.* **40**, 438–66.

14. PLAYS AND PLAYERS

Gasset, J. Ortega y (1969). *The Dehumanisation of Art.* Princeton.

Goffman, E. (1970). *Where the Action is.* Allen Lane, Penguin Press.

Gregory, R. (1970). *The Intelligent Eye.* New York and London; McGraw-Hill.

Ryle, G. (1949). *The Concept of Mind.* London; Hutchinson.

Strawson, P. F. (1950). On referring. *Mind* 59, 320–44.

Wain, J. (1968). *Shakespeare in the Theatre.* Harmondsworth; Penguin.

Willett, J. (1964) (ed.). *Brecht on Theatre.* London; Methuen.

Wittgenstein, L. (1968). *Philosophical Investigations.* Oxford; Blackwell.

15. ACTION AND EXPRESSION IN WESTERN ART

Alberti, L. B. (*c.* 1435). *Della Pittura.* Ed. H. Janitschek, 1877, Vienna; W. Braumüller.

Brandt, E. (1965). *Gruss und Gebet. Eine Studie zu Gebärden in der minoisch-mykenischen und frühgriechischen Kunst,* Waldsassen, Bavaria; Stiftland Verlag.

Brilliant, R. (1963). *Gesture and Rank in Roman Art. Mem. Conn. Acad. Arts Sci.* **14.**

Buehler, K. (1933). *Ausdruckstheorie.* Jena; Fischer.

Buschor, E. (1921). *Griechische Vasenmalerei.* Munich; R. Piper.

Clark, K. (1963). Motives. In *Acts of the Twentieth International Congress of the History of Art at New York,* 1961, vol. IV, Princeton University Press.

Deonna, W. (1914). *L'Expression des sentiments dans l'Art grec.* Paris; H. Laurens.

Engel, J. J. (1785–6). *Ideen zu einer Mimik.* Berlin; Mylius.

Filarete (Antonio Averlino) (*c.* 1460). *Treatise on Architecture.* Ed. J. R. Spencer, 1965, New Haven and London; Yale University Press.

Goethe, J. W. (1817). Joseph Bossi über Leonardo da Vincis Abendmahl. *Über Kunst und Alterthum,* **1,** 3.

Gombrich, E. H. (1935). Review of J. Bodonyi, Entstehung und Bedeutung des Goldgrundes, in *Kritische Berichte zur kunstgeschichtlichen Literatur* **5,** 66–75.

Gombrich, E. H. (1945). Botticelli's Mythologies. *J. Warburg and Courtauld Insts* **8.**

Gombrich, E. H. (1950). *The Story of Art.* London; Phaidon.

Gombrich, E. H. (1960). *Art and Illusion.* New York; Pantheon, London; Phaidon.

Gombrich, E. H. (1962). Expression and Communication. Reprinted in *Meditations on a Hobby Horse,* 1963, London; Phaidon.

Gombrich, E. H. (1964). Moment and Movement in Art. *J. Warburg and Courtauld Insts* **27.**

Gombrich, E. H. (1965). Visual Discovery through Art. Reprinted in *Psychology and the Visual Arts.* Ed. James Hogg, 1969, Harmondsworth; Penguin.

Gombrich, E. H. (1966). Ritualized gesture and expression in art. *Phil. Trans. Roy. Soc.* B, No. 772, **251.**

Gombrich, E. H. (1968). The Leaven of Criticism in Renaissance Art. In *Art, Science and History in the Renaissance.* Ed. C. S. Singleton, Baltimore; Johns Hopkins.

Gombrich, E. H. (1969). The Evidence of Images: The Priority of Context over Expression. In *Interpretation.* Ed. C. S. Singleton, Baltimore; Johns Hopkins.

Groenewegen-Frankfort, H. A. (1951). *Arrest and Movement.* London; Faber.

Janitschek, H. (ed.) (1877). *Alberti Della Pittura.* Vienna; Braumüller.

Leonardo da Vinci (*c.* 1500). *Treatise on Painting.* Ed. A. P. McMahon, 1956, Princeton University Press.

Lomazzo, G. P. (1584). *Trattato dell'Arte della Pittura, Scultura ed Architettura.* Milan; P. G. Pontio.

Le Brun, Charles (1689). *Conférence sur l'Expression générale et particulière.* Paris; E. Picart.

Lister, R. (1966). *Victorian narrative paintings.* New York; Clarkson N. Potter.

Mâle, E. (1932). *L'Art religieux après le Concile de Trente.* Paris; Librairie Armand Colin.

Montagu, J. (1960). *Charles Le Brun's Conference sur l'expression.* Unpublished Ph.D. Thesis, University of London.

Murray, H. A. (1943). *Thematic Apperception Test.* Harvard University Press.

Neumann, G. (1965). *Gesten und Gebärden in der griechischen Kunst.* Berlin; W. de Gruyter.

Nordenfalk, C. (1950). Tizians Darstellung des Schauens. In *Nationalmusei Årsbok,* 1947–8.

Pfuhl, E. (1923). *Malerei und Zeichnung der Griechen.* Munich; F. Bruckmann.

Reisner, Bob and Kapplow, H. (1958). *Captions Outrageous.* London and New York; Abelard-Schuman.

Ringbom, S. (1965). *Icon to Narrative.* Åbo (Turku); Åbo Akademi.

Schäfer, H. (1930). *Von ägyptischer Kunst,* 3rd edition. Leipzig; J. C. Heinrichs.

Shaftesbury, Anthony, the Earl of (1714). *Characteristicks.* London.

Testelin, H. (1696). *Sentimens des plus habiles peintres.* Paris; Mabre-Cramoisy.

Xenophon (*c.* 400 B.C.). *Memorabilia,* trans. E. C. Marchant, 1923, The Loeb Classical Library, London; Heinemann.

EPILOGUE

Chomsky, N. (1967). The general properties of language. In *Brain Mechanisms Underlying Speech and Language.* Ed. F. L. Darley, New York; Grune and Stratton.

Sebeok, T. A. (ed.) (1968). *Animal Communication.* Bloomington, Indiana and London; Indiana University Press.

Tavolga, W. N. (1968). Fishes. In Sebeok, 1968.

AUTHOR INDEX

Note: page numbers in **bold** type relate to the lists of references

INDEX

28